24 Advances in Polymer Science

Fortschritte der Hochpolymeren-Forschung

Edited by H.-J. CANTOW, Freiburg i. Br. · G. DALL'ASTA, Cesano Maderno
K. DUŠEK, Prague · J. D. FERRY, Madison · H. FUJITA, Osaka · M. GORDON,
Colchester · W. KERN, Mainz · G. NATTA, Milano · S. OKAMURA, Kyoto
C. G. OVERBERGER, Ann Arbor · T. SAEGUSA, Kyoto · G. V. SCHULZ,
Mainz · W. P. SLICHTER, Murray Hill · J. K. STILLE, Iowa City

W0079479

With 130 Figures

Springer-Verlag Berlin Heidelberg GmbH 1977

Editors

Prof. Dr. Hans-Joachim Cantow, Institut für Makromolekulare Chemie der Universität, Stefan-Meier-Str. 31, 7800 Freiburg i. Br., BRD

Prof. Dr. Gino Dall'Asta, SNIA VISCOSA – Centro Sperimentale, Cesano Maderno (MI), Italia

Prof. Dr. Karel Dušek, Institute of Macromolecular Chemistry, Czechoslovak Academy of Sciences, 162 06 Prague 616, ČSSR

Prof. Dr. John D. Ferry, Department of Chemistry, The University of Wisconsin, Madison 6, Wisconsin 53706, U.S.A.

Prof. Dr. Hiroshi Fujita, Osaka University, Department of Polymer Science, Toyonaka, Osaka, Japan

Prof. Dr. Manfred Gordon, University of Essex, Department of Chemistry, Wivenhoe Park, Colchester C04 3 SQ, England

Prof. Dr. Werner Kern, Institut für Organische Chemie der Universität, 6500 Mainz, BRD

Prof. Dr. Giulio Natta, Istituto di Chimica Industriale del Politecnico, Milano, Italia

Prof. Dr. Seizo Okamura, Department of Polymer Chemistry, Kyoto University, Kyoto, Japan

Prof. Dr. Charles G. Overberger, The University of Michigan, Department of Chemistry, Ann Arbor, Michigan 48 104, U.S.A.

Prof. Takeo Saegusa, Kyoto University, Department of Synthetic Chemistry, Faculty of Engineering, Kyoto, Japan

Prof. Dr. Günter Victor Schulz, Institut für Physikalische Chemie der Universität, 6500 Mainz, BRD

Dr. William P. Slichter, Bell Telephone Laboratories Incorporated, Chemical Physics Research Department, Murray Hill, New Jersey 07 971, U.S.A.

Prof. Dr. John K. Stille, Colorado State University, Department of Chemistry, Fort Collins, CO 805 23, U.S.A.

ISBN 978-3-662-15476-2   ISBN 978-3-540-37469-5 (eBook)

DOI 10.1007/978-3-540-37469-5

Library of Congress Catalog Card Number 61-642

© by Springer-Verlag Berlin Heidelberg 1977
Originally published by Springer-Verlag Berlin Heidelberg New York in 1977.
Softcover reprint of the hardcover 1st edition 1977

2152/3140 – 543210

Contents

Polymer-Metal Complexes and Their Catalytic Activity

Eishun Tsuchida and Hiroyuki Nishide

Department of Polymer Chemistry, Waseda University, Tokyo 160, Japan

Table of Contents

1. Introduction

A polymer-metal complex is composed of a synthetic polymer and metal ions. Its
synthesis represents an attempt to give an organic polymer inorganic functions.
Catalytically active polymers can be obtained by inducing a metal complex to catalyze a reaction with a polymer backbone, and it is reasonable to assume that the
metal complex bound to the polymer will show a specific type of catalytic behavior, reflecting the properties of the polymer chain. Indeed, many synthetic polymer-
metal complexes have been found to exhibit high catalytic efficiency. On the other
hand, it has been shown that in metalloenzymes such as oxidase and hemoglobin,
where a metal complex is the active site, the macromolecular protein part plays a
significant role, or even controls the reactivity of the metal complex. Thus, research
on the catalytic activity of polymer-metal complexes has attracted considerable
interest, which has increased in recent years.

A wide variety of investigations have been carried out on polymeric metal complexes; these include studies of semiconductivity, thermostability, redox reactions,
collection of metal ions, biomedical effects, and so on. The polymeric metal complexes are classified into the following groups:

(A) polymer-metal complexes

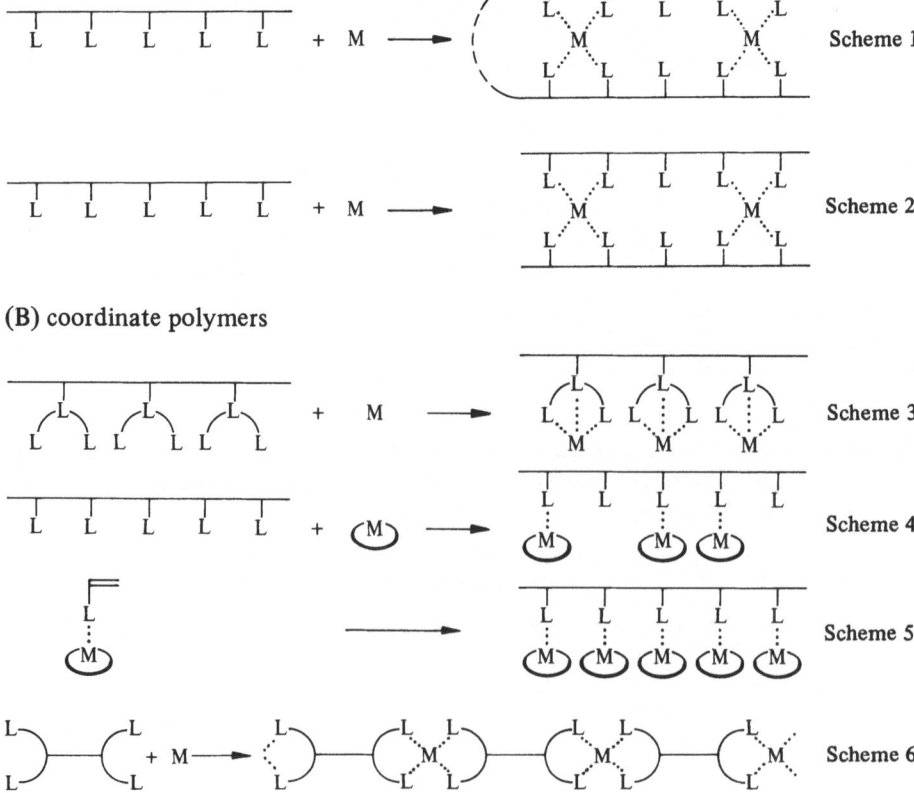

Scheme 1

Scheme 2

(B) coordinate polymers

Scheme 3

Scheme 4

Scheme 5

Scheme 6

(C) poly(metal-phthalocyanine) type

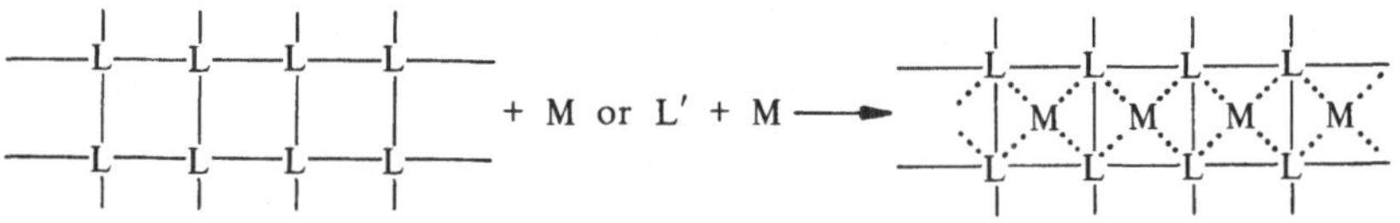

Scheme 7

L = coordinating atom or group; M = metal ion

Polymer-metal complexes, represented by Schemes 1 to 5, are defined as complexes composed of a polymer ligand and metal ions in which the metal ions are attached to the polymer ligand by a coordinate bond. Here a polymer ligand is understood to be a polymeric substance that contains coordinating groups or atoms (mainly N, O, and S), obtained by the polymerization of monomers containing coordinating sites, or by the chemical reaction between a polymer and a low-molecular-weight compound having coordinating ability. Typical polymer ligands previously reported are listed in Table 1. When a polymer ligand is mixed directly with a metal ion, which generally has four or six coordinate bonding hands, a polymer-metal complex is formed. This may be of the intra-polymer chelate type (Scheme 1) or of the inter-polymer chelate type (Scheme 2). Complex formation proceeds via Scheme 3, where the polymer backbone contains multidentate ligands, such as the iminodiacetic acid group, or acts as a carrier for low-molecular-weight multidentate ligands; many so-called chelating resins fit this scheme. The polymer-metal complexes represented by Schemes 1 to 3 have chelating structures in their polymer ligands and are therefore called polymer chelates. The pendant-type polymer-metal complex (Scheme 4) is formed by the reaction of a polymer ligand with a stable metal complex, the central metal ion of which has already been masked with low-molecular-weight ligands except for one coordinate site that remains vacant, *e.g.* metalloporphyrins, or cobaltic chelates. A polymer-metal complex is also obtained by polymerizing a monomeric metal complex (Scheme 5).

Scheme 6 represents *coordinate polymers*. A low-molecular-weight compound with multidentate groups on both ends of the molecule grows into a linear polymer with metal ions, and the polymer chain is composed of coordinate bonds. The parquetlike polymer complexes, poly(metal-phthalocyanine) and poly(metal-tetracyanoethylene), are classified into Scheme 7. They are formed by inserting metal ions into planar-network polymers or by causing a low-molecular-weight ligand derivative to react with a metal salt and a condensation reagent.

This article deals with the polymer-metal complexes (Schemes 1–5), because they have the following merits in comparison with other polymeric metal complexes. (i) Metal ion and ligand site can be chosen for study without restrictions. (ii) It is not difficult to control the molecular weight of a polymer complex and to modify the structure of a polymer ligand. (iii) The polymer complex is soluble in both aqueous and nonaqueous solvent. (iv) It is possible to change the ratio of the organic polymer part to the inorganic metal complex part. This explains why the polymer often affects the behavior of the metal complex.

Table 1. Typical polymer ligands

Coordinating groups	Repeating units of polymer ligands		
Amines $-NH_2$, $\rangle NH$, $\rangle N$	$-CH_2-CH-$ $\quad\ \ \|$ $\quad\ \ NH_2$ $-NHCH_2CH_2-$	$-CH_2-CH-$ (with phenyl–$CH_2NHCH_2CH_2NR_2$)	$-CH_2CHO-$ $-CH_2-$ $\ \|$ NR_2	$-NH-CH-CO-$ $\qquad\|$ $\qquad (CH_2)_4$ $\qquad\ \|$ $\qquad NH_2$

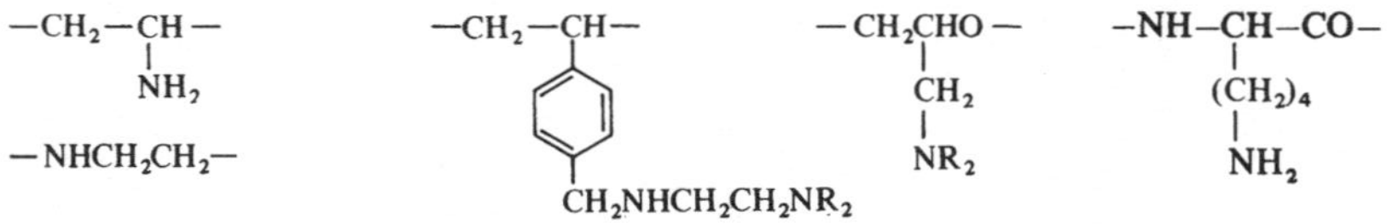

Nitrogen of heterocyclic compounds

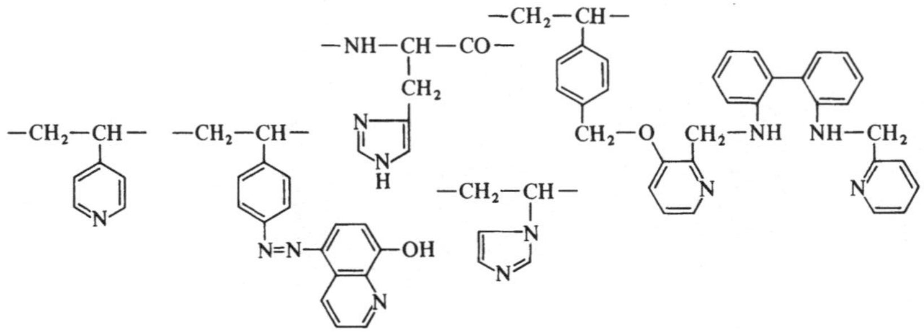

Schiff base
$\rangle C=N-$

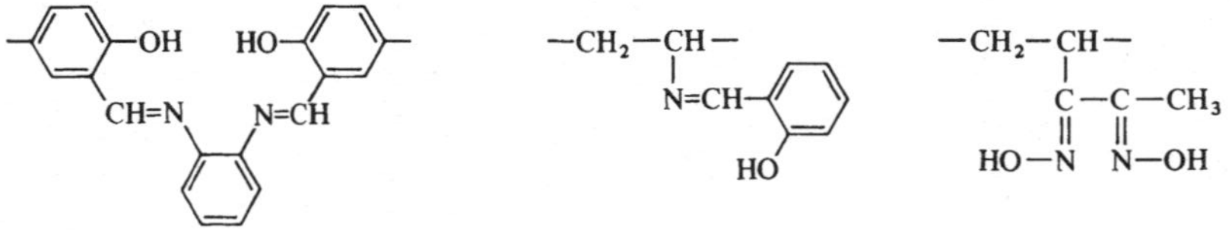

$-CH_2-CH-$
$\ \ \ \|$
$\ \ \ OH$

Alcohols $-OH$
carboxylic acids
$-COOH$

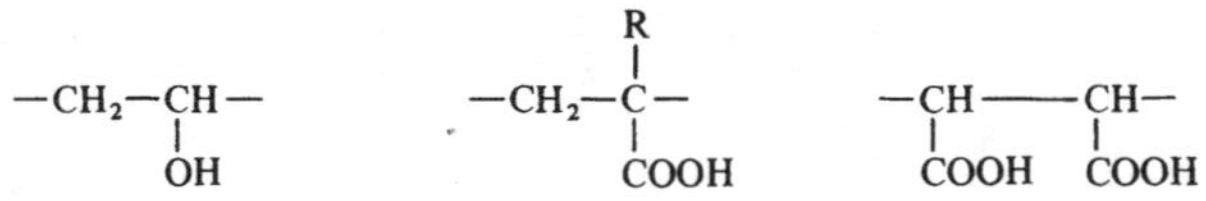

$-CH_2-CH-$
$\qquad\|$
$\qquad OH$

Ketones $>C=O$ esters, amides	$-CH_2-\overset{\displaystyle R}{\underset{\displaystyle C}{\vert\,\vert}}-\overset{\displaystyle O}{\underset{\displaystyle C}{\vert\vert}}-CH_2-\overset{\displaystyle O}{\underset{\displaystyle C}{\vert\vert}}-CH_3$	$-R-\overset{O}{\overset{\vert\vert}{C}}-CH_2-\overset{O}{\overset{\vert\vert}{C}}-R-$	$-CH_2-CH-O-\overset{O}{\overset{\vert\vert}{C}}-$ (phenol) $-CH_2-CH-\overset{O}{\overset{\vert\vert}{C}}-NH-OH$
Aminopoly- carboxylic acids	$-CH_2-CH-$ (phenyl) $CH_2N(CH_2COOH)_2$	$N(CH_2COOH)_2$ (benzene ring) $N(CH_2COOH)_2$, $-CH_2$, CH_2-	$-N-CH_2-CH_2-$ $\quad CH_2COOH$
Phosphonic acids $-PO(OH)_2$	$-CH_2-CH-$ (phenyl) $O=P(OH)_2$	$O-CH_2PO(OH)_2$ (benzene ring) $-CH_2$, CH_2-	
Thiols $-SH$	$-CH_2-CH-$ $\quad\quad SH$	$-CH_2-CH-$ (phenyl) CH_2-SH	$-CH_2-CH-O-O-CH_2-CH-CH_2$ $\quad\quad\quad\quad SH\ SH$ $-NH-N-$ $\quad\quad\quad C=S$ $\quad\quad\quad -NH-N-$

It is the purpose of this article to discuss whether or not there are any differences between the chemical reactivity of a polymer-metal complex and that of the corresponding monomeric complex. Although various extensive investigations on polymer-metal complexes have been reported, most of these complexes are too complicated to be discussed quantitatively due to the nonuniformity of their structure. These compounds include not only "complexes of macromolecules" but also the structurally labile "metal complex". Before detailed information can be obtained about the properties of polymer-metal complexes, and especially about the reactivity and catalytic activity of polymer-metal complexes, their structure must be elucidated. A polymer-metal complex having a uniform structure may be defined as follows;
(i) the structure within the coordination sphere is uniform, *i.e.* the species and the composition of the ligand and its configuration are identical in any complex unit existing in the polymer-metal complex;
(ii) the primary structure of the polymer ligand is known.

If the structure within the coordination sphere is identical in a polymer complex and a monomeric complex, their reactivity ought to be the same even though the complex is bound to a polymer chain. However, it is clear that the reactivity is sometimes strongly affected by the polymer ligand that exists outside the coordination sphere and surrounds the metal complex. The effects of polymer ligands will be discussed in the text under the following two terms: (1) the steric effect, which is determined by the conformation and density of the polymer-ligand chain, and (2) the special environment constituted by a polymer-ligand domain. To illustrate these effects, the following questions will be discussed:

(i) When metal complexes attach to a polymer ligand at high concentrations, are the stability and the reactivity of the complexes changed?

(ii) In a rigid polymer do the attached metal complexes, at low concentrations, behave virtually as in a solution at infinite dilution?

(iii) A polymer ligand in solution may be extended or densely coiled up, or sometimes helically structured. Does the conformation of the ligand chain influence the reactivity?

(iv) Are there any differences between the chelate formation of a polymer-ligand chain and that of the individual monomeric ligand?

(v) Because polymer-metal complexes are poly(electrolyte)s, an electrostatic effect is predominant in the reactions of ionic species.

(vi) A metal complex in an aqueous solution may be occluded by a hydrophobic polymer matrix, and this hydrophobic microenvironment affects reactivity.

These discussions will embrace homogeneous solutions of polymer-metal complexes. Of course one of the important advantages offered by the use of a polymer ligand, especially a crosslinked polymer ligand, in catalysis is the insolubilization of the attached complexes; the insolubility of the polymer catalyst makes it very easy to separate from the other components of the reaction mixture. Several polymer-metal complexes have been used for this purpose, although such applications are not covered in this article. The aim here is (1) to characterize polymer-metal complexes and their behavior in such simple but important elementary reactions as complex formation, ligand substitution, and electron transfer, and (2) to describe their catalytic activity.

II. Formation and Structure of Pendant-Type Polymer-Metal Complexes

Examples of the pendant-type polymer-metal complexes represented by Schemes 4 and 5 are shown in Table 2. They are obtained by the reaction in Scheme 8: a polymer ligand is made to coordinate to a vacant site of a previously prepared, stable, low-molecular-weight metal complex. The structure and hence the functions of the metal complex are hardly changed by attachment to the polymer ligand. The metal complexes attached to the polymer ligands 1 to 8 are the classic cobaltic

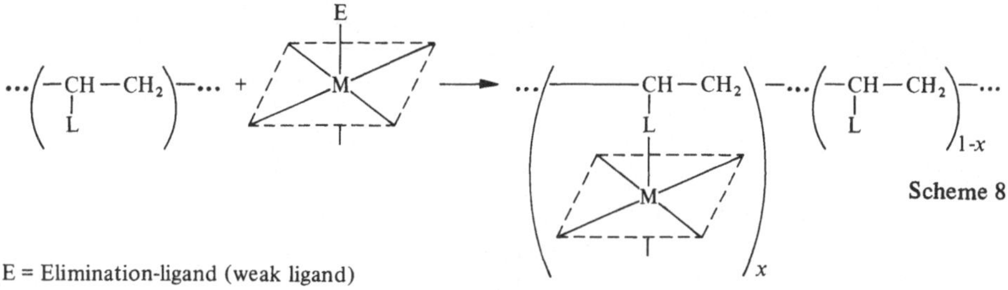

Scheme 8

E = Elimination-ligand (weak ligand)

chelates $9-13$, paramagnetic chromic chelate 14, metalloporphyrin 15, which reacts with molecular oxygen, metal-chlorophyllin 16, which is a derivative of chlorophyll, and so on. In Scheme 8, the degree of coordination (x) means the molar ratio [metal complex]/[repeating unit of polymer ligand]. If all the ligands in the polymer are coordinated to M, the value of x becomes unity.

A. Pendant-type Polymer-Co(III) and -Cr(III) Complexes

We prepared a series of pendant-type polymer-metal complexes having a uniform structure by the substitution reaction between a polymer ligand and a Co(III) or Cr(III) chelate, the chelate being inert in ligand-substitution reactions[1, 2]. A polymer-Co(III) complex, e.g. cis-[Co(en)$_2$(PVP)Cl]Cl$_2$ (en=ethylenediamine, PVP= poly(4-vinylpyridine)) 17, was prepared as follows[1]:

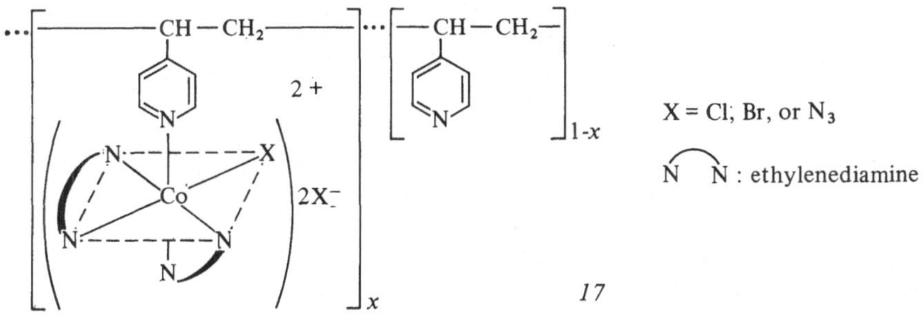

X = Cl, Br, or N$_3$

N   N : ethylenediamine

17

An aqueous solution of cis-[Co(en)$_2$Cl$_2$]Cl was added to an ethanolic solution of PVP and the mixture heated. After a few hours the mixture was filtered and dia-

Table 2. Pendant-type polymer-metal complexes

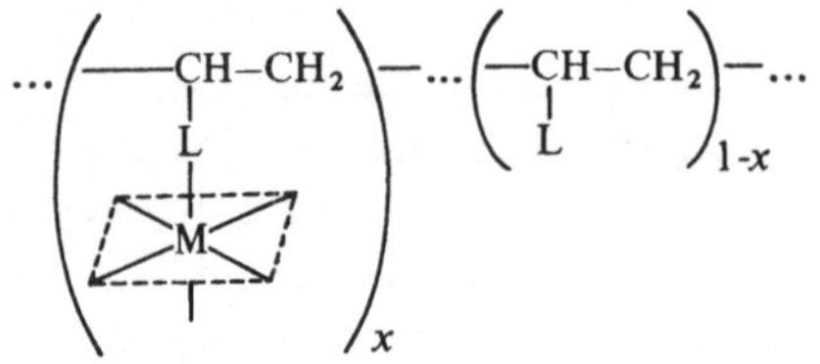

Polymer ligand	$\overset{|}{\underset{L}{|}}$	Metal complex	

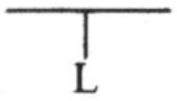

$-CH_2-CH-$
PVP *1*

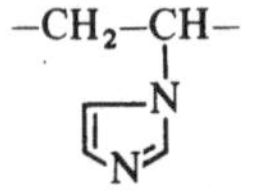

$-CH_2-CH-$
PNVI *2*

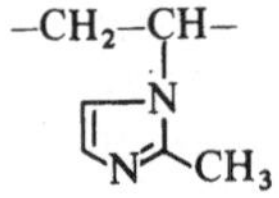

$-CH_2-CH-$
PMVI *3*

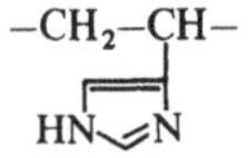

$-CH_2-CH-$
PVI *4*

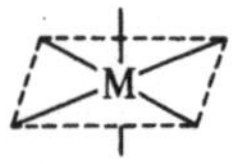

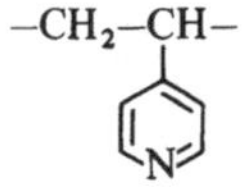

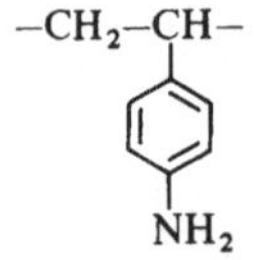

$-CH_2-CH-$
PAS *5*

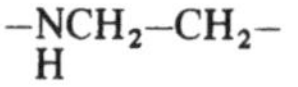

$-NCH_2-CH_2-$
H LPEI *6*

$-NCH_2-CH_2-NCH_2-CH_2-$
CH_2H
CH_2
NH_2 BPEI *7*

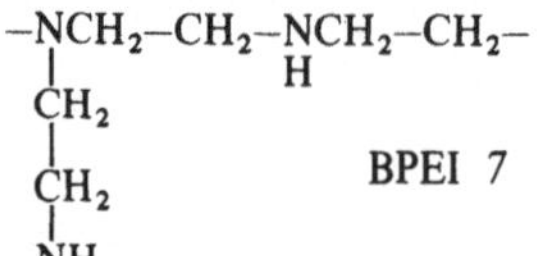

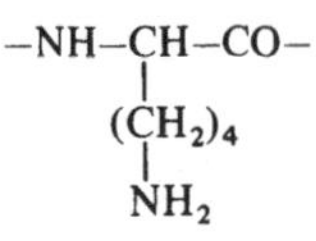

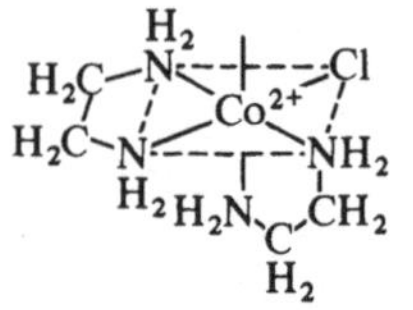

$-NH-CH-CO-$
$(CH_2)_4$ PLL *8*
NH_2

9

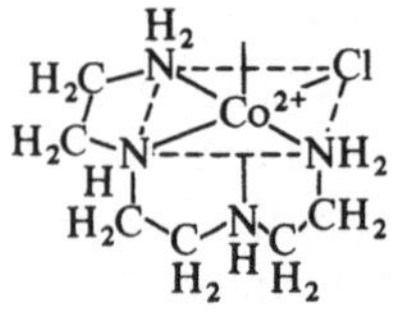

10

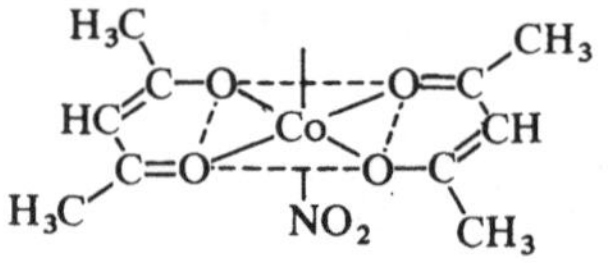

11

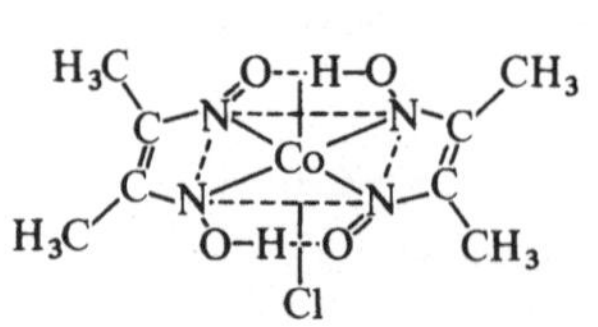

12

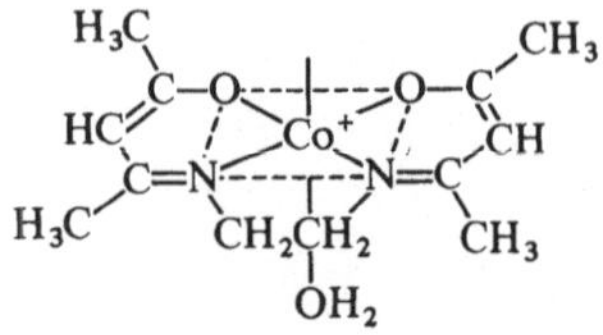

13

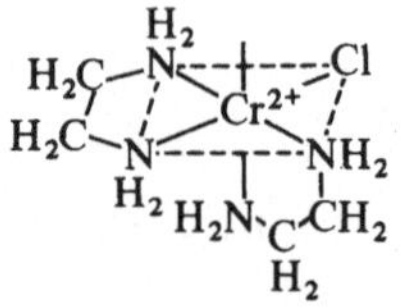

14

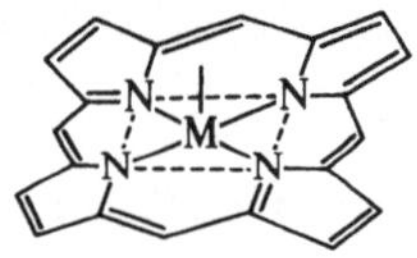

15

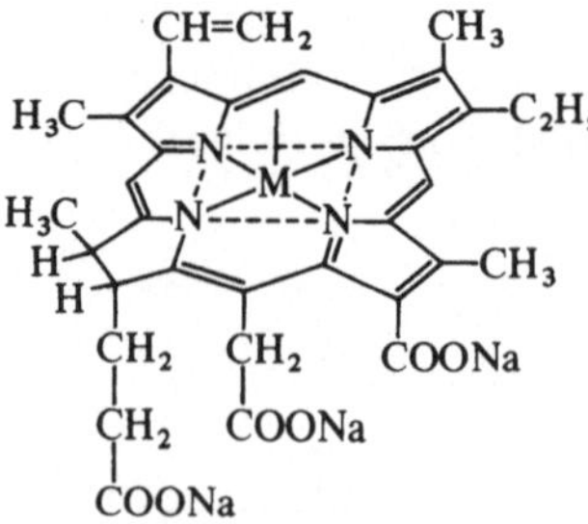

16

lyzed in cold water, then, after the water had evaporated, the PVP-Co(III) complex was obtained as a transparent, dark-reddish filmy substance. The polymer complex, [Co(en)$_2$(PVP)Cl]Cl$_2$ 17, had one bivalent cation per complex repeating unit and was water-soluble, regardless of the solubility of the polymer ligand.

Co(III) and Cr(III) chelates are inert in ligand-substitution reactions, so we considered that complex formation between a polymer ligand and the Co(III) or Cr(III) chelate would be irreversible under the applied experimental conditions[2]. This assumption was supported by the fact that the degree of coordination (x) did not decrease either during the reaction time or during the purification of the polymer complexes and that no degradation product from the polymer-Co(III) complexes could be found. Figure 1 shows the formation curves of the polymer-Co(III)

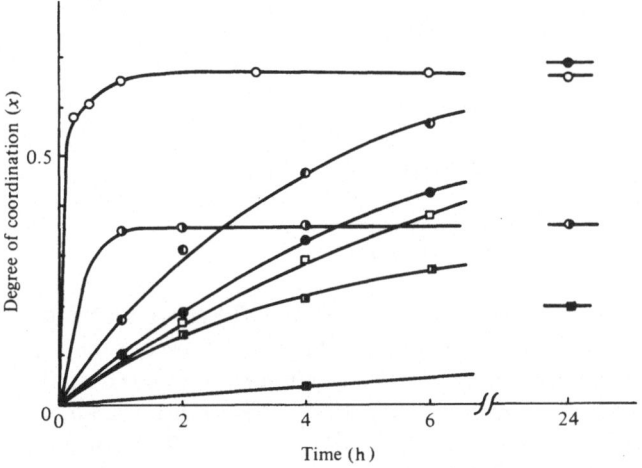

Fig. 1. Complex formation curves for polymer-Co(III) and -Cr(III) complexes[1−5]
○: [Co(acac)$_2$(PVP)(NO$_2$)], ◑: [Co(en)$_2$(BPEI)Cl]Cl$_2$, ◓: [Cr(en)$_2$(PVP)Cl]Cl$_2$,
●: [Co(en)$_2$(PVP)Cl]Cl$_2$, □: [Co(en)$_2$(QPVP)Cl]Cl$_2$($-R = -CH_3$), ◪: [Co(en)$_2$(QPVP)Cl]Cl$_2$
($-R = -CH_2C_6H_5$), ■: [Co(en)$_2$(PAS)Cl]Cl$_2$, PVP = poly(4-vinylpyridine), BPEI = branched
poly(ethyleneimine), QPVP = partially quaternized PVP, PAS = poly(p-aminostyrene),
acac = acetylacetone, en = ethylenediamine, 80 °C

and polymer-Cr(III) complexes, $i.e.$ the variation of x with time[1−5]. The initial rates of complexation for [Co(acac)$_2$(PVP)(NO$_2$)] (acac=acetylacetone) 11, for [Cr(en)$_2$(PVP)Cl]Cl$_2$ 14, and for [Co(en)$_2$(BPEI)Cl]Cl$_2$ (BPEI = branched poly(ethyleneimine) = 7), 9 are larger than those for the other complexes under the same conditions. These differences are attributed to the relatively high ligand substitution activity of the acac chelate and the Cr(III) chelate and to the higher basicity of BPEI as compared to PVP and PAS(pols(p-aminostyrene) = 5). Complexation of PVP of a lower degree of polymerization proceeds more easily[1], and the complexation rate for partially alkylated poly(4-vinylpyridine) (QPVP) 18 is $-R = -H > -CH_3 > -CH_2C_6H_5$, as shown in Fig. 1. These findings are interpreted as due to the steric hindrance of the polymer ligands[3].

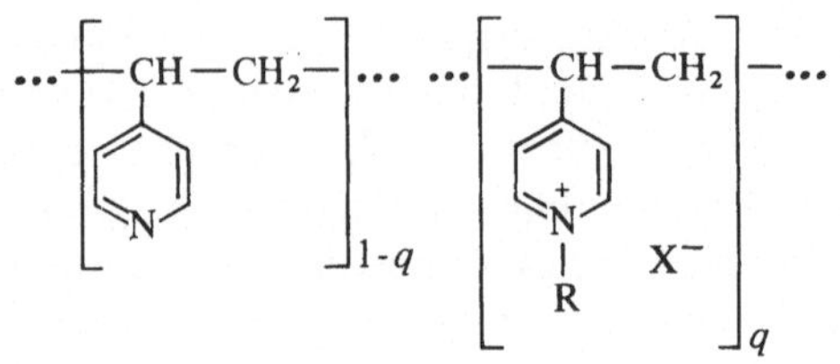

18

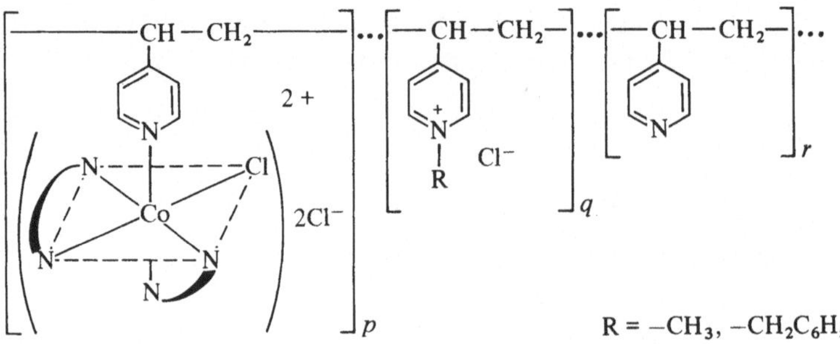

19

$$R = -CH_3, -CH_2C_6H_5$$

$p + p + r = 1$ ($q \times 100$) indicates the percentage of quaternization
Degree of coordination is defined as $x = p/(p + r)$ in this case.

In Fig. 1, the degree of coordination (x) of [Co(acac)$_2$(PVP)(NO$_2$)] seems to reach a constant level of about 0.65. Our experiment showed that, even if the Co(III) chelate was added in a large excess, or if the reaction mixture was heated for more than a day, the value of x remained below 0.7[2]. The limiting value of x was found to be also about 0.65 for the other PVP-Co(III) chelates. The conformational profile of the PVP-Co(III) complex with a higher degree of coordination is illustrated schematically in Fig. 2(a). The bulky Co(III) chelates coordinate along a PVP chain at high density, so, in the closest packing, some of the pyridine units must remain uncoordinated to avoid steric collision between the Co(III) chelates. A statistical calculation was carried out and the calculated value of x at saturation (0.63) was nearly equal to the experimental value of x[2]. The profile for the BPEI-Co(III) complex was considered to be as in Fig. 2(b), with the Co(III) chelates coordinated preferentially to the end primary amine groups of the branched poly(ethyleneimine)[4].

The suggested rod like structure of the pendant-type PVP-Co(III) complex is supported by the viscosity behavior of the polymer-complex solution (Fig. 3)[2]. The PVP-Co(III) complexes have higher viscosity than PVP; this suggests that the polymer complex has a linear structure and that intra-polymer chelation does not occur. The dependence of the reduced viscosity on dilution and the effect of ionic strength further show that [Co(en)$_2$(PVP)Cl]Cl$_2$ is a poly(electrolyte). The polymer complexes with higher x values have a rodlike structure due to electrostatic repulsion or the steric bulkiness of the Co(III) chelate. On the other hand, the solubility and solution behavior of the polymer complex with a lower x value is similar to that of the polymer ligand itself.

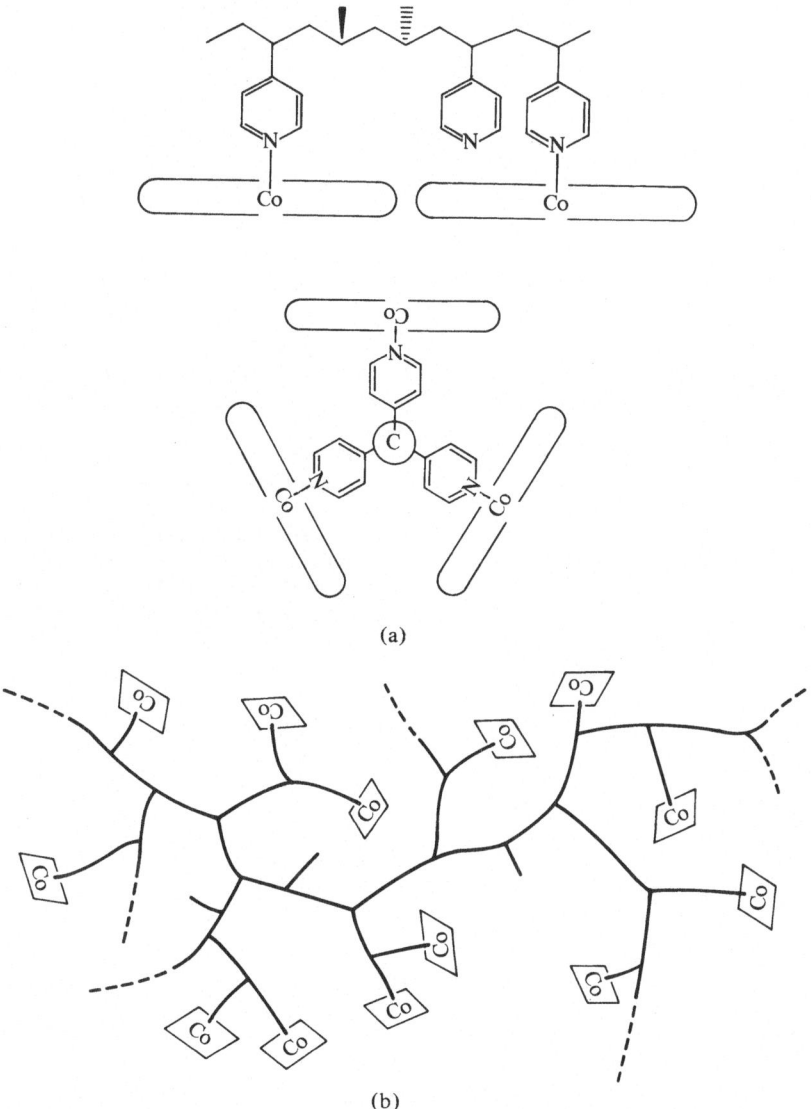

Fig. 2. Conformational profiles of polymer-Co(III) complex: poly(vinylpyridine)-Co(III) complex (a) and branched poly(ethyleneimine)-Co(III) complex (b)

When complex formation was carried out by simultaneously mixing a polymer ligand, a Co salt, and low-molecular-weight ligands[6], or when the formation conditions were not moderate[7], the polymer ligands were crosslinked by the Co(III) chelate as shown in *20* and *21*. Attention must be paid to the synthetic procedure in order to obtain a soluble pendant-type polymer complex having a uniform structure[8].

The inner-sphere structures of the polymer-Co(III) and -Cr(III) complexes were determined from their characteristic absorption bands in the visible, infrared, and

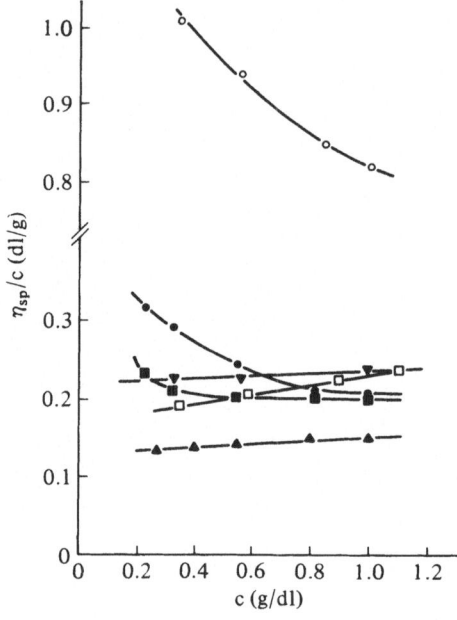

Fig. 3. Viscosity behavior of poly(4-vinyl-pyridine)-Co(III) complexes and poly-(4-vinylpyridine) ligand[2]

○: [Co(en)$_2$(PVP)Cl]Cl$_2$ in H$_2$O,
□: [Co(en)$_2$(PVP)Cl]Cl$_2$ in H$_2$O + 0.2 M Na Cl,
●: [Co(en)$_2$(PVP)Cl]Cl$_2$ in DMSO,
■: [Co(en)$_2$(PVP)Cl]Cl$_2$ in DMSO + 0.1 M NaClO$_4$, ▲: [Co(acac)$_2$(PVP)(NO$_2$)] in DMSO, ▲: PVP in DMSO

20

21

far-infrared spectra, as listed in Table 3 together with those of the corresponding monomeric complexes[1-5]. For example, the visible spectra of cis-[Co(en)$_2$(PVP)Cl]Cl$_2$ (λ_1 = 518 nm, ϵ_1 = 72, λ_2 = 367 nm in Table 3) are similar to those of the pyridine complex, cis-[Co(en)$_2$(Py)Cl]Cl$_2$, which has the cis config-uration. The IR absorption peak at 1600 cm^{-1}, based on the stretching of the pyri-dine ring, shifts by about 20 cm^{-1} to a higher wave number, corresponding to the coordination of the pyridine unit of PVP to the Co(III) chelate. However, the absorption peak at 1600 cm^{-1} is also observed in the PVP complexes, suggesting that uncoordinated repeating units of PVP remain in the PVP-Co(III) complexes[1].

Table 3. Characteristic absorption bands of visible, infrared, and far-infrared spectra and magnetic susceptibility of pendant-type polymer-Co(III) and -Cr(III) complexes

Complexes	X^a	DP^b	Vis. (nm) λ_1	λ_2	IR $\nu_{C=C}$ (cm^{-1}) coord.	uncoord.	FIR (cm^{-1})[5] ν_{Co-N}(arom)	$\chi_M \cdot 10^{3\ c}$ (1/mol)[5]	Ref.
[Co(en)$_2$(Py)Cl]Cl$_2$	9	–	510	365	1610	–	236, 218	–	1)
[Co(en)$_2$(PVP)Cl]Cl$_2$	9	0.68	98	518	367	1620	1600	232, 210	–	1)
[Co(en)$_2$(PVP)Cl]Cl$_2$	9	0.32	98	518	367	1620	1600	233, 213	–	3)
[Co(en)$_2$(PVP)Cl]Cl$_2$	9	0.66	19	518	367	1619	1600	232, 211	–	3)
[Co(en)$_2$(BuNH$_2$)ClCl$_2$	9	–	–	510	–	–	–	–	–	4)
[Co(en)$_2$(BPEI)Cl]Cl$_2$	9	0.39	1600	510	–	–	–	–	–	4)
[Co(acac)$_2$(Py)(NO$_2$)]	11	–	–	538	424	1615	–	–	–	2)
[Co(acac)$_2$(PVP)(NO$_2$)]	11	0.66	98	543	421	1640	1600	–	–	2)
[Co(acac)$_2$(An)(NO$_2$)]	11	–	–			1565	–	–	–	2)
[Co(acac)$_2$(PAS)(NO$_2$)]	11	0.65	42			1560^d	1619^d	–	–	2)
[Cr(en)$_2$(Py)Cl]Cl$_2$	14	–	–	522	387	1612	–		6.4	5)
[Cr(en)$_2$(PVP)Cl]Cl$_2$	14	0.70	98	524	388	1620	1600		6.6	5)
[Cr(en)$_2$(PVP)Cl]Cl$_2$	14	0.20	98	524	388	1620	1600		6.1	5)

^{a} Degree of coordination.

^{b} Degree of polymerization of polymer ligand.

^{c} Molar magnetic susceptibility. Py = pyridine, PVP = poly(4-vinylpyridine) 1, BuNH$_2$ = n-butylamine, BPEI = branched poly(ethyleneimine) 7, An = aniline, PAS = poly(p-aminostyrene) 5, en = ethylenediamine, acac = acetylacetone.

^{d} The shift of the absorption of the Py ring to a higher wave number due to the coordination is considered to be caused by the overlap between the d-orbital of Co(III) and π-conjugated system of the Py ring. Conversely, the shift of the absorption of the An ring to a lower wave number in the complex is considered to result from the localization of the lone electron pair of the An nitrogen due to the coordination.

The PVP-Cr(III) complex is paramagnetic because a Cr(III) ion possesses three spins $(3d^3)$. The molar magnetic susceptibility of the PVP-Cr(III) complexes given in Table 3 is the same as that of the pyridine-Cr(III) complex. The ESR spectra of PVP-Cr(III) are similar to that of pyridine-Cr(III)[5]. Thus, we considered that the Cr(III) complexes on the PVP chain were magnetically dilute and the Cr(III) ions did not interact with each other, although they coordinated along the PVP chain at high concentration.

There still remains the question whether the coordinate bond between a polymer ligand and a metal ion is stronger than that of the monomeric complex or not. The far-infrared absorptions assigned to the stretching bands of the coordinate bond between Co and N of the polymer ligand are also shown in Table 3[5]. The peaks of ν_{Co-N} shift by about 3–8 cm^{-1} to a lower wave number in the PVP-Co(III) complexes, and the shifts are larger in the PVP complex with the higher x value. It is suggested that the coordinate bond between Co and the pyridine unit in PVP is weaker than the bond of the monomeric pyridine. This suggestion is also supported by the fact that the d-d absorption band of the Co(III) complex is somewhat red-shifted in the PVP complex[1].

Differential thermal analysis of the PVP-Co(III) complexes also supplied information about the strength of the coordinate bond between Co and PVP[5]. The dissociation temperature of the coordinate bond increases with degree of coordination (x): $198°$ $(x = 0.17) < 204°$ (monomeric pyridine complex) $< 224°$ $(x = 0.32) < 242°$ $(x = 0.67)$. The PVP-Co(III) complex probably has a rigid rod-like structure due to the steric bulkiness of the Co(III) chelate, as shown in Fig. 2(a), and the degree of freedom of the PVP complex itself or of the Co(III) chelates in the PVP complex is considered to be much smaller than that of the monomeric pyridine complex. Thus it is presumed that more energy is required to break the coordinate bond between Co and PVP and that the dissociation temperature increases with x[5].

B. Structures of Polymer-Heme Complexes

The metalloporphyrin complexes and their derivatives have been studied from the standpoint of model compounds of hemoglobin. The polymer-metalloporphyrin complexes are also formed by the reaction in Scheme 8, and a few qualitative investigations have been made with poly(L-lysine)[9, 10], poly(L-histidine)[11], and poly(vinylimidazole)[12] as the polymer ligand. Blauer[9] has studied the complex formation of heme with poly(L-lysine) and has discussed the effects of the molecular weight and secondary structure of poly(L-lysine) on complex formation. Shibata et al.[11] have reported on the heme complexes of poly(L-histidine) and the copolymer of L-histidine and L-glutamic acid. Poly(L-histidine) and the copolymer formed the six-coordinate and five-coordinate heme complex, respectively. Hatano[12] found that the formation constant of the complex of heme with poly(N-vinyl-2-methylimidazole) 3 was larger by a factor of 10^2 or 10^3 than that of the corresponding monomeric heme complex.

In this section, the complex formations and the coordination structures of ferri- and ferro-protoporphyrine IX (ferriheme and ferrohem, respectively) 22 with

CH=CH₂ H   CH₃

H₃C   C   CH=CH₂

=N   N-

HC   Fe   CH

N   N

H₃C   C   CH₃

CH₂   H   CH₂

CH₂   CH₂

COOH   COOH

22

various polymer ligands are summarized quantitatively[13–16]. The reactions of these polymer-heme complexes with molecular oxygen will be described in Chapter V.

We determined the coordination number($\bar{n}$), which is the molar ratio of the axial ligand to the heme ion, and the formation constants (K) of these complexes by spectrometric titration, using Miller-Drough's method[17], and list them in Table 4[13, 14]. The $\bar{n}$ values of the ferri- and ferroheme complexes of the low-molecular-weight ligands, $i.e.$ pyridine, imidazole and butylamine, are all equal to one, which is in agreement with the $\bar{n}$ values of the other polymer ligands, poly(4-vinyl-pyridine)(PVP, 1), poly(N-vinyl-4-methylimidazole)(PVMI, 3) and branched poly-(ethyleneimine)(BPEI, 7), except for poly(L-lysine)(PLL, 8). The absorption spectra and ESR spectra of polymer-ferriheme and -ferroheme are identical with those of the monomeric complexes ($e.g.$ ESR parameters: $g_{\parallel} = 2, g_{\perp} = 6$ for the ferriheme complex at high spin state)[15], and this indicates that the coordinate structures are the same in both the polymer complex and the monomeric complex, as represented in Fig. 4(a). The formation of the five-coordinate iron-porphyrin complexes with $\bar{n} = 1$ is explained by the $trans$ effect and the steric hindrance of the polymer chain[14]. Table 4 shows, furthermore, that the formation constants(K) of the polymer-ferriheme and -ferroheme complexes are about 10^2 times as large as those of the monomeric complexes. This result is thought to be caused by the apparent higher concentration of ligands in the polymer domain[14].

The $\bar{n}$ values of the PLL-ferriheme and -ferroheme complexes in Table 4 are equal to 2, and the six-coordinate heme structure is formed preferentially; this may be due to an inherent factor in which PLL differs from the other polymer ligands. It is well known that PLL can assume an α-helical conformation above pH 10. The complex formation between PLL and heme was carried out at pH 12, where the ε-amino groups of the side chain of PLL were not protonated and PLL formed a complete α-helix[14]. The helix parameter (molar ellipticity at 222 nm) was the same before and after complex formation with heme. The helix content, $i.e.$ the overall conformation of PLL, was not affected. The viscosity of the PLL solution also remained constant on the addition of ferriheme[14]. These results mean that complex formation occurs with a structure complementary to the α-helical conformation and that bending of the PLL helix does not occur. The inter-polymer sandwich structure proposed by Blauer [Fig. 4(c)][9] is thus ruled out. The structure

Table 4. Axial coordination number ($\bar{n}$) and formation constant (K) for polymer-ferriheme and -ferroheme complexes[13, 14]

Ligand	Solvent	$\bar{n}$		K (1/mol for $n=1$; 1^2/mol^2 for $n=2$)	
		Ferriheme	Ferroheme	Ferriheme	Ferroheme
PVP	DMF−MeOH(7/3)	0.95	1.3	3.8×10^2	1.1×10^3
Pyridine	DMF−MeOH(7/3)	1.0	1.1	6.0	5.1×10
PVMI	DMF−MeOH(1/3)	1.1	1.2	1.8×10^2	2.6×10^3
Imidazole	DMF−MeOH(1/3)	1.1	1.2	9.5	4.6×10
QPVP	H$_2$O−DMF(9/1)	1.3	1.3	2.6×10^2	2.8×10^4
PLL	H$_2$O(pH 12)	2.0	2.2	4.2×10^4	3.5×10^9
BPEI	H$_2$O(pH 12)	1.0		10^2	
Butylamine	H$_2$O(pH 12)	1 ~ 2		–	
PBLG · Im	DMF−MeOH(7/3)	1.0		3.9×10^2	

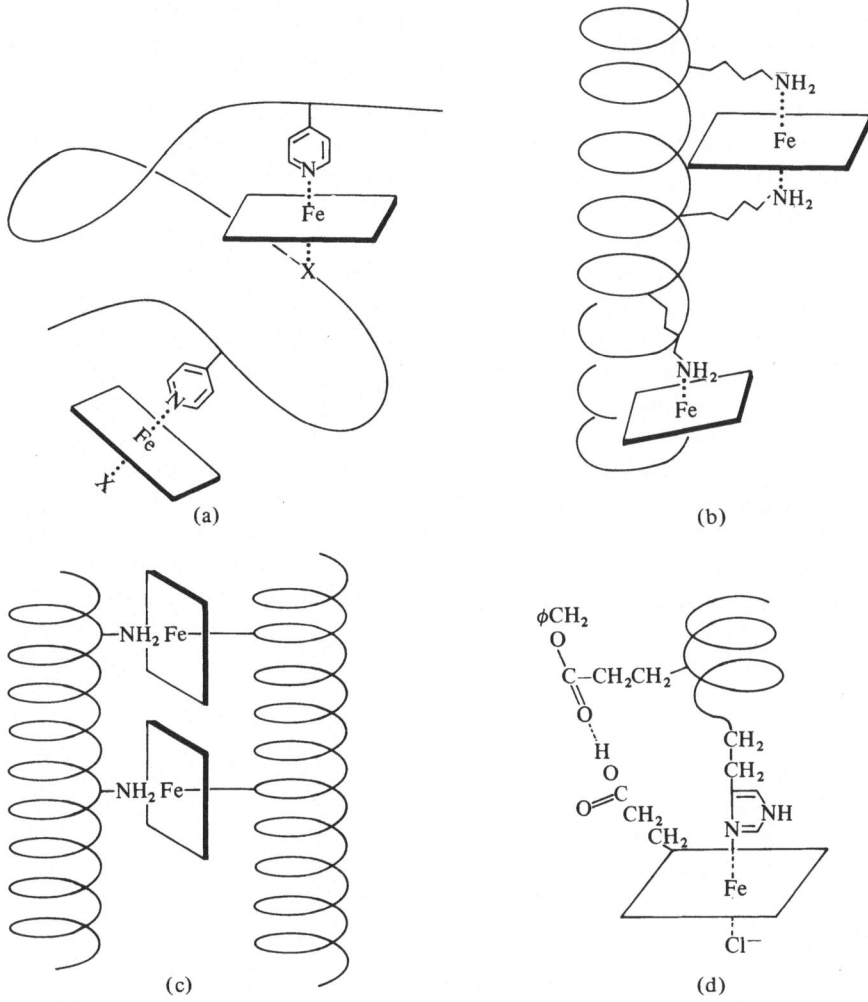

Fig. 4. Schematic representation of the structure of polymer-heme complexes: the ferriheme complex with poly(4-vinylpyridine) (a), poly(L-lysine) (b), (c), or poly(γ-benzyl-L-glutamate) with a pendant imidazole (d)

should be an intra-molecular chelating complex in which two ε-amino groups of an α-helical PLL in the same polymer chain coordinate to the heme iron from the opposite side of the porphyrin plane, as illustrated in Fig. 4(b). We further observed the induced Cotton effect for the PLL-heme system. This optical activity was consistent with the asymmetric structure formed by the chelation of two ε-amino groups of PLL to a central iron ion. However, the helix content of PLL decreased slightly when the ratio $[PLL]_0/[heme]_0$ was lower than $10^{18)}$. This indicates that there is some steric hindrance in the structure for PLL to coordinate a bulky heme.

When poly(γ-benzyl-L-glutamate) containing a pendant imidazole at the chain end (PBLG·Im)[19] was used as the polymer ligand, the coordination number in

Table 4 was unity, which indicated the five-coordinate structure as in the PVMI-heme complex. PBLG·Im forms an α-helix, and the helix content and intrinsic viscosity were unchanged in the PBLG·Im ferriheme complex. The formation constant of the ferriheme complex with PBLG·Im was not so different from that of the imidazole complex (Table 4). The strong coordination was thought to be due to an additional hydrogen bond between a propionic residue of ferriheme and a carbonyl residue in the side chain of PBLG·Im, as shown in Fig. 4(d)[14].

Complex formation with metallochlorophyllins(Chn, *23*), which are more readily water-soluble than heme, was carried out with partially quaternized poly(4-vinylpyridine)(QPVP, *18*) in an aqueous solution of pH $10^{16)}$. The coordination number was unity for all metallo-Chn, having Co(III), Fe(III), Ni(II), Cu(II), or Zn(II) as the central metal ion. The formation constants of metallo-Chn with QPVP were about $10^3 1/mol$, much larger than those of the monomeric complexes. Furthermore, the difference in K values was slight among the different metallo-Chn. It has been suggested that the electrostatic bonding between the carboxylate anion of Chn and the quaternized pyridinium cation on the QPVP chain governs the stability of QPVP-metallo-Chn complexes. Wang[20] had reported that the negatively charged hemes were combined to PLL by salt linkages, although there was no evidence to support his structure. Apart from the coordinate bond, the electrostatic or hydrophobic interaction between heme and a polymer in an aqueous solution cannot be neglected.

23

a) R = CH$_3$
b) R = CHO

The dispersion effect of water-soluble polymers on a heme aggregate is mentioned below. However, these water-soluble polymers have no coordinating ability and are not polymer ligands. It is well known that heme and its complexes easily aggregate in aqueous solution by stacking interaction, and that stable monomeric dispersed heme complexes are formed under restricted conditions such as extremely low concentration and low ionic strength[21, 22].

(heme complex) ⇌ (heme complex)$_2$ ⇌ (heme complex)$_n$ Scheme 9

That aggregation is due to the hydrophobic interaction between porphyrin planes is supported by two facts: (1) the aggregate is dissociated by the addition of alcohol, and (2) the aggregate becomes tighter with increasing temperature and ionic strength. When we added a water-soluble polymer, such as poly(ethyleneoxide)(PEO), poly-(vinylalcohol)(PVA), and poly(N-vinylpyrrolidone)(PVD), to an aqueous solution of the aggregated heme, equilibrium (Scheme 9) was established on the left, dispersed side [23]. The heme complexes were completely dispersed to the monomeric state at a high concentration of polymer. It was concluded that the heme complex is dispersed on the polymer by hydrophobic interaction because the aggregate is much reduced with increasing hydrophobicity of the polymer: PVD > PEO > PVA[23] and because we found an increase in fluorescence intensity due to the hydrophobic interaction between porphyrin ring and polymer[24]. The aggregated ferroheme was most effectively dispersed, of course, by the copolymers of N-vinylpyrrolidone with vinylimidazol or vinylpyridine which were water-soluble but hydrophobic polymer ligands[24]. Ferroheme is presumed to be dispersed by the hydrophobic effect of the polymer ligands and/or be tightly coordinated on the polymer-ligand matrix to prevent the aggregation sterically.

We describe some recent work in which we have studied how the magnetic and electronic properties of heme iron vary when it is occluded by a polymer ligand.

The spin state of the Fe(III) of methemoglobin is considered to be at thermal equilibrium between the high-spin state and the low-spin state[25]. This phenomenon is often observed for chelate compounds such as porphyrin having a strong quadridentate ligand. We studied the spin equilibrium of polymer-bound heme complexes[26]. On the basis of studies of the temperature dependence of the magnetic susceptibility, as Fig. 5 indicates, it was found that the high-spin complex and the low-spin com-

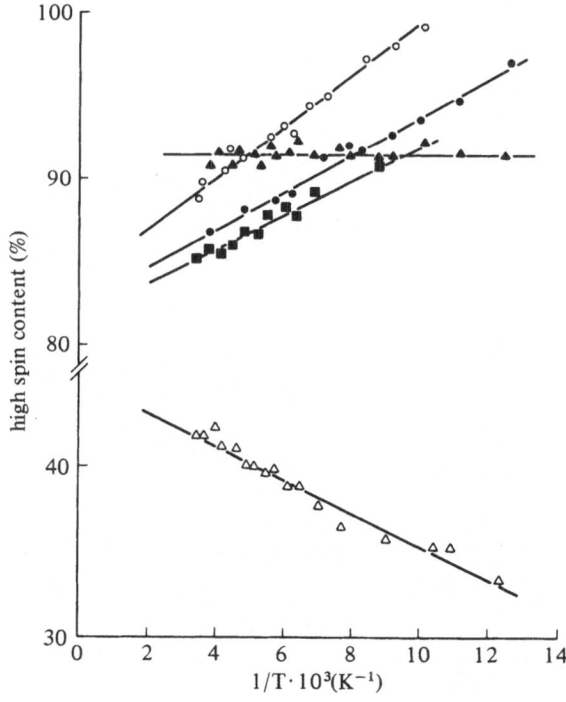

Fig. 5. Content of the high-spin heme complex[28]
●: pyridine ferriheme, ○: PVP-ferriheme, △: partially quaternized PVP-ferriheme, ▲: QPVP-ferroheme, ■: pyridine-ferriheme imbedded in poly(styrene)

plex were in thermal equilibrium at the Fe(III) state. Spin equilibrium was assumed from the result that the values for molar magnetic susceptibility of the ferriheme complexes were intermediate between the calculated values for the high-spin state and the low-spin state[27]. The slopes and intercepts of the straight lines in Fig. 5 are transformed into entropic and enthalpic parameters, respectively. As may be seèn in Fig. 5, these thermodynamic parameters are affected by the polymer-ligand species, although the coordination structure is a five-coordinate one for all complexes. On the other hand, the spin state is not changed when the pyridine-ferriheme complex is imbedded in non-coordinative poly(styrene), as also shown in Fig. 5. These results indicate that the primary structure of the polymer ligand influences the magnetic property, i.e. the electronic configuration of heme iron.

We recently studied the ESR spectra of the mixed heme complex 24, where the nitrogen monoxide radical is inserted into the sixth coordinate site of heme-iron as a probe[28], in order to examine the electronic structure of the heme complex[29]. Figure 6(a) shows the ESR spectrum of the NO radical which coordinates to the imidazole-ferriheme complex. The hyperfine structure (hfs) due to the N of the

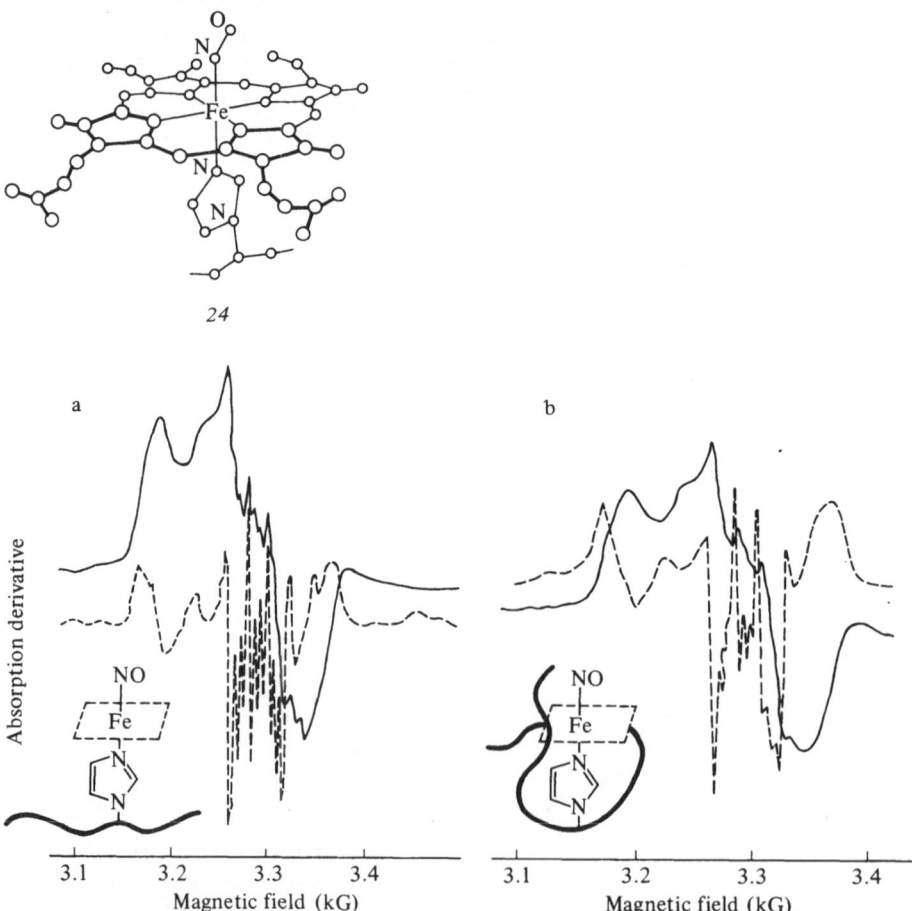

Fig. 6. Hyperfine structure of poly(N-vinylimidazole)-ferriheme-NO complex: in methanol solvent (a) and in dimethylformamide solvent (b)[29]

fifth coordinated imidazole is clearly split into 9 lines; this indicates strong electronic interaction between NO and the imidazole ligand, *i.e.* delocalization of the electron along the axis of the heme. Similar ESR spectra were observed for the poly(N-vinylimidazole)(PNVI)-ferriheme-NO complex in methanol, and it was concluded that PNVI coordinated strongly to heme iron[29]. However, as shown in Fig. 6(b), the hfs lines become broad in the spectrum of PNVI-ferriheme-NO in dimethylformamide(DMF) solvent. Even when the solvent was varied, such broadening of the hfs lines was never observed for the monomeric imidazole-heme complex; however, broadening was seen for the 2-phenylimidazole-heme complex. This was interpreted as meaning that the steric hindrance of the 2-phenyl substituent hindered electronic interaction between the N of the imidazole and the NO radical at the *trans* position. Therefore, the broad spectrum of the PNVI complex in DMF indicated that the interaction between the PNVI ligand and NO, *i.e.* the coordinate bond between PNVI and heme, was weakened by the steric hindrance of the PNVI chain, which was densely contracted in DMF. DMF is a poor solvent for PNVI, whereas methanol is a good solvent in which the strong coordinate bond of PNVI can be observed. This result confirms that the conformation of the polymer-ligand chain is much influenced by the strength of the coordinate bond.

The above phenomenon was also studied for the PVP-heme complex with the NO radical inserted[30]. The hfs due to electronic interaction between the NO radical and the N of PVP was clearly observed for the PVP complex in methanol, which is a good solvent of PVP. This result indicates a strong coordinate bond between PVP and heme iron. Conversely, in DMF the coordinate bond is weakened by the steric hindrance of the contracted PVP chain. The ESR spectrum was also measured for the heme-NO complex of the copolymer of styrene and 4-vinylpyridine in toluene solvent[29]. The electronic interaction between NO and the N of the pyridine unit of the copolymer increases with the styrene content of the copolymer. The strength of the coordinate bond of the polymer ligand is also influenced by the copolymer composition of the polymer ligand. These results can only be interpreted by invoking the conformational effect of the polymer ligand, and this is determined by the solvent species and the primary structure of the polymer ligand.

C. Polymerization of Coordinated Monomer

If a monomeric metal complex containing a vinyl group is polymerized without side reactions, a polymer-metal complex having uniform structure can be obtained.

Both typical and exceptional examples of the polymerization of a vinyl monomer containing a transition-metal ion are provided by the radical polymerization of vinylferrocene[31]. Vinylferrocene and its derivatives are polymerized by a radical or a cationic initiator to form a polymer of high molecular weight. The high polymerizability is based on the property that the ferrocene compounds are extraordinarily stable against chemical reactions.

It is well known that transition-metal salts and metal complexes, unlike non-Werner-type ferrocene compounds, act as inhibitors in the polymerization of vinyl monomers. For example, the radical polymerization of vinylpyridine is strongly inhibited in the presence of Cu(II) or Fe(III)[32]. However, vinylpyridine with Cu(I) or Co(III), *i.e.* its vinylpyridine complex, is polymerized to form poly(vinylpyridine) containing the metal ion[33]. Cu(I) and Co(III) ions have a completely filled d orbital and are relatively inert to redox reactions with organic radicals and ions. Hence, the metal ion species is selected to avoid redox reactions and to polymerize the vinyl monomer.

We have reported the radical polymerization of Cu complexes with the Schiff-base ligand containing a vinyl group 25[34]. The polymer was obtained when the

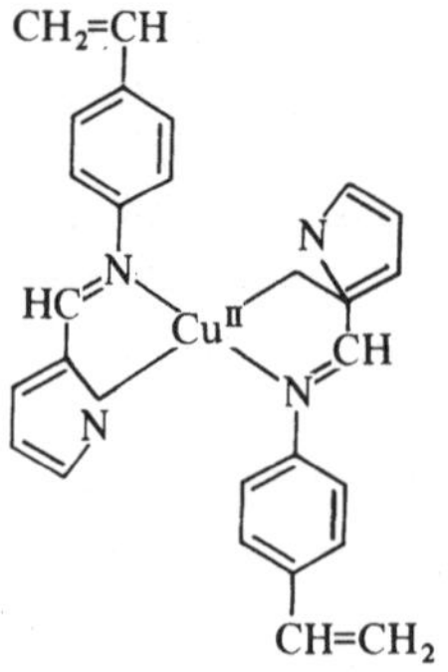

CH₂=CH

25

concentration of the primary radical produced from the initiator was higher than that of the Cu(II) ion of the monomer 25. The polymerization mechanism was studied by spectroscopic measurement of the Cu complex and from the redox of the complex monomer with the DPPH radical. The primary radical first attacks the vinyl group of the complex monomer, then the radical is transferred from the vinyl group to the central Cu(II) ion through the π-conjugated ligand. Only after the central Cu(II) ion has been reduced does the propagation reaction of the complex monomer occur. It is necessary for the polymerization of the complex monomer that the central metal ion and a vinyl group do not conjugate.

Polymerization of the complex monomer is also interesting as a method for controlling the reactivity of the vinyl monomer and the steric configuration of the resulting polymer. Osada[35, 36] has reported the radical polymerization of methacrylate(MA) monomers coordinated to Co(III) complexes: $[Co(NH_3)(MA)](ClO_4)_2$ 26 and *cis*-$[Co(NH_3)_4(MA)_2]ClO_4$ 27. It was found that the complex monomers 26

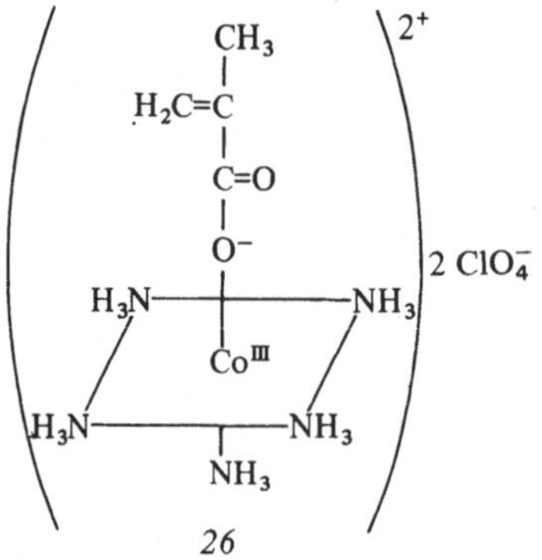

26

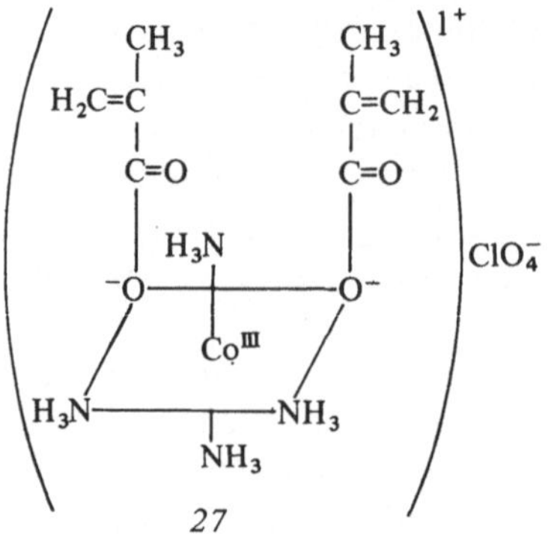

27

and *27* were incorporated into the polymers in an aqueous solution at pH 7. $Q = 1.4$ and $e = 0.46$ were determined from the results of the copolymerization of the complex monomer *26* with methacrylic acid. The reactivity of the MA anion ($Q = 0.9$, $e = -1.0$) was affected by coordination to the Co(III) complex. In other words coordination, decreased the electron density of the vinyl group.

The microtacticity of the polymers obtained by the radical polymerization of the complex monomer *26* with methacrylic acid was studied[36] because a complex monomer of this type would be expected to exert a significant influence on the tacticity of the polymers through its extraordinarily bulky side group and its positive charge. It was reported that the introduction of *26* into the macromolecules at pH 7 resulted in a significant increase in isotactic configuration, and that the tacticity was slightly influenced by the copolymerization of *27*. The deviation from unity of P_{dd1}/P_{d1} was thought to mean that the penultimate effects played an important role in determining steric configuration.

We recently synthesized the organocobalt compounds containing a vinyl group represented in *28, 29,* and *30.* They were not homopolymerized by the redox

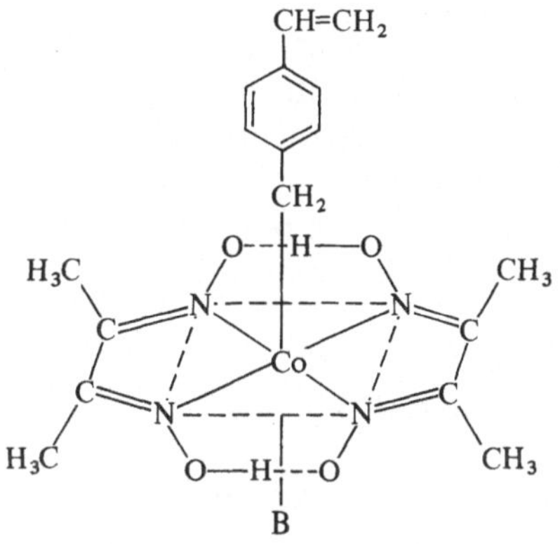

28

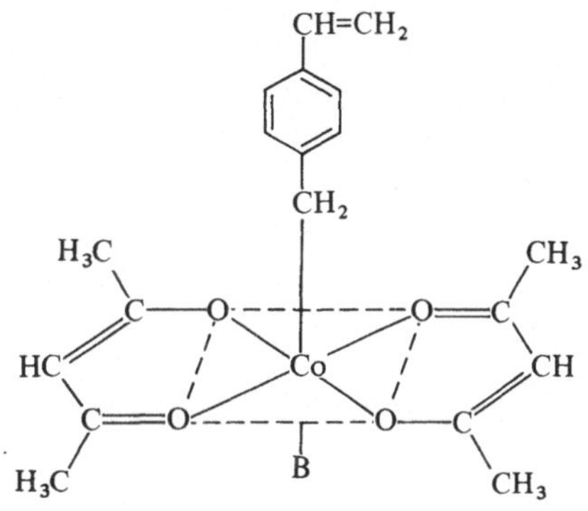

29

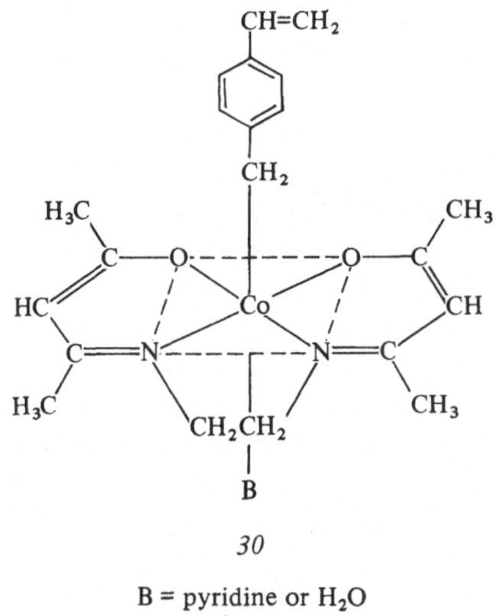

30

B = pyridine or H₂O

reaction between a radical and the organocobaltic σ-bond. However, the complex monomers *29* and *30* were copolymerized with styrene. The reactions and photo-decomposition of these polymers were also studied and compared with those of the monomeric organocobalt complexes and of vitamin B_{12}[37].

III. Formation and Stability of Polymer Chelates

When a "naked" metal ion is added to a solution of a polymer ligand such as poly-(acrylic acid), poly(vinylalcohol), poly(ethyleneimine), or poly(vinylpyridine), a polymer chelate is rapidly formed, as described in Schemes 1 to 3. The equilibrium situation of complex formation between a polymer ligand and a labile metal ion, *e.g.* Cu(II), Ni(II), Zn(II), or Co(II) ions, gives information about the character-istics of the polymer ligand, although the formation constant of the complexation involving a polymer ligand is not easily estimated. Recently, the methods for calcu-lating the formation constants of polymer chelates have been further refined. Gregor[38] has modified Bjerrum's method, taking into account the electrostatic effect of a complexed polymer chain, because a polymer ligand and a complexed polymer are both poly(electrolyte) s. Marinsky[39] has evaluated the distribution of mobile metal ions neighboring a polymer and the difference in potential for a metal ion between the surface of a polymer and a bulk solution. The stability and shape of polymer chelates are characterized in this chapter by using the results obtained for the equilibrium of complexation of a polymer with labile metal ions.

A. Formation and Stability of Intra-Polymer Chelates of Cu(II)

Studies on the formation of polymer chelates have been almost exclusively con-cerned with the Cu(II) ion. The reasons are (i) the stability of Cu(II) chelates in an aqueous solution is sufficient to make them difficult to hydrolyze under exper-imental conditions; (ii) Cu(II) ions can be readily determined by chelate titration, atomic absorption spectrophotometry, etc.; (iii) Cu(II) chelates are active in both visible and ESR spectra.

Immediately after a solution of Cu(II) ions has been mixed with a solution of a polymer ligand, a polymer-Cu chelate is rapidly formed. We have used methods for rapid-reaction analysis to follow the kinetic behavior of chelate formation[40, 41]. Rate constants for the complexation of poly(acrylic acid)(PAA) with Cu(II) have been determined by the temperature-jump method[40]; the magnitudes of the for-mation-rate constant (k_f) and of the dissociation-rate constant (k_d) were 10^8 $1/mol \cdot sec$ and 10^4 $1/sec$, respectively, or nearly equal to those of the corresponding monomeric dicarboxylic acid, glutaric acid. The larger k_f value and smaller k_d value in the PAA-Cu system, as compared with those in the glutaric acid-Cu system, can be explained by electrostatic attraction between PAA and Cu(II) and by the con-formational change in PAA accompanied by the dissociation of Cu from the PAA chain. The stopped-flow spectroscopic method has also been used to monitor the complex formation of poly(ethyleneimine)(PEI) with Cu(II)[41]. For the PEI system

k_f was about 10^8 l/mol sec. The electrostatic repulsion of the protonated PEI with Cu(II) influenced the rate constants. The kinetics of the Cu complex formation was studied under restricted conditions; therefore further discussion is avoided here.

The formation of a Cu(II) chelate can be recognized by the color change of the solution. The visible and ESR spectra of Cu complexes with poly(4-vinylpyridine) (PVP) and with pyridine are shown in Fig. 7(a, b), and (c, d), respectively[42-45]. As can ben seen from Fig. 7(b), the $d-d$ absorption maximum of the Cu complex shifts progressively to shorter wavelengths as the Cu concentration in an aqueous solution decreases (for different ligand ratios[ligand]/[Cu]); this suggests stepwise formation of the complex: $Cu(pyridine)_1 \rightarrow Cu(pyridine)_4$. However, the absorption maximum of the Cu complex of PVP does not shift [Fig. 7(a)], which means that complex formation between Cu and PVP is not a step-by-step mechanism and that the composition of the Cu complex in PVP remains almost constant throughout the course of the reaction. The ESR spectrum of the pyridine-Cu complex [Fig. 7(d)] changes

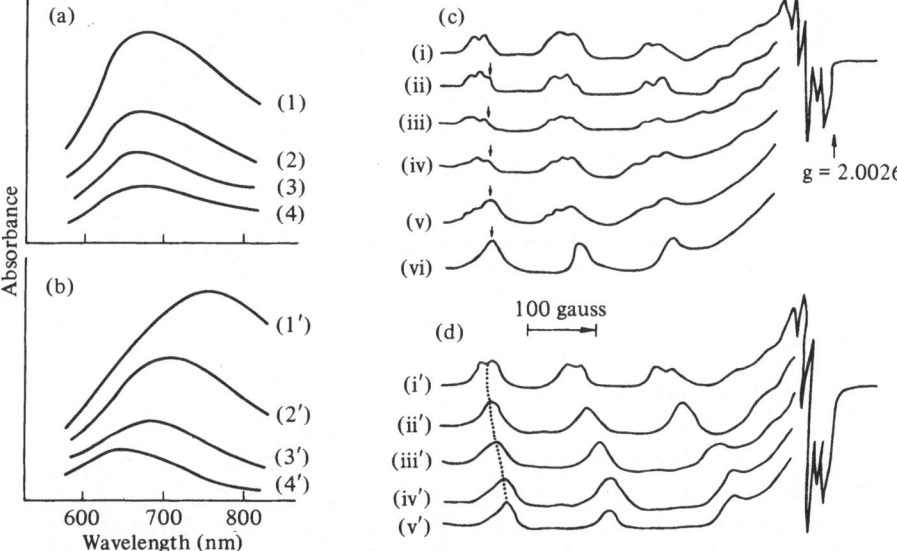

Fig. 7. Visible spectra of the Cu complex of poly(4-vinylpyridine) (a) and of pyridine (b)[42, 43], and ESR spectra of the Cu complex of poly(4-vinylpyridine) (c) and of pyridine (d)[44, 45]
(a) [pyridine unit in PVP]/[Cu] = 1 (1), 2.5 (2), 5 (3), 20 (4),
(b) [pyridine]/[Cu] = 1 (1'), 2.5 (2'), 5 (3'), 100 (4'),
(c) [pyridine unit in PVP]/[Cu] = 0 (i), 4 (ii), 6 (iii), 10 (iv), 20 (v), 100 (vi),
(d) [pyridine]/[Cu] = 0 (i'), 4 (ii'), 6 (iii'), 10 (iv'), 100 (v'). Aqueous solution, pH 5.5, $CH_3COOH-CH_3COONa$ buffer

with the ligand ratio, corresponding to the stepwise formation of the $Cu(pyridine)_4$ complex. Figure 7(c) shows that in the PVP-Cu system, on the contrary, the completely coordinated Cu complex is already formed at the lower ligand ratio, i.e. completely complexed Cu ions coexist with free, uncomplexed Cu ions even at the low ligand ratio. This result also suggests that the PVP-Cu chelate is formed in a one-step reaction.

In order to study the shape of a polymer chelate, viscometric measurements of a homogeneous solution of PVP were carried out[46]. The reduction in viscosity of the PVP solution with increasing concentration of metal ions is shown in Fig. 8(a). At constant PVP concentration, an increase in the amount of metal ions added causes a decrease in viscosity, which reveals that the polymer-ligand chain is markedly contracted due to intra-polymer chelation. An intra-polymer chelate takes a very compact form and the metal ions are crowded within the contracted polymer chain, as illustrated in Fig. 8(b). The contracted shape of an intra-polymer chelate was also reported for PAA[47], poly(vinylamine)[48], and poly(vinylalcohol)[49].

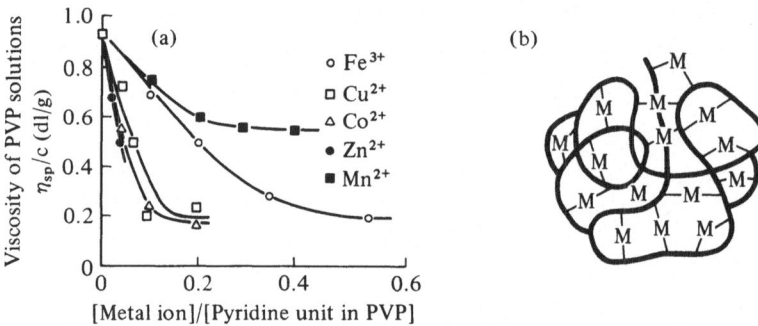

Fig. 8. Viscosity of poly(4-vinylpyridine) solutions in the presence of metal ions (a)[46] and schematic profile of an intra-polymer chelate (b)

Figure 8(a) also shows that the curves for decrease in viscosity, *i.e.* the contracted structure of PVP chelated with a metal ion, are different for the various metal-ion species[46]. It is an important finding that the structure of an intra-polymer chelate depends on the metal-ion species.

The composition of the Cu chelates involving polymer ligands has been determined by spectroscopic, pH, and polarographic titration, and the four-coordinate Cu complexes represented in Table 5 were estimated as the main species[50-56]. The overall formation constants (β_4) of the four-coordinate Cu complexes with polymer ligands are summarized in Table 5 together with those of the monomeric Cu complexes[50-56]. The K values are from one to five orders of magnitude greater in the polymer-chelate systems than in the monomeric Cu-complex systems, which indicates that the Cu chelate forms very readily in a polymer ligand. The polymer-Cu chelate is very stable so that the Cu ions are occluded in a "cage" of contracted polymer chains, and Cu ions are hard to elute from the polymer domain to the bulk solution, although labile Cu ions repeatedly attach and detach on ligand units of the polymer. This phenomenon appears to be general for polymer-chelate systems, and it can be explained by assuming that the concentration of the ligand is locally higher in the polymer domain. This assumption is supported by the stepwise formation constants for a polymer-chelate system, as mentioned below.

Table 5. Overall formation constants of Cu(II) complexes with various polymer ligands

Ligand	$c \cdot 10^2$ [a] (mol/l)	μ [b]	Composition of Cu(II) complex	β_4 [c]	Ref.
Poly(acrylic acid)	0.58	0.1	$Cu(-COO^-)_2(-COOH)_2$	1.8×10^6	50)
Poly(acrylic acid)	3.4	0.1	$Cu(-COO^-)_2(-COOH)_2$	2.2×10^5	50)
Poly(acrylic acid)	1.0	0.2	$Cu(-COO^-)_2(-COOH)_2$	3.0×10^7	51)
Poly(methacrylic acid)	1.0	0.1	$Cu(-COO^-)_2(-COOH)_2$	7.6×10^4	52)
Acetic acid	–	0.2	$Cu(CH_3COO^-)_2(CH\ COOH)_2$	5.8×10^2	51)
Poly(vinylalcohol)	1.0	0.1	$Cu(-O^-)_2(-OH)_2$	8.5×10^{15}	53)
Ethylalcohol	–	–	–	0	
Poly(methacryl acetone)	1.0	0.2	$Cu(-COCH_2COCH_3)_2$	5.5×10^4	54)
Acetyl acetone	–	0.2	$Cu(CH_3COCH_2COCH_3)_2$	10	54)
Poly(vinylpyridine)	1.0	0.1	$Cu(-C_5H_4N)_4$	3.2×10^{10}	55)
QPVP [d] (Bz, 10%)	1.0	0.1	$Cu(-C_5H_4N)_4$	5.0×10^{10}	55)
(Bz, 28%)	1.0	0.1	$Cu(-C_5H_4N)_4$	6.3×10^{10}	55)
(Bz, 43%)	1.0	0.1	$Cu(-C_5H_4N)_4$	3.2×10^{10}	55)
(Et, 13%)	1.0	0.1	$Cu(-C_5H_4N)_4$	2.5×10^{10}	55)
(Et, 25%)	1.0	0.1	$Cu(-C_5H_4N)_4$	3.2×10^9	55)
(Et, 35%)	1.0	0.1	$Cu(-C_5H_4N)_4$	2.0×10^9	55)
Pyridine	–	0.5	$Cu(C_5H_5N)_4$	3.2×10^6	55)
Poly(N-vinylimidazole)	1.0	0.1	$Cu(-C_3H_3N)_4$	5.8×10^{12}	56)
Imidazole	1.0	0.16	$Cu(C_3H_4N)_4$	4.0×10^{12}	56)

[a] Concentration of ligand.
[b] Ionic strength.
[c] Overall formation constant.
[d] Partially quaternized poly(4-vinylpyridine), quaternizing reagent: Bz = benzyl chloride, Et = ethyl bromide.

The stepwise formation constants, K_1 to K_4 in Eq. (1), of Cu complexes with

$$Cu + L \overset{K_1}{\rightleftharpoons} CuL$$

$$CuL + L \overset{K_2}{\rightleftharpoons} CuL_2$$

$$CuL_2 + L \overset{K_3}{\rightleftharpoons} CuL_3 \qquad (1)$$

$$CuL_3 + L \overset{K_4}{\rightleftharpoons} CuL_4$$

$$\beta_4 = \prod_{n=1}^{4} K_n \qquad (2)$$

PVP were calculated by computer, using the modified method of Bjerrum, and the constants obtained are plotted in Fig. 9 together with those of a pyridine-Cu system[55]. In the monomeric pyridine system, the stepwise formation constant decreases with the number of coordinated ligands because the number of vacant sites on the Cu ion, or the ligand-accepting ability of the Cu ion, decreases with succes-

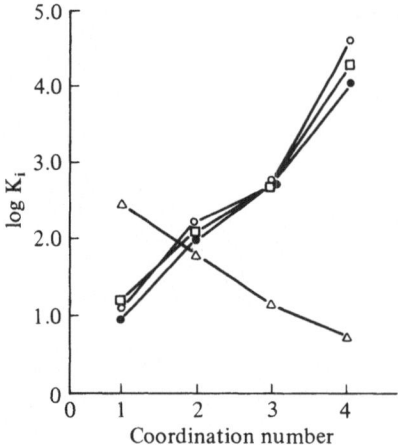

Fig. 9. Stepwise formation constants of poly(4-vinylpyridine)-Cu complexes[55] degree of polymerization of PVP; ○, □: 19, ●: 49, [N]/[Cu]. ○, ●: 5, □: 10, △: stepwise formation constant of pyridine-Cu complex, 25 °C

sive complex formation. In contrast to the monomeric pyridine-Cu system, the stepwise formation constants increase in the PVP system. This is explained by the high local concentration of ligands in the polymer domain, *i.e.* once a Cu ion is attached to one ligand group of the PVP chain, the other ligands coordinate more readily. As a result, complex formation appears to proceed in one step. A similar result has been reported in the poly(vinylamine)-Cu system[57] and the poly(N-vinyl-imidazole)-Cu system[58].

Thermodynamic parameters have been reported for Cu chelate formation with PAA[59], the copolymer of ethylene and maleic acid[60], poly(vinylalcohol)[61], and PVP[42], and it was pointed out that the entropic factor was important in the chelate formation of polymer ligands. The large change in entropy observed for polymer chelate formation was considered to be associated with the effect of a polymer ligand and the high local concentration of ligands.

Some factors that influence the stability of polymer chelates should be mentioned. Hojo *et al.*[61] have reported the effect of the ligand ratio [ligand]/[metal ion] on the formation of the Cu chelate of poly(vinylalcohol)(PVA). Figure 10 shows the relationship between the formation constant of the Cu complex, the viscosity of an aqueous solution of PVA, and the ligand ratio. The viscosity diminishes very sharply at about [PVA]/[Cu] = 32; this corresponds to an increase in the formation constant. A tightly packed conformation of PVA, caused by intra-polymer chelation with Cu, facilitates more and more chelate formation.

An enhancement of stability due to the contracted conformation of a polymer-ligand chain may also be deduced from Table 5. The β_4 value of a PAA-Cu chelate is larger in the system with higher ionic strength. The interpretation is that the local concentration of ligands is increased by the contraction of the PAA chain on the addition of the neutral salt[50].

Copolymers have been used to study the effect of neighboring groups of ligand units in a polymer chain. Hojo *et al.*[53, 62] reported the formation constants of the copolymers of PVA. The K value of partially acetalized PVA *31* with Cu decreased with degree of acetalization and also decreased for the various aldehyde groups

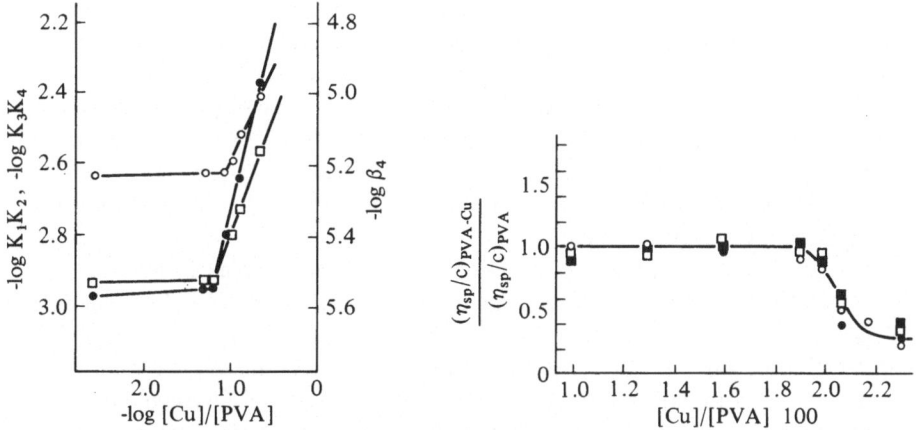

Fig. 10. Relationship between ligand ratio and formation constant (a), and between ligand ratio and viscosity (b) for the poly(vinylalcohol)-Cu system[61]
(a) ○: K_1K_2, □: K_3K_4, ●: β_4, (b) concentration of PVA; ○: 0.21, □: 0.31, ■: 0.50, ●: 0.84, μ 0.1

(R– in *31*) in the order: H– > CH$_3$– > n-C$_3$H$_2$– > iso-C$_4$H$_9$– > n-C$_4$H$_7$–[53].
This was taken to mean that complex formation might be sterically hindered by the introduction of a rigid acetal ring into the main chain. The complex formation of partially acetylated PVA (*32*) was also reported[62]. The formation of four-coordinated Cu chelate was hindered in PVA derivatives with a degree of acetylation above 16%. Furthermore, the chelate formation constants with Cu were reported for partially acetylated PVA with different distribution of acetyl groups. The formation constants were smaller for acetylated PVA with the acetyl groups randomly distributed on the chain. It was considered that the PVA-Cu chelate consisted of vinylalcohol groups adjacent to each other on the PVA chain.

31   *32*

The complex formation of partially alkylated PVP was reported by Kabanov *et al.*[42] and by us[55]. The K value fell by about two orders of magnitude with the degree of alkylation for the PVP derivatives that were alkylated or quaternized by methyl bromide[42]. This reduction of K was due to electrostatic repulsion on the quaternized PVP chain. On the other hand, Table 5 shows no effect of alkylation on the stability of the polymer chelate when benzyl chloride is used as the alkylating reagent, while the formation constant falls slightly for PVP alkylated by ethyl bromide. Since the benzyl group has a high hydrophobicity, the affinity among benzyl groups may be able to cancel the electrostatic strain of the polymer chain, so allowing the polymer chain to bend easily and form the Cu chelate[55].

The formation of polymer chelates is independent of the degree of polymerization of the polymer ligand for PVA ligands (DP 400–1400)[61] and for PVP ligands (DP 19–108)[55]. Kavanov et al.[42] reported that the sedimentation coefficient of the PVP-Cu solution increased from 1 S to 6 S and suggested that polymer ligands associated by Cu complex formation.

The configuration of polymer ligands also influences complex formation. Morawetz et al.[63] reported that isotactic PAA formed a more stable Cu chelate. Nakagawa et al.[64] estimated the formation constants of poly(methacrylic acid) with metal ions. While the stepwise formation constants K_1 and K_2 were hardly affected by the configuration of the polymer ligand, K_3 and K_4 were greater for isotactic than for syndiotactic poly(methacrylic acid). The configuration of the PVA ligands also influenced the third and fourth steps of Cu chelate formation[61].

A reaction of protein with the Cu(II) ion, which forms a violet Cu chelate, is well known as a biuret reaction. Studies of this reaction[65] show that Cu chelate formation was hindered for polypeptides which assumed a helical or β-structure.

B. Complexation of Metal Ions on Crosslinked Polymer Ligands

Intra-polymer chelates show the following features, as mentioned in the previous section: (i) the polymer ligand is markedly contracted because of intra-polymer chelation and (ii) the formation constants of a polymer chelate are much larger than those of a monomeric complex. It is also an important feature that the shape and stability of the polymer chelate depend on the metal ion species.

Thus, when a crosslinked polymer is used as a polymer ligand, the crosslinked polymer ligand can be expected to form stable complexes with a specific metal ion because the polymer-ligand chain has been previously contracted and immobilized. The present section describes studies of the adsorption behavior of metal ions on crosslinked polymer ligands.

Gregor et al.[66] reported the Cu-complex formation of poly(methacrylic acid) (PMA) resins crosslinked with 1% or 9% divinylbenzene. The formation constants of the Cu complexes with the resins were smaller than that of noncrosslinked PMA. The stepwise formation constants decreased from K_1 to K_4 in the resin system, which was the opposite of the noncrosslinked PMA system. The rigidity of the polymer-ligand chain was considered to hinder chelate formation. Formation constants of the PMA resin were also reported by Gustafson et al.[67]. The formation constant of the noncrosslinked PMA decreased for various metal ions: Cu > Zn > Ni > Co; however, for the PMA resin the order was Cu > Ni > Zn. The difference in the order was explained by the tetrahedral structure of Zn.

Poly(ethyleneimine)(PEI) bridged by alkylene dihalide has been used as a chelating resin for Cu and Co[68]. Dingman et al.[69] studied the adsorption of metal ions on PEI resins crosslinked with toluene diisocyanate. The amount of metal ions adsorbed decreases with the degree of crosslinking. These crosslinked PEI resins

bind metal ions with high efficiency, and never before has the attempt been made
to produce a resin with a selective binding ability for a metal ion.

We used partially crosslinked poly(4-vinylpyridine)(DBQP) as the polymer
ligand. PVP was crosslinked by alkylation of the pyridine groups in PVP with 1,4-
dibromobutane to yield insoluble DBQP resins. The free and unquaternized pyridine
groups in DBQP could be used to form the polymer chelate, as represented in *33*.

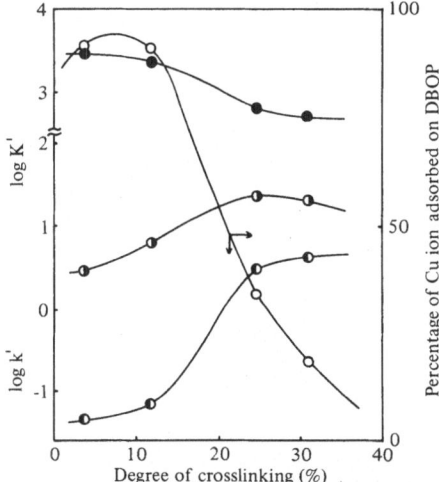

33

Fig. 11. Effect of crosslinking in binding
ability of DBQP resins for Cu(II) Ions[43]
●: percentage of Cu ions adsorbed on
DBQP resins (%),
○: stability constant of DBQP-Cu complexes
(determined by Langmir's eq.),
◑: apparent formation rate constant of
DBQP-Cu (k'_f),
◐: apparent dissociation rate constant of
DBQP-Cu (k'_d), DBQP = partially crosslinked poly(4-vinylpyridine) with
1,4-dibromobutane

These DBQP resins readily adsorbed metal ions[46]. The stability constant of the
the DBQP-Cu chelate was about one order of magnitude greater than that of the
noncrosslinked PVP system, which indicated that the DBQP resin took up Cu ions
from solution with high efficiency. This result was ascribed to the locally higher
concentration of ligands in the polymer domain, especially in the domain of the
crosslinked and strongly contracted resin system. The effect of crosslinking of
DBQP on Cu ion adsorption may be seen in Fig. 11[46]. The amount of Cu adsorbed,
i.e. the binding ability for Cu ions, decreases with degree of crosslinking (C%). The
stability constant K' and the rate constants k'_f and k'_d are plotted against C% in

Fig. 11, where k_f' is the apparent rate constant of Cu complex formation [Eq. (3)], and k_d' is the apparent dissociation rate constant.

$$DBQO + Cu \xrightarrow{k_f'} DBQO\text{-}Cu \tag{3}$$

$$DBQP\text{-}Cu + edta \xrightarrow{k_d'} DBQP + Cu\text{-}edta \tag{4}$$

Equation (4) shows the rate constant for the reaction in which the chelating agent, edta, takes Cu ions from the stable resin complex. K' is determined by the adsorption behavior based on Langmuir's type. The K' value decreases rapidly with increasing number of crosslinking points. Corresponding to the decrease in the stability constant, the k_d' value increases with the crosslinking. That is, the Cu ion is more easily dissociated in the highly crosslinked DBQP because of the instability of the Cu complex.

We observed that the d—d absorption maxima of the DBQP-Cu chelates were red-shifted from 660 nm toward 700 nm with C%, which could be caused by the relatively weak ligand field introduced by the crosslinking[46]. The ESR spectra of DBQP-Cu are shown in Fig. 12[46]. The parameters, $g_{\parallel} = 2.3$, $g_{\perp} = 2.1$ and

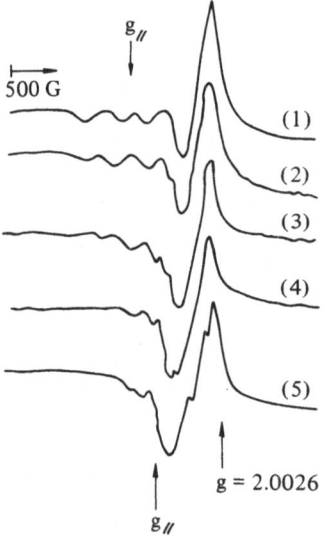

Fig. 12. ESR spectra of Cu complexed DBQP[46]
(1) partially quaternized PVP-Cu, (2) DBQP(C%4)-Cu,
(3) DBQP(C%12)-Cu, (4) DBQP(C%25)-Cu,
(5) DBQP(C%31)-Cu, C% = degree of crosslinking,
aqueous solution at pH 5.5

$A_{\parallel} = 150 \pm 10$ gauss, for the DBQP-Cu chelates with the lower C% are in agreement with those of the monomeric Cu(pyridine)$_4$ complex whose structure is square-planar. In the highly crosslinked chelate, however, the signal due to the parallel orientation is shifted to the higher magnetic field where the perpendicular signal is observed. This result is due to the fact that the anisotropy arising from the square-planar structure of the Cu complex has disappeared in the highly crosslinked resins.

The square-planar structure is probably distorted toward a tetrahedral structure. In the highly crosslinked polymer ligands, the ligand chain is thought to be unable to form a stable square-planar Cu complex because the steric hindrance of the tetramethylene bridges is strong and because the conformation of the ligand chain cannot easily be mobilized due to the number of crosslinking points.

The adsorption of various metal ions on DBQP is shown in Fig. 13[46]. For Cu and Fe ions an adsorption maximum is observed at about 10% crosslinking. The

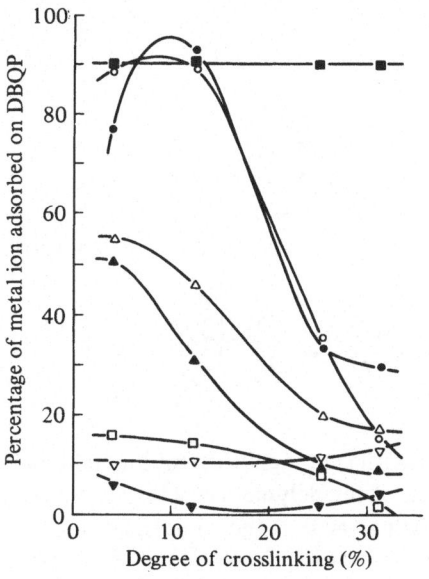

Fig. 13. Metal ion adsorption on DBQP resins[46]
$\bigcirc$: Cu^{2+}, $\bullet$: Fe^{3+}, $\triangle$: Co^{2+}, $\triangledown$: Ni^{2+}, $\square$: Zn^{2+}, $\blacksquare$: Hg^{2+}, $\blacktriangledown$: Cd^{2+}, $\blacktriangle$: Cr^{2+}, concentration of unquaternized pyridine groups in DBQP is kept constant at 0.01 mol/l. aqueous solution, pH 5.5

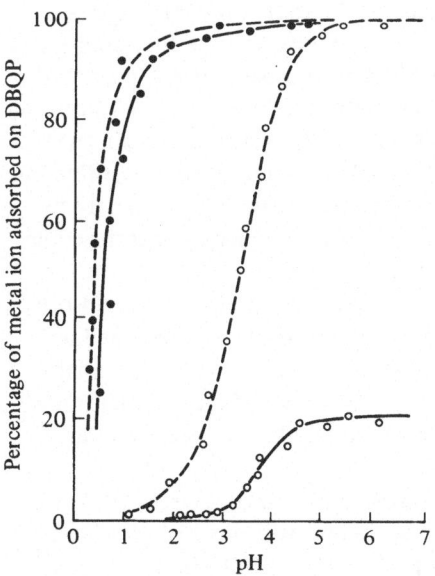

Fig. 14. Adsorption on DBQP resins from aqueous solution containing equivalent amounts of Cu and Hg[46]
$\bigcirc$: Cu ion adsorption, $\bullet$: Hg ion adsorption, $-$: DBQP resin (C%31), $\cdots$: DBQP resin (C%12), C% = degree of crosslinking in DBQP resins

adsorption of Ni, Cr, or Zn ions decreases with increasing C%. Hg ion adsorption, on the other hand, is independent of C%. From these results one can conclude that DBQP resins exhibiting a crosslinking degree of 30% selectively adsorb Hg. This conclusion is confirmed by Fig. 14[46]. The 31% crosslinked DBQP selectively binds Hg from a solution containing equivalent amounts of Cu and Hg, although both ions, Cu and Hg, are adsorbed on the DBQP resin with 12% crosslinking. The effect of crosslinking differs among the metal ion species, and it is possible by controlling the crosslinking reaction to produce resins that selectively adsorb a given metal ion.

We recently proposed a new procedure for synthesizing a chelate resin, as illustrated in the following Scheme[70]:

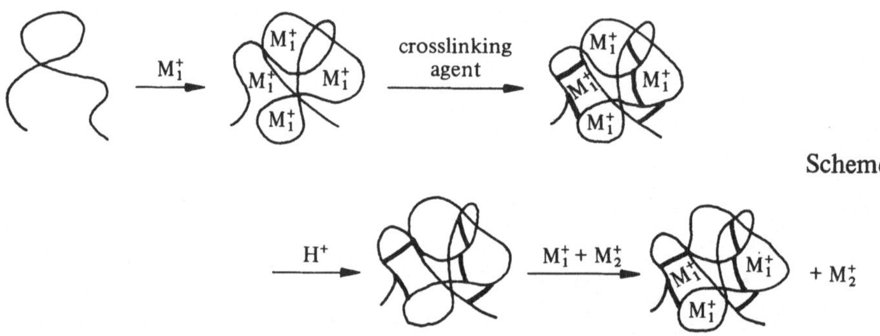

Scheme 10

First a complex is prepared between a polymer ligand, *e.g.* PVP, and a metal ion(M_1) and a crosslinking agent is added to the solution of the polymer complex. The metal ion M_1 is then removed by treating the resin with an acid. If the conformation of the polymer-ligand chains in this resin remains the best for the metal ion M_1, used as the template, then the resin should preferentially form complexes with the metal ion M_1 when dipped into a solution containing various metal ions.

We selected four metal ions as template ions[70]: the four-coordinate square-planar Cu ion, the six-coordinate octahedral Fe and Co ions, and the four-coordinate tetrahedral Zn ion. The solutions of complexed PVP were green, brown, pink, and white, respectively. The characteristics of the template reactions involving transition-metal ions are as follows: (i) the coordination bond energy is strong enough; (ii) the template metal ion is readily removed by treating with an acid; (iii) the coordination number and stereostructure of the metal complex is specific for a given metal-ion species; (iv) the template reaction is recognized without difficulty by spectroscopic and magnetic measurements Table 6 shows the adsorption of metal ions on the DBQP resins synthesized by the template reaction: the resins preferentially adsorb the metal ion used as template. For example, the amount of Cu ions adsorbed on the Cu-template resin is higher than that on the other resins. It is significant that the resins differ in their binding ability for metal ions in spite of the fact that they have the same chemical composition (the degree of crosslinking of each resin in Table 6 was about 40%).

The stability constant of the Cu complex was found[71] to be largest for the Cu template resin ($K' = 3860$ l/mol), which adsorbed Cu ions with the highest efficiency, the stability constants for the other resins being $K' = 100-1400$ l/mol. The changes

Table 6. Adsorption of metal ions on template resins[46, 70]

Template ion	Percentage of adsorbed metal ions			
	Cu^{2+}	Fe^{3+}	Co^{2+}	Zn^{2+}
none	15		7	6
Cu^{2+}	52	4	6	8
Fe^{3+}	13	25	7	7
Co^{2+}	16		10	8
Zn^{2+}	0.2		6	14

$CH_3COOH–CH_3COONa$ buffer, pH 5.5, degree of cross-linking of DBQP resin = 38–42%. The concentration of un-quaternized and coordinative pyridine groups in DBQP was kept constant for each resin.

in enthalpy and entropy for Cu complex formation were $\Delta H = -1.2$ kcal/mol, $\Delta S = 13$ e.u. for the Cu-template resin, and $\Delta H = -0.8$, $\Delta S = 9{,}8$ ($K' = 540$) for the resin synthesized without any template ion. The larger change in entropy observed in the complexation of the Cu-template resin indicated that the Cu-template resin selectively adsorbed Cu ions by entropic effect. Furthermore, the absorption spectrum of the Cu complex of the Cu template resin was located at a wavelength 10–20 nm shorter than those of the other resins[70] and the ESR parameters of the Cu complex of the Cu-template resin were similar to those of the non-distorted planar Cu complex[71]. From these results, it was suggested that the conformation of the polymer-ligand chain in the Cu template resin remained the best one for the Cu ion.

An important use of chelating resins is to remove transition-metal ions from aqueous solutions. For this purpose, various types of multidentate ligands have been introduced into a network polymer[72]. Although such chelating resins take up almost all transition-metal ions in high yield, they do not meet the requirements of an important industrial process, that is, to remove a particular metal ion from aqueous solutions containing several metal ions. Such chelating resins do not have sufficient selectivity for metal ions. For example, the adsorption behavior of imino-diacetic acid resin is shown in Fig. 15(a). The adsorption capacity is plotted as ordinate and the pH of the aqueous solution as abscissa. The complex formation constant with metal ions of this chelating resin is so large that the resin adsorbs as many ions as there are chelating sites, i.e. iminodiacetic acid groups. In other words, the efficiency of metal ion adsorption is near 100%. One is able to remove particular metal ions by controlling the pH of the aqueous solution. For example, when the pH range is set at 1 to 2, Hg and Cu ions are taken up by the resin, but Zn and Cd ions are not adsorbed. However, this method is not practical because of the difficulty of controlling the pH of the bulk aqueous phase and that of drainage in a large quantity. Furthermore, whereas the metal ions on the left side of Fig. 15(a) are preferentially adsorbed at lower pH values, it is impossible to remove only the metal ions situated on the right side.

Such adsorption behavior of a chelating resin is schematically illustrated in Fig. 15(b), where the binding ability of the resin varies with pH. In contrast to Fig. 15(b), if the adsorption capacity is adjusted for each metal ion, i.e. the ordinate

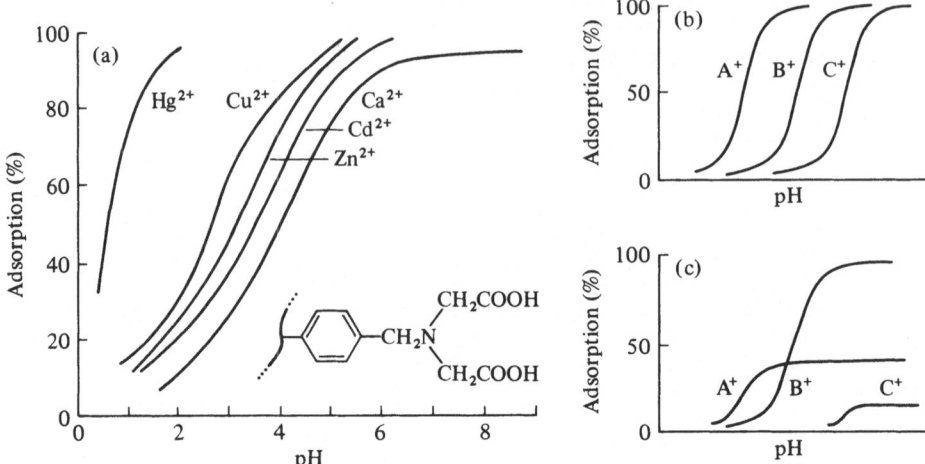

Fig. 15. Adsorption of metal ions on iminodiacetic acid resin (a) and schematic representation of adsorption behavior of polymers (b), (c)

value is changed for the respective metal ion species, selective binding ability for metal ions can be accomplished as shown in Fig. 15(c). From this point of view, it is proposed as a trial method for producing a resin with selective binding ability for a metal ion that a polymer-ligand chain be brought to the best conformation for the coordination sphere of a metal ion by crosslinking a polymer ligand, as described above.

C. Resolution of α-Amino Acids by Chiral Polymer Complexes

Stereoselective complex formation of a labile metal ion with α-amino acids is well known[73]. Since the equilibrium of a labile metal complex is established very rapidly, it seems possible that the stereoselectivity of the metal complex formation could be used to resolve optically active amino acids.

Davankov et al.[74] prepared the polymer containing L-proline units *34* by the

34

reaction of a chloromethylated styrene-divinylbenzene copolymer with the amino group of L-proline. The L-proline resin in the presence of Cu(II) ions displayed a much higher affinity to the D-amino acids than to the L-enantiomers, *i.e.* the L-isomers of alanine, valine, leucine, proline and some other amino acids moved much faster along the column packed with the L-proline-Cu resin than did their D-enantiomer, thus affording their separation. This result was considered to be due to the fact that the mixed Cu complexes of ligands of the N-benzyl-L-proline type were more stable with the D-amino acids than were the corresponding complexes with the L-isomers.

They further observed the stereoselective effect in several other systems containing other amino acids as the polymer ligands[74], and identified certain prerequisites for effective resolution; (i) mixed complexes are more stable and it is better to use Cu(II) or Ni(II), (ii) the sum of the dentation of the polymer ligands and the mobile ligands (amino acids) should be equal to the coordination number of the central metal ion. The difference in free energy of formation of the two diastereomeric complexes with L- and D-proline was estimated to be 0.4 to 0.5 kcal/mol for the L-proline-Cu resin. It was emphasized that the stereoselectivity of this process was sufficient for the complete resolution of racemic α-amino acids[74].

Snyder et al.[75] bound N-carboxymethyl-L-valine to a styrene-divinylbenzene copolymer 35 and used its Cu complex to resolve racemic amino acids. With all

$$
\begin{array}{l}
| \\
CH_2 \\
| \\
CH
\end{array}
\!\!-\!\!\bigcirc\!\!-\!\!CH_2-N\!\!<\!\!\begin{array}{l}
CH(CH_3)_2 \\
| \\
CHCOOH \\
\\
CH_2COOH
\end{array}
$$

35

amino acids, the D-isomer eluted first, which was consistent with the equilibrium studies showing that the Cu-complex formation of N-carboxymethyl-L-valine with L-amino acid was favored over that of the D-enantiomer. The degree of resolution increased with increasing bulkiness of the side chain on the amino acid. These results suggested that there was steric repulsion between the side chain of the amino acid and the isopropyl side chain of the L-valine polymer.

In the above-mentioned systems the poly(styrene) matrices act only as carriers of optically active ligands, although the resolution is made possible by the insolubility of the polymer ligands. In order to study the steric effect of the main chain of a polymer on the stereoselectivity of Cu complex formation, we synthesized the polymer containing, N,N-dibenzylleucine, as shown in Fig. 16[76]. Resolutions of amino acids were attempted by using the column with the Cu complex of the polymer, and L-isomer eluted ahead of D-isomer for all amino acids. This stereoselectivity was, however, opposite to the monomeric analog of the Cu complex where the complex of L-isomer was more stable. As illustrated in Fig. 16, it was considered that one side of the coordination plane was crowded with the isobutyl group, the benzyl group of the polymer, and the main chain of the polymer, so that when, for instance, the isopropyl group of valine was on the same side of the plane, steric repulsion among the side chains and the polymer ligand was large and the complex was more unstable. Therefore, the Cu complex of D-valine, which held the isopropyl group on the other side of the crowded space, was more stable than the complex of L-enantiomer.

The stereoselectivity observed above must be of thermodynamic origin. On the other hand, stereoselectivity of kinetic origin, which is well known for the kinetically

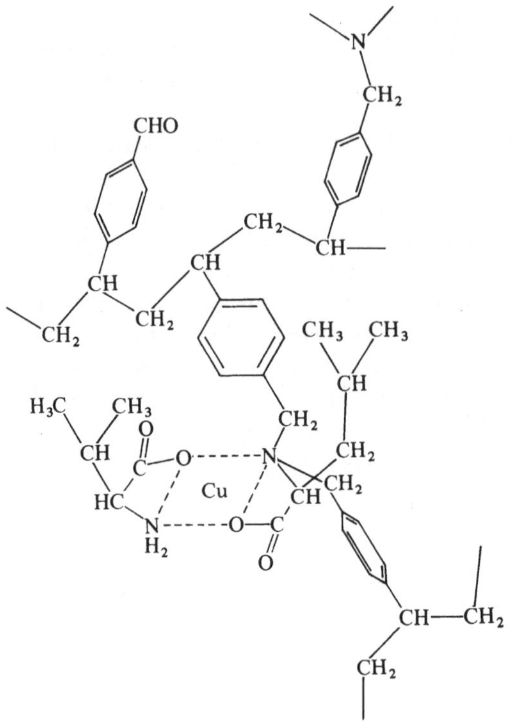

Fig. 16. Crowded structure of the Cu complex of N,N'-dibenzylleucine polymer and valine[76]

inert Co(III) and Cr(III) complexes[77], has not yet been reported for a polymeric system.

IV. Reactivity of Polymer-Co(III) Complexes

A Co(III) complex is inert in ligand-substitution reactions, and its uniform structure is thus maintained even in an aqueous solution. The reaction mechanism of a Co(III) complex in solution is well known, so that a pendant-type polymer-Co(III) complex, e.g. 17, 19, is one of the most suitable compounds for a quantitative study of the effects of a polymer ligand on the reactivity of a metal complex. The reactivities of the polymer-Co(III) complexes are discussed here kinetically and compared with those of the monomeric Co(III) complexes in the following reactions: electron-transfer reactions between the polymer complexes and Fe(II) [Eqs. (5) and (6)], and the ligand-substitution reaction of the polymer-Co(III) complex with hydroxy ions or water [Eqs. (7) and (8)]. One of the electron-transfer reactions proceeds via

$$[Co(III)(en)_2L\,Cl]^{2+} + Fe(II) \rightarrow [L(en)_2Co(III)-Cl-Fe(II)]^{\ddagger}$$
$$\rightarrow Co(II) + Fe(III) \tag{5}$$

$$[Co(III)(en)_2L\,Cl]^{2+} + [Fe(II)(phen)_3]^{2+} \rightarrow Co(II) + Fe(III) \tag{6}$$

$$[Co(III)(en)_2 L Cl]^{2+} + OH^- \qquad\qquad [Co(III)(en)_2 L(OH)]^{2+} + Cl^- \qquad (7)$$

$$[Co(III)(en)_2 L Cl]^{2+} + OH_2 \qquad\qquad [Co(III)(en)_2 L(OH_2)]^{3+} + Cl^- \qquad (8)$$

L = polymer-ligand, e.g. PVP = poly(4-vinylpyridine), PVI = poly(N-vinylimidazole); en = ethylenediamine, phen = phenantroline

an inner-sphere mechanism [Eq. (5)] where the Cl ion acts as the bridging ligand and the bridged inner-sphere complex is formed in the intermediate state, while the other has an outer-sphere mechanism [Eq. (6)] in which electron transfer occurs when oxidant and reductant move into sufficient proximity. These reactions are simple but important elementary reactions of a metal complex. As expected, it was confirmed that the reaction mechanisms were the same in both the polymer complexes and the corresponding monomeric complexes[3].

A. Electrostatic Effect of Polymer-Co(III) Complexes

A polymer-Co(III) complex behaves as a poly(electrolyte) in an aqueous solution, and the reaction is expected to be sensitive to the charge of reactant species. Rate constants (k_{PVP}) for the reactions of the poly(4-vinylpyridine) (PVP)-Co(III) complexes, $[Co(III)(en)_2(PVP)Cl]Cl_2$ [Eqs. (5), (7), and (8)] and those (k_{Py}) of the monomeric pyridine complex, $[Co(III)(en)_2(Py)Cl]Cl_2$, were determined[78-80]. The reactivity ratios of the PVP complex and the pyridine complex (k_{PVP}/k_{Py}) are summarized in Table 7.

Table 7 shows that the reactivity ratio depends essentially on the charge of the reactants. Not only are the rates for the reaction with the negatively charged Fe(II)-chelate^{n-} very high, as the reaction occurs between a cation (or a polycation) and an anion, but also the reaction of the PVP complex is markedly accelerated compared to that of the pyridine complex. The reactivity ratios (k_{PVP}/k_{Py}) for the Fe(II)-chelates^{2-} are about 20. Moreover, the ratio is affected by the magnitude of the negative charge when different Fe(II)-chelates, produced by changing the chelating agents, are used as reductants. On the other hand, the reactivity ratio for reduction with a ferrous aquo ion Fe^{2+} with two positive charges falls to about one-third. The reactions with noncharged $FeSO_4$ or H_2O were hardly affected by electrostatic factors.

This phenomenon may be explained as follows[79]: the PVP-Co(III) complex is a polycation and forms a "domain" with a greater charge density in the aqueous solution. The electrostatic effect due to this polycation domain is predominant in the reactions of the Co(III) complex with charged species. The polycation domain tends to concentrate the oppositely charged Fe(II)-chelate^{n-} species due to electrostatic attraction, and an increase in the local concentration of Fe(II)-chelate enhances the reaction rate [Fig. 17(a)]. Conversely, the polymer complex inhibits the positively charged Fe^{2+} species from entering the domain of the polymer complex due to electrostatic repulsion. This explanation is further supported by the influence of ionic strength on reactivity, as shown also in the far-right column of

Table 7. Reactivity of poly(4-vinylpyridine)-Co(III) complex[79-81]

Reactant	Charge of reactant	Reaction	k_{PVP} $(1 \cdot mol^{-1} \cdot sec^{-1})$	k_{Py} $(1 \cdot mol^{-1} \cdot sec^{-1})$	(k_{PVP}/k_{Py})	(k_{PVP}/k_{Py})[f]	Ref.
$Fe(OH_2)_6^{2+}$	+2	Eq. (5)[a]	8.82×10^{-5}	37.1×10^{-5}	0.24	0.35	79)
H_2O	0	Eq. (8)[b]	8.1×10^{-6}	10×10^{-6}	0.81	–	81)
$FeSO_4$	0	Eq. (5)[c]	2.42×10^{-3}	4.10×10^{-3}	0.59	–	78)
OH^-	–1	Eq. (3)[d]	7.7×10^2	4.5×10^2	1.7	–	81)
$Fe\text{-}edtaoh^-$	–1	Eq. (5)[e]	8.8×10^4	2.4×10^4	3.7	–	80)
$Fe\text{-}edta^{2-}$	–2	Eq. (5)[e]	28×10^4	1.3×10^4	21	3.2	79)
$Fe\text{-}cydta^{2-}$	–2	Eq. (5)[e]	19×10^4	1.2×10^4	16	–	80)
$Fe\text{-}hdtpa^{2-}$	–2	Eq. (5)[e]	28×10^4	1.3×10^4	22	3.5	80)

Chelating agents: edta = ethylenediaminetetraacetic acid, edtaoh = hydroxyethylethylenediaminetriacetic acid, cydta = cyclohexanediaminetetraacetic acid, hdtpa = protonated diethylenetriaminepentaacetic acid, $[Co(III)(en)_2(PVP)Cl]Cl: x$ ethyl 0.45, DP of PVP 98.

[a] μ 0.14, $HClO_4$ 0.1 mol $\cdot$ 1^{-1}, 25°. Under these conditions, $[Fe(OH_2)_6]$ $2 ClO_4$ is almost dissociated to $Fe(OH_2)_6^{2+}$.

[b] $HClO_4$ 0.1 mol $\cdot$ 1^{-1}, 50 °C.

[c] $\Sigma [SO_4^{2-}]$ 0.25 mol $\cdot$ 1^{-1}, H_2SO_4 0.5 mol $\cdot$ 1^{-1}, 25°. Under such conditions, the dissolving species as a reductant is assumed to be totally non-charged $FeSO_4$, since the formation constant for the $FeSO_4$ has been estimated to be about 10^{2-3}.

[d] NaOH 0.02 mol $\cdot$ 1^{-1}, μ 0.06, 25°.

[e] μ 0.018, 25°.

[f] Neutral salt $NaClO_4$ added system μ 0.12.

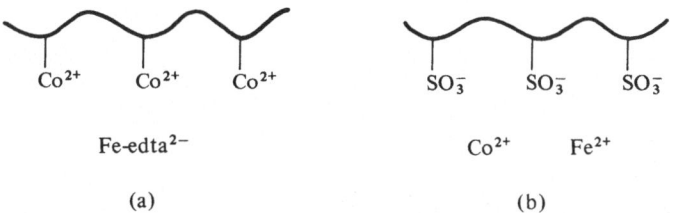

Fe-edta^{2-}

(a)   (b)

Fig. 17. Electrostatic domain of polymer-Co(III) complex (a) and of poly(styrene sulfonate) (b)

Table 7. When the neutral salt sodium perchlorate is added, the electrostatic domain of the PVP complex is shielded and the reactivity ratio falls to unity.

This reaction profile of the polymer complex has some similarities with the phenomenon of the polyelectrolyte-catalyzed reactions. It has been reported that the reactions between two positively charged species in aqueous solution are drastically accelerated in the presence of polyanions[82−84]. For example, the electron-transfer reaction between $[Co(III)(en)_2(Py)Cl]^{2+}$ and $[Fe(II)(OH_2)_6]^{2+}$ is very slow because the reaction occurs between two cations; however, the addition of a small amount of poly(styrenesulfonate) accelerates the reaction by a factor of 10^{3} [84]. This result is also interpreted as indicating that the two positively charged reactants are both concentrated in the polyanion domain, so that they encounter each other more frequently [Fig. 17(b)].

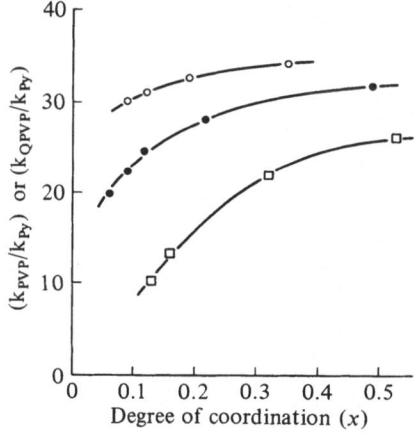

Fig. 18. Effect of degree of coordination (x) and of quaternization (q) on the reactivity of $[Co(en)_2(PVP)Cl]^{2+}$ or $[Co(en)_2(QPVP)Cl]^{2+}$ with Fe(II)edta^{2-} [79, 85]
□: PVP complex, ●: QPVP complex ($q = 0.35$), ○: QPVP complex ($q = 0.55$), PVP = poly(4-vinylpyridine), QPVP = partially quaternized PVP

Figure 18 shows the relationship between the degree of coordination (x in 17), i.e. the number of positive charges on a polymer chain, and the reactivity ratio (k_{PVP}/k_{Py}) for the reaction with Fe(II)-edta$^{2−}$ [79]. The reactivity ratio increases with x value. Hence, the PVP complex with a higher degree of coordination is considered to have a stronger electrostatic domain because it has more positive charges on the PVP chain. The reactivity ratios (k_{QPVP}/k_{Py}) for the partially quaternized PVP(QPVP)-Co(III) complexes 19 are also plotted in Fig. 18. The reactivity ratio

Table 8. Activation parameters for reactions of the Co(III) complex with Fe(II) (25 °C) [79, 85]

Co(III) complex	Fe(II)	Additives (mol·l⁻¹)	k (l·mol⁻¹·sec⁻¹)	$\Delta H^{\ddagger}$ (kcal·mol⁻¹)	$\Delta S^{\ddagger}$ (e.u.)
a) $[Co(III)(en)_2(Py)Cl]^{2+}$	Fe-edta^{2-}	—	1.3×10^4	8.7	−10
b) $[Co(III)(en)_2(PVPy)Cl]^{2+}$	Fe-edta^{2-}	—	28×10^4	7.7	− 7.7
c) $[Co(III)(en)_2(QPVP)Cl]^{2+}$	Fe-edta^{2-}	—	65×10^4	7.0	− 8.3
d) $[Co(III)(en)_2(Py)Cl]^{2+}$	Fe^{2+}	—	3.7×10^{-4}	13	−30
e) $[Co(III)(en)_2(Py)Cl]^{2+}$	Fe^{2+}	PSS 4	55×10^{-4}	16	−16
f) $[Co(III)(en)_2(PVP)Cl]^{2+}$	Fe^{2+}	—	1.1×10^{-4}	16	−23

PVP = poly(4-vinylpyridine): DP 98, QPVP = partially-quaternized PVP: DP 98, degree of quaternization (q) 0.55, PSS = poly(styrenesulfonate): DP 320, Py = pyridine, en = ethylenediamine, edta = ethylenediaminetetraacetic acid; x of Co(III)-PVP 0.65, x of Co(III)-QPVP 0.13. Conditions for (a) to (c): Co(III) 0.01 mmol·l⁻¹, Fe(II) 0.05 mmol·l⁻¹, pH 5.0, μ 0.018; conditions for (d) to (f): Co(III) 4 mmol·l⁻¹, Fe(II) 33.5 mmol·l⁻¹, H⁺ 30.5 mmol·l⁻¹, H⁺ 30.4 mmol·l⁻¹.

is higher in the QPVP-Co(III) complex with the larger q value. For the QPVP complex the electrostatic attraction of $Fe(II)\text{-edta}^{2-}$ is considered to be enhanced by pyridinium cations situated on the QPVP chain[85]. However, the charge density does not much influence reactivity for the polymer complex with a sufficiently high x or q value.

The activation parameters are presented in Table 8[79]. For the reactions be between the Co(III) complex^{2+} and $Fe\text{-edta}^{2-}$, (a) to (c) in Table 8, the activation enthalpy is smaller and the activation entropy larger than for the reduction by Fe^{2+}, (d) to (f), which is a reaction of two cations. A comparison of the parameters for the polymer complex, (b) or (c), with those for the pyridine complex, (a) shows that the acceleration for the PVP or QPVP complex is based on a decrease in activation enthalpy and an increase in activation entropy. This is the opposite of the polyelectrolyte-catalyzed reaction, in which the acceleration is due to an increase in activation entropy (compare(e) with (d)). In the polyelectrolyte-catalyzed system the acceleration and increase in activation entropy are attributed to the increase in the local concentration of the two reactants, the Co(III)-Py complex^{2+} and Fe^{2+} [84], whereas in the reaction of the polymer complex the large activation entropy and small activation enthalpy are held to be due to the increase in the local concentration of the reactant $Fe(II)\text{-edta}^{2-}$ and the electrostatic attraction between the reactant and the Co(III) complex, which is fixed to the polycation chain.

B. Hydrophobic Interaction

In the reaction between the Co(III) complex and $Fe(II)(phen)_3^{2+}$ [Eq. (6)] we observed higher reactivity of the polymer complexes (Table 9[85]). This reaction was carried out at very high ionic strength so that the electrostatic effect on the reaction was neglible. Therefore, other attractive interactions caused by the essential properties of the polymers must be considered in the reaction with $Fe(II)(phen)_3$. On the spectroscopic measurement, the hyperchromicity observed at the d–d first absorption band of $Fe(II)(phen)_3$ (510 nm) was caused by the hydrophobic interaction between polymer ligands and the bulky phenantroline ring. The magnitude of hyperchromicity decreases in the order: $Fe(II)(phen)_3$ mixed with poly(N-vinyl-

Table 9. Outer-sphere electron-transfer reaction between $[Co(III)(en)_2LCl]$ and $[Fe(II)(phen)_3]$ [86]

L	DP of L	x	$k_p \cdot 10^2$ $(1\ mol^{-1} \cdot sec^{-1})$	k_p/k_m	Hyperchromicity (Abs.O.D. $\cdot$ 10^2)
PVMI	250	0.32	31	23	2.2
PVP	98	0.32	7.8	5.3	1.2
PVP	19	0.39	4.3	2.8	0.5

PVMI = poly(N-vinyl-2-methylimidazole), PVP = poly(4-vinylpyridine), DP = degree of polymerization, x = degree of coordination, k_p = rate constant for the Co(III)-polymer complex, k_m = rate constant for the monomeric Co(III)-N-ethylimidazole or Co(III)-pyridine complex, Co(III) 3.5 mmol $\cdot$ 1^{-1}, Fe(II) 0.09 mmol $\cdot$ 1^{-1}, pH 5.0, μ 5.0, μ 2, 30°.

2-methylimidazole)(PVMI)(DP = degree of polymerization 250)-Co(III) $\gg$
PVP(DP 98)-Co(III) > PVP(DP 19)-Co(III) $\gg$ N-ethylimidazole-Co(III) $\simeq$ pyridine-
Co(III), and this order agrees with that of the reactivity. The hydrophobic interaction between polymer ligands and Fe(II)(phen)$_3$ is considered to account for the
higher reactivity of the polymer complexes.

In general, the polymer ligand surrounding the metal complex constitutes an
electrostatic or hydrophobic domain in aqueous solution. This polymer domain
governs the chemical reactivity of the metal complex. This phenomenon is said to
be an "environmental effect of polymer" on a chemical reaction.

C. Influence of Conformational Change of Ligand Chain

It is thus expected that the conformation of the polymer-ligand chain will influence
the reactivity of a metal complex. The influence of the conformational change of a
poly(N-vinyl-2-methylimidazole)(PVMI 3) ligand has been studied in the electron-
transfer reaction of a Co(III) complex in aqueous-alcoholic solvents[87].

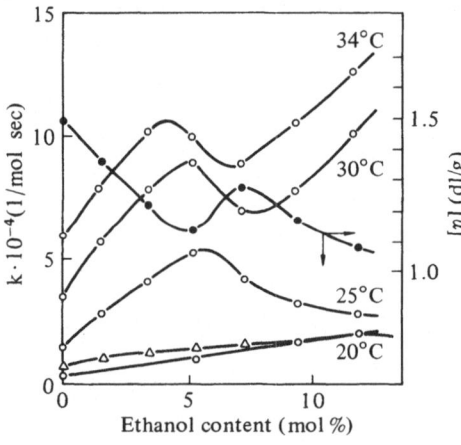

Fig. 19. Relationship between ethanol
content, rate constant (k), and
viscosity of solution in the electron-
transfer reaction for poly(N-vinyl-
2-methylimidazole)-Co(III)~Fe-edta^{2-}[87]
○: rate constant for reaction between
[Co(III)(en)$_2$(PNVI)Cl]Cl$_2$ and
Fe(II)edta, △: rate constant for
reaction between monomeric
[Co(III)(en)$_2$(NEI)Cl]Cl$_2$ and
Fe(II)edta, ●: intrinsic viscosity of
PVMI-Co complex solution, PVNI =
poly(N-vinyl-2-methylimidazole),
NEI = N-ethylimidazole

Figure 19 shows the rate-constant curves for the reaction of [Co(III)(en)$_2$L Cl]$^{2+}$
(L = PVMI or N-ethylimidazole(NEI)) with Fe-edta^{2-} in H$_2$O-alcohol mixed solvents. An increase in the alcohol content brings about the enhancement of the
reactivity of the Co(III) complex; however, maxima and minima appeared in the
reaction between PVMI-Co(III) and Fe-edta^{2-}. The viscosity behavior of the PVMI-
Co(III) complex solution is also shown as a function of ethanol concentration in
Fig. 19. There is good correlation between the reactivity curve and the viscosity
curve, with the maximum (minimum) point of the reactivity precisely coincident
with the minimum (maximum) point of the viscosity. The decrease in viscosity
reflects a contraction in the PVMI chain caused by the suppression of charge dissociation of the Co(III) complexes or of electrostatic repulsion due to the nonpolar
ethanol solvent. At intermediate ethanol concentration (5–7 mol%) the hydro-

phobic interaction between the polymers is cancelled; the polymer chain is temporarily extended and again contracts on further addition of ethanol. At higher reaction temperatures the polymer chain expands even at lower ethanol concentration, so that the maximum point of the rate constant shifts to a lower ethanol concentration. Similarly, the maxima and minima shift to a lower alcohol concentration in the order: isopropylalcohol > ethanol > methanol[87].

The higher reactivity of the PVMI-Co(III) complex is attributed to the electrostatic domain of the polymer complex, as in the above PVP system. When the PVMI chain contracts, the charge density in the polymer domain increases and the reaction rate also increases. On the other hand, when the polymer chain expands, the electrostatic domain is weakened, which produces a fall in reactivity. These results confirm that the conformation of the polymer complex is closely related to the strength of its electrostatic domain and to the reaction rate. The effects of the polymer chain on reactivity are to be understood not only in terms of "static" chemical environment but also as "dynamic" effects which vary with the solution conditions, e.g. pH, ionic strength, solvent composition, temperature, and so on.

V. Reaction of Polymer-Heme Complexes with Molecular Oxygen

Biological oxygen supply depends on the rapidly reversible combination of molecular oxygen with hemoglobin and myoglobin. The oxygen-binding site is the Fe(II)-protoporphyrin IX (ferroheme) complex with the imidazole group of the globin protein, and the ferroheme complex is included within the hydrophobic domain of globin. If ferroheme is isolated from the protein and situated with imidazole in a

$$(Im)Fe^{II}(P) + O_2 \rightleftharpoons (Im)(P)Fe-O_2 \rightarrow (Im)Fe^{III}(P) + O_2^- \qquad (9)$$

P = porphyrin, Im = imidazole

solution, the protein-free ferroheme complex is immediately oxidized to ferriheme upon exposure to air, although the ferroheme complex is quite resistant to oxidation in the strict absence of water and other acids at low temperature.

Cobalt(II)-Schiff-base complexes with a nitrogenous base are also well known to bind molecular oxygen in an organic solvent at room temperature.

$$(B)Co^{II}(S) + O_2 \rightleftharpoons (B)(S)Co-O_2 \xrightleftharpoons{(B)Co^{II}(S)} (B)(S)Co-O_2-Co(S)(B)$$
$$\rightarrow 2(B)Co^{III}(S) + O_2^{2-} \qquad (10)$$

S = Schiff-base, B = nitrogenous base

But most of the Co(II) complexes react with molecular oxygen to yield binuclear dioxo-bridged species of the type $(B)(S)Co-O_2-Co(S)(B)$, and these may be readily oxidized to the corresponding super dioxo-bridged complexes, $Co(III)-O_2^{2-}-Co(III)$.

It is important to form a stable 1:1 dioxygen-metal complex, because oxyhemoglobin and oxymyoglobin are 1:1 dioxygen-iron complexes which can reversibly desorb molecular oxygen. The following conditions are considered to be required for reversible oxygenation (M(II) = lower valence metal ion, M(III) = higher valence metal ion): An oxygenated metal complex reacts with another M(II) and the binuclear dioxo-bridged complex, $M-O_2-M$ is formed. Thus the first condition is that the oxygenated complex must be isolated each other and here the steric facter of the metal complex is important in order to inhibit oxidation via dimerization. An oxygenated complex is in the resonance between the non-bonding structure, $M(II)-O_2$ and the dative structure, $M(III)-O_2^-$. So the second condition is that the oxygenated complex must be surrounded by a hydrophobic environment in order to suppress charge separation and to exclude water molecules from the vicinity of the oxygenated complex.

In hemoglobin and myoglobin the globin protein protects the ferroheme, which is tucked into a hydrophobic crevice of the protein, known as the "heme pocket". A polymer ligand might be expected to protect the dioxygen-metal complex against autoxidation in much the same way as the globin protein does. In this chapter, we describe how polymer-metal complexes react with molecular oxygen and introduce attempts to construct synthetic oxygen carriers.

A. Oxygenation of Polymer-Heme Complexes

Nitrogenous base-ferroheme complexes in aqueous solution are immediately oxidized to the corresponding ferriheme complexes on exposure to oxygen. Irreversible oxidation was also observed for the ferroheme complexes of polymer ligands in aqueous solution[88, 89], and stable oxygen carriers composed of polymer-heme complexes have not recently been synthesized. Thus, we began our studies on the reaction of polymer-metal complexes with molecular oxygen with the oxygenation of Co(II) complexes in organic solvents.

Co(II)-Schiff-base chelates have a greater tendency to form oxygenated complexes at the temperature 0 °C than do metalloporphyrins, possibly because the aromatic Schiff-base ligand more readily donates π-electron density to a central metal ion than does the porphyrin ligand. We found that the polymer complexes of Co(II)bis(salicylaldehyde)ethylenediimine(salen) 36 and of Co(II)bis(acetylacetonato)ethylenediimine(acacen) 37 were reversibly oxygenated at room temperature[90]. For example, Fig. 20 shows the absorption and desorption of molecular oxygen by the Co(II)salen complex with the copolymer of styrene and 4-vinylpyridine (PSP). The PSP-Co(II)salen complex effectively takes up molecular oxygen

36

37

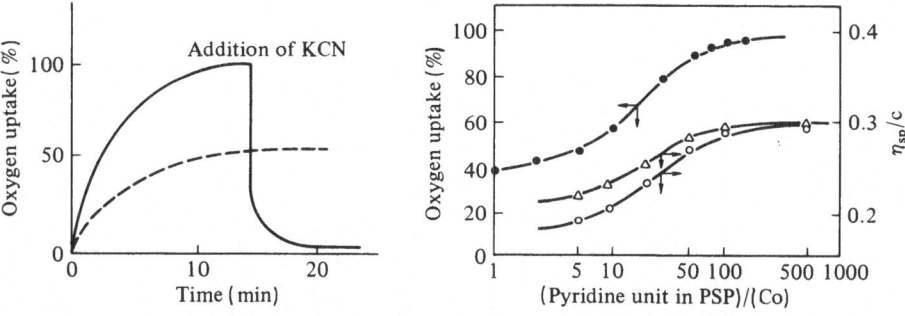

Fig. 20. Absorption and desorption of molecular oxygen by polymer-Co(II) salen complex (a) and effect of added amount of polymer ligand on oxygen absorption (b)[90]
(a) ——: Co(II)salen complex with copolymer of styrene and 4-vinylpyridine(PSP), −−−: Co(II)salen complex with pyridine, in toluene, at 15 °C, ●: percentage of oxygen absorption by Co(II)salen with PSP, △: viscosity of PSP-Co(II)salen solution before oxygenation, ○: viscosity after oxygenation

(the equilibrium constant for the oxygen binding: $K_{O_2} \simeq 4 \times 10^{-3}$ mmHg, 20 °C) and releases it by degassing or on the addition of KCN, which coordinates to Co(II) more strongly than does molecular oxygen. The degree of oxygen absorption is 1.0 for the PSP system, while that for the pyridine-Co(II)-salen complex is 0.5, corresponding to the reversible formation of 1:1 dioxygen-Co complex and of the 1:2 dioxygen-Co complex which is rapidly oxidized, respectively. The effect to the added amount of PSP ligand on oxygen absorption is shown in Fig. 20(b). For the ligand ratio [PSP ligand]/[Co] > 80, the 1:1 dioxygen-Co complex is preferentially formed, although the binuclear dioxygen complex is formed in the region where a small amount of PSP is added. This result is interpreted as follows. The formation constant of PSP-Co(II)salen is so large that Co(III)salen molecules are fixed on the PSP chain at low concentration when the ratio [PSP]/[Co] is large enough. This inhibits formation of the binuclear complex, (-vinylpyridine unit) (salen)Co−O$_2$−Co(salen)(vinyl-pyridine unit-). This interpretation is further supported by the fact that the viscosity of the PSP-Co(salen) solution, see Fig. 20(b), is the same before and after oxygenation in the region [PSP]/[Co] > 80.

Parallel studies were carried out on the Co(II)porphyrin complexes[91]. We chose two porphyrin derivatives soluble in organic solvents: protoporphyrin IX dimethyl-ester (PPME) *38*, and tetraphenylporphyrin (TPP) *39*.

Oxygenation of these polymer-Co(II)porphyrin complexes took place when the solution (*e.g.* toluene) was cooled and exposed to oxygen. The oxygenation was monitored by either ESR or visible spectroscopy. For example, the visible absorption of the oxygenated complex of PSP-Co(II)PPME was observed at 420, 543, and 575 nm in toluene at −90 °C and immediately after bubbling nitrogen gas was changed to the absorption (404 and 555 nm) of the five-coordinated PSP-Co(II)-PPME complex through the isosbestic points[92]. The absorption band could be rapidly changed back to the oxygenated type by exposing the solution to oxygen. The ESR spectra clearly demonstrate the formation of the 1:1 dioxygen-Co complex. While the spectrum of PSP-Co(II)PPME was similar to that of the five-coordinate Co(II)PPME complex under nitrogen atmosphere, a new signal with 8 shf lines

38   *39*

developed in the vicinity of g = 2.0 when the solution was exposed to oxygen at
−90 °C. The ESR parameters of the oxygenated PSP-Co(II)PPME complex were
the same as those of the monomeric pyridine-Co(II)PPME complex with oxygen[93],
and this suggests that the polymer ligand has no major effect on the electronic
structure of the oxygenated complex.

Table 10. Thermodynamic data for reversible oxygen binding to Co(II) PPME complexes and
to myoglobin

Co(II) complexes	Solvent	Temp.	ΔF kcal/mol	ΔH kcal/mol	ΔS e.u.	Ref.
MeIm-CoPPME	Toluene	−45°	1.8	−11.8	−59	94)
BuPy-CoPPME	Toluene	−45°	2.9	−10.0	−57	94)
Py-CoPPME	Toluene	−45°	3.0	−9.2	−53	94)
Py-CoPPME	Toluene	−45°	3.3	−9.0	−54	91)
PSP-CoPPME	Toluene	−45°	3.2	−6.8	−43	91)
Coboglobin[a]	Aq. pH 7	20°	2.5	−13.3	−53	95)
Myoglobin[a]	Aq. pH 7	20°	−0.2	−18.1	−60	96)

PPME = protoporphyrin IX dimethylester, MeIm = 1-methylimidazole, BuPy = butyl-pyridine,
Py = pyridine, PSP = copolymer of styrene and 4-vinylpyridine (vinylpyridine content: 23%,
molecular weight: 1.1 x 10^4).

[a] Sperm whale.

Thermodynamic data for the reversible oxygenation of the Co(II)PPME complexes are given in Table 10[91, 94−96]. The PSP ligand shows an effect on the
thermodynamic parameters for the oxygen binding of Co(II)PPME. One notices
that the polymer ligand makes a significant favorable entropic contribution to
oxygen binding. Binding to the PSP-Co(II)PPME complex is disfavored enthalpically
compared with the monomeric pyridine-Co(II)PPME complex, but this is compensated by a further favorable change in the entropy. The corresponding values for
myoglobin (Mb) and cobalt-myoglobin (CoMb) are also shown in Table 10. A simi-

lar entropic effect is found for CoMb, if it is compared with imidazole-Co(II)-PPME. The protein or polymer environment is considered to produce a large effect on the stability of the oxygen complex through its entropic contribution.

At room temperature these polymer-Co(II)porphyrin complex, except CoMb, were irreversibly oxidized to the Co(III) complexes. Oxidation of the Fe(II)-porphyrin complexes proceeds more rapidly than that of the Co(II)porphyrin complexes. The next section deals with attempts to make polymer ligands act as inhibitors to the irreversible oxidation of metalloporphyrin complexes.

B. Immobilization of Heme on Polymer Matrixes

One of the major difficulties encountered in attempts to prepare 1 : 1 dioxygeniron complexes which can desorb molecular oxygen, particularly oxygenated complexes of ferroheme, is the strong driving force toward the irreversible formation of the stable μ-oxo ferriheme dimer, as represented in Eq. (11). The oxygenated

$$\text{Fe(II)} + \text{O}_2 \rightleftharpoons \text{Fe--O}_2 \xrightarrow{\text{Fe(II)}} \text{Fe--O}_2\text{--Fe} \rightarrow \text{Fe(III)--O--Fe(III)} \tag{11}$$

complex rapidly reacts with another ferroheme and the binuclear dioxygen-bridged complex is formed. The binuclear complex is irreversibly oxidized to produce the oxo-bridged ferriheme complex. In other words, the first problem of reversible oxygenation is how to inhibit the dimerization.

Much recent work has been aimed at overcoming this problem, and two approaches have been partially successful. The first is the elegant steric approach: porphyrins have been substituted in a fashin that inhibits dimerization. The second approach is to attach the heme complexes to a rigid polymer chain at low concentration so as to prevent two heme complexes from approaching each other in such a manner as to lead to dimerization. By these means a reversible reaction between molecular oxygen and metalloporphyrins has been achieved.

In a classic experiment Wang[97] reported the first synthetic oxygen carrier of Fe(II)porphyrin. He embedded the diethyl ester of ferroheme in a hydrophobic

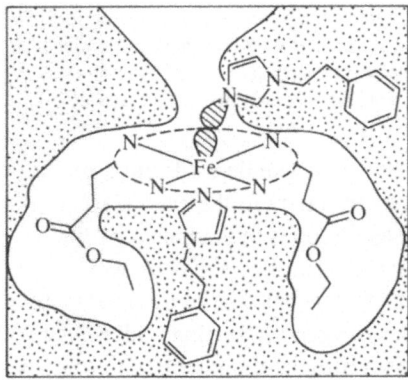

Fig. 21. Schematic diagram of ferroheme derivate embedded in matrix of poly-(styrene) and 1-(2-phenylethyl)-imidazole[97]

matrix of poly(styrene) and 1-(2-phenyl-ethyl)-imidazole. The Fe(II) ion of each embedded heme molecule was firmly coordinated to a substituted imidazole on one side and only partially coordinated to a second imidazole group on the other side because of the constraint of the matrix, as illustrated in Fig. 21. The embedded ferroheme was indeed found to combine reversibly with molecular oxygen. Spectral evidence was as follows: 528, 556 nm in deoxy-state and 537, 563 nm in oxy-state. The change was reversible with respect to oxygen exposure. The oxygenated heme complex embedded in the poly(styrene) film remained stable for a few days without oxidation even if water was present. Wang considered that the low dielectric constant of the hydrophobic poly(styrene) matrix excluded water molecules and suppressed charge separation in the oxygenated complex. Oxidation via dimerization was also thought to be prevented since the embedded hemes were isolated from each other by the matrix.

Hatano[12] dispersed the ferroheme complex of poly(N-vinyl-2-methylimidazole) in fluid paraffin. He reported a reversible change in the visible spectra with respect to oxygen exposure.

We have studied the oxygenation of ferroheme bound to a polymer ligand in the solid state[98]. The profiles of oxygen uptake by powders of polymer-heme complexes were measured by volumetry, as shown in Fig. 22. The heme complex embedded in the porous polymer matrix or in the poly(electrolyte) aggregate takes up

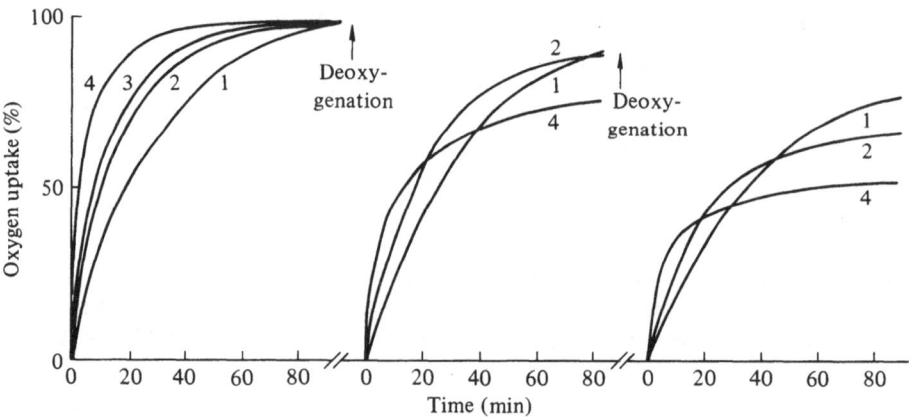

Fig. 22. Oxygen uptake by polymer-heme complexes (cycle of oxygenation-deoxygenation)[98]
1 : poly(4-vinylpyridine)(PVP)-ferroheme, 2 : partially quaternized PVP-ferroheme to which poly(styrene sulfonate) was added to form the polyelectrolyte aggregate, 3 : PVP-ferroheme with porous surface, 4 : PVP-ferroheme immersed in water; deoxygenation: 1 h *in vacuo*

oxygen more rapidly than others. The rate of oxygen uptake depends upon the surface condition or gas permeability of the polymer matrix in the case of solid-state oxygen binding, although the degree of oxygenation is independent of the physical conditions of the polymer matrix. Figure 22 also shows that uptake by the polymer complex immersed in solvents is much faster than that of the solid system.

The oxygenated complex desorbed molecular oxygen on heating under reduced oxygen pressure. Desorption of oxygen, *i.e.* regeneration of the ferroheme complex, was confirmed by measuring an increase in magnetic susceptibility; the spin state of Fe(II) of the ferroheme complex changed from the low spin state to the high-spin state with deoxygenation. Figure 22 also shows that oxygen-bindung ability gradually declines with the cycle of oxygenation-deoxygenation. This degradation, due to irreversible oxidation of the ferroheme complex, is more marked for the polymer complex immersed in solvents and for the polymer complexes with porous surface.

Other results reported concern the reactivity of Fe(II)porphyrin attached to the surface of a solid. Collman et al.[99] prepared Fe(II)(TPP) 39 coordinated to an imidazole group bonded to crosslinked poly(styrene), as represented in 40. Treatment of 40 with oxygen in benzene caused oxidation and formed the μ-oxo dimer

H₂C

HC—⟨benzene ring⟩—CH₂—N⟨imidazole⟩N—[Fe]

40

O[Fe(III)TPP]₂. It was considered that the crosslinked poly(styrene) ligand was not rigid enough to prevent dimerization upon treatment with oxygen.

Basolo et al.[100] found that the attachment of Fe(II)(TPP) to a rigid modified silica gel support produced an efficient oxygen carrier. The silica gel used contained a 3-imidazolylpropyl group bonded to the surface atoms of silicon. This then reacted with Fe(II)(TPP)(B)₂, and the axial base was removed by heating the silica gel. The five-coordinate Fe(II)(TPP) complex 41 was prepared. The open coordination site could reversibly bind molecular oxygen. It was concluded that

—O   OCH₃
    Si
—O   CH₂CH₂CH₂—N⟨imidazole⟩N—[Fe]

41

resistance to the formation of the μ-oxo dimer was due to motional hindrance occasioned by the attachment of Fe(TPP) to the rigid surface of the silica.

As mentioned above, the dimerization of Fe(II)porphyrin was considerably inhibited by embedding the porphyrin complex in polymer matrices or by attaching the porphyrin complex to the surface of rigid polymers, and oxygen carriers composed of Fe(II)porphyrin were produced by using solid state polymers. However, the rate of oxygenation and deoxygenation of these polymer-Fe(II)porphyrin complexes was much lower than with complexes in homogeneous solutions, because ligand-exchange reactions of metal complexes in the solid state occur very slowly. Thus, the next attempt should be to synthesize a polymer-metalloporphyrin complex which will form a stable oxygenated complex even in a homogeneous solution.

From this point of view, some interesting metalloporphyrins have been produced by clever synthetic techniques. Collman et al.[101] synthesized the Fe(II) complex of "picket-fence porphyrin" 42 to favor the five-coordinate structure and inhibit dimerization. Their porphyrin derivative had steric bulkiness on one side of

42

43

44

the porphyrin plane and yet left the other side unencumbered. They reasoned that the use of a suitable bulky axial base ligand such as N-alkyl imidazole would allow coordination of the imidazole on the unhindered side of the porphyrin, with the other side of the porphyrin remaining as a pocket for binding molecular oxygen. Moreover, the picket fence would discourage the dimerization of the oxygenated complex. It was reported that the Fe(II)porphyrin complex *42* could bind molecular oxygen reversibly in benzene or tetrahydrofuran at room temperature when the picket fence was constructed with the pivalamidophenyl group. They succeeded in obtaining the crystal of the oxygen adduct of *42*. Their result means that the steric bulkiness and the special environment around the oxygen binding site are important to form the reversible oxygen complex.

Baldwin *et al.* [102] synthesized the Fe(II) complex of "capped porphyrin" *43*. This Fe(II) porphyrin complex was reported to bind molecular oxygen in pyridine at 25 °C, the life of the oxygenated complex being about 20 h.

The complex of porphyrin derivative *44* was synthesized by Traylor *et al.* [103]. Fe(II)porphyrin of this type formed preferentially the five-coordinate complex. The life time of the oxygenated complex was several hours in dimethylformamide at −45 °C but only a few minutes at room temperature.

The above three studies on the synthesis of oxygen carriers are of great significance because they have demonstrated steric and environmental effects on reversible oxygenation. Recently the porphyrin ring was perfectly bonded to a polymer by a covalent bond in order to inhibit the dimerization of the porphyrin complex.

Ikeda *et al.* [104] polymerized tetra(p-styryl)porphyrin with styrene and obtained the polymer-Fe(II)(TPP derivative) complex *45* in which the porphyrin complex

45

was covalently bonded to the poly(styrene) chain at low concentration. The visible spectrum of the polymer complex *45* with imidazole was reported to change reversibly at room temperature from 431, 536, and 560 nm in the deoxy state to 421, 550, and 590 nm in the oxy state. As mentioned, the oxidation of Fe(II)porphyrin via dimerization was inhibited by immobilization of the porphyrin ring on the polymer matrix.

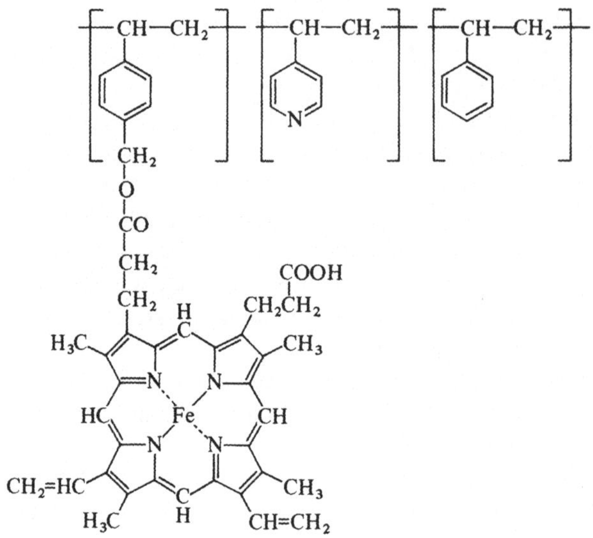

46

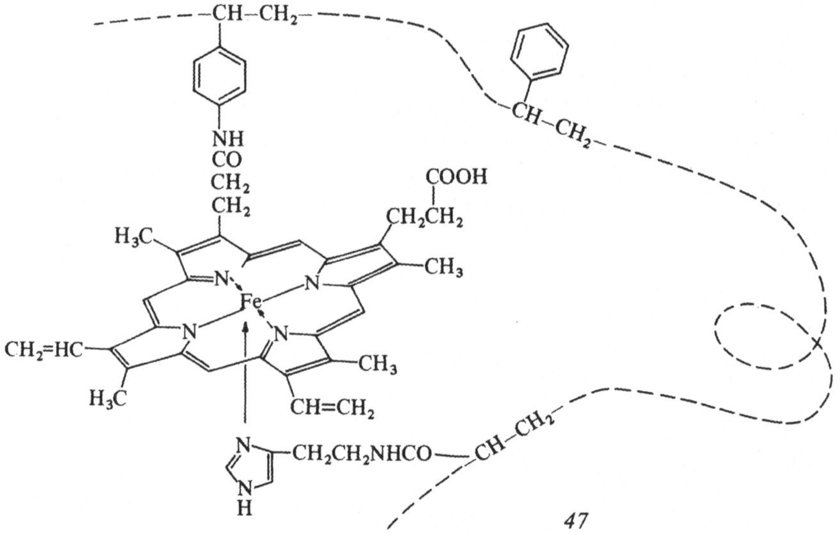

47

We also synthesized polymer-metalloporphyrin complexes in which both the porphyrin ring and a nitrogenous axial ligand were fixed on a polymer chain by covalent bonds, as represented in 46 and 47[105]. ESR and visible measurements indicated that the Co(II)- and Fe(II)porphyrin complexes of 46 and 47 were reversibly oxygenated in toluene or dimethylformamide. These results confirmed that the metalloporphyrin complex attached on a polymer at low concentration behaves virtually as in a solution at infinite dilution.

Although attempts to prepare stable and reversible oxygenated complexes were successful in organic solvents, it is not yet possible to predict which complex will perform the reversible oxygenation in an aqueous solution. Further research on oxygen binding by hemoglobin and myoglobin will provide more insight into the possibilities and limitations of attempts to synthesize oxygen carriers.

C. Pseudo-Allosteric Effect of Poly(L-lysine)-Heme Complex

It is well known that hemoglobin effectively absorbs molecular oxygen in the lungs and desorbs it in the terminal tissue although there is only a small difference in the partial pressure of oxygen. This phenomenon is clearly demonstrated by Fig. 23.

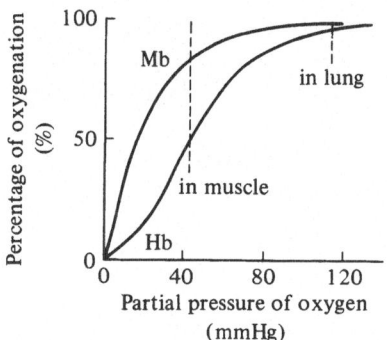

Fig. 23. Oxygen binding curves of hemoglobin (Hb) and myoglobin (Mb)

The equilibrium curve of oxygen binding by hemoglobin is sigmoidal, while the curve for myoglobin is hyperbolic. The oxygen-binding behavior of hemoglobin, called "allosteric effect", is believed to be caused by heme-heme interaction. Hemoglobin contains four heme units and myoglobin only one. The four heme units, which are situated separately in the hemoglobin molecule, are indirectly connected with each other through the globin chain, and they bind four oxygen molecules cooperatively. The conformational change of the globin protein plays an important role in the cooperative binding of oxygen and basic theoretical treatments of the cooperativity have been presented[106]. Such a "cooperative phenomenon", though often observed *in vivio,* had not been found in any synthetic system. However, we found a cooperative phenomenon in interactions of the poly(L-lysine)-heme complex with molecular oxygen or carbon monoxide.

As described in Section IIB, the heme complex illustrated in Fig. 4(b) is formed with poly(L-lysine)(PLL). The fifth and sixth coordination positions of the heme iron are occupied by the primary amines on the side chain of the PLL ligand, which

$$\text{PLL} \underset{\text{NH}_2}{\overset{\text{NH}_2}{\diagup\text{Fe}\diagup}} + O_2 \rightleftharpoons \text{PLL} ---\text{NH}_2\cdots|\,\text{Fe}\cdots\cdots O_2 \longrightarrow \text{Fe(III)} + O_2^- \qquad (12)$$

$$\text{PLL} \underset{\text{NH}_2}{\overset{\text{NH}_2}{\diagup\text{Fe}\diagup}} + CO \rightleftharpoons \text{PLL} ---\text{NH}_2\cdots|\,\text{Fe}\cdots\cdots CO \qquad (13)$$

gives an α-helical conformation. The PLL-ferroheme complex reacts with molecular oxygen, and one of the ε-amino side groups is out of the coordination position of the heme iron, as shown in Eq. (12). The PLL-ferroheme complex also combines with carbon monoxide and forms the bright red PLL-heme-CO mixed complex which is thermodynamically stable [Eq. (13)]. Figure 24(a) shows the relationship between oxygen pressure and rate of oxygen absorption by the PLL-heme complex in aqueous solution (pH 12.0)[107]. The oxygen absorption curve of the PLL-heme complex is sigmoidal, while those of the heme complex of poly(4-vinylpyridine), of poly(ethyleneimine), and of course, of a low-molecular-weight nitrogenous ligand are hyperbolic. The sigmoidal response is also observed for the CO-binding equilibrium curve of the PLL-heme complex [Fig. 24(b)][108].

The cooperative parameter based on Hill's equation[a] was determined and the results are listed in Table 11[107, 108]. The cooperative parameter (n) of the PLL-

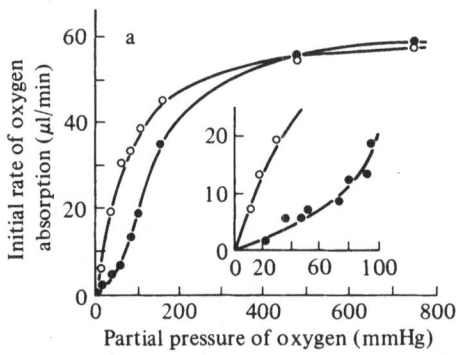
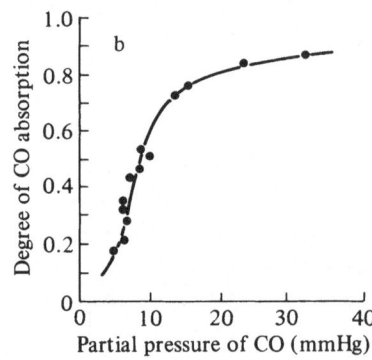

Fig. 24. Relationship between partial pressure of oxygen and oxygen absorption rate by poly(L-lysine)-heme complex (a) and equilibrium curve for CO binding by poly(L-lysine)-heme complex (b)[107, 108]

(a) ●: poly(L-lysine)(PLL)-ferroheme, ○: PLL-ferroheme with poly(ethyleneoxide) which interacts with PLL and destabilizes the helical structure of PLL, in aq. solution pH 12

---

[a] Hill's equation[109] is represented as

$$Y = KP^n/(1 + KP^n) \tag{14}$$
$$\log(Y/(1 - Y)) = n \log P + \log K \tag{15}$$

where Y and P represent reaction rate or degree of saturation for an equilibrium reaction, and concentration of substrate or partial pressure of oxygen. n is the cooperative parameter or Hill's coefficient. The relationship between P and Y with $n = 1$ corresponds to a hyperbolic curve, and that for $n > 1$ is sigmoidal. The reactions with $n > 1$ are understood as follows: (i) a macromolecular compound contains several reaction or adsorption sites; (ii) these sites are connected indirectly with each other; (iii) the first of these sites does not interact easily with the substrate. However, once the first site has interacted, reactivity increases successively for the second, the third, and so on. Thus, the overall reaction is completed at a stroke. The cooperative parameter (n) corresponds to the strength of the cooperative interaction between the sites. The value of n for oxygen binding by hemoglobin is 2.8, whereas $n = 1.0$ is obtained for myoglobin. There are believed to be three requisities for the cooperative binding of oxygen by hemoglobin (allosteric oxygenation of hemoglobin)[110]: (i) the hemoglobin molecule contains four hemes; (ii) these hemes are in equilibrium between two states that differ in their affinity for molecular oxygen; (iii) the four hemes interact with each other indirectly.

heme complex is quite large ($n = 2.1$), while those of the other synthetic polymer-heme complexes are small ($n \simeq 1$). The cooperative parameter for the CO equilibrium curve of PLL-heme was 2.0, as large as that in the oxygen system[108]. This indicates that coorperative interaction does exist in the PLL system and that the contribution to the cooperativity of irreversible oxidation, which takes place after the oxygen binding in Eq. (12), is negligible.

Table 11. Hill's coefficient for reaction of polymer-heme complexes with molecular oxygen

System	Solvent	Additives	Hill's coefficient (n)	$P_{1/2}$[a] mmHg
PVP-heme	DMF MeOH (1/1)	–	1.1	58
BPEI-heme	Phosphate buff. (pH 12)	–	1.0	100
PLL-heme	Phosphate buff. (pH 12)	–	2.0	130
PLL-heme[b]	Phosphate buff. (pH 12)	–	2.0	130
PLL-heme	Phosphate buff. (pH 12)	0.1M NaCl	1.7	140
PLL-heme	Phosphate buff. (pH 12)	0.4M NaCl	1.3	270
PLL-heme	Phosphate buff. (pH 12)	Poly(ethyleneoxide)	1.3	75
PLL-heme	Phosphate buff. (pH 12)	Poly(glutamic acid)	1.8	67
Hemoglobin	Phosphate buff. (pH 7)	–	2.8	10
Myoglobin	Phosphate buff. (pH 7)	–	1.0	0.5

PVP = poly(4-vinylpyridine), BPEI = branched poly(ethyleneimine), PLL = poly(L-lysine).
[a] $P_{1/2}$ is the pressure at which one-half of the complex has reacted with molecular oxygen.
[b] The [PLL]/[heme] ratio is changed from 7 to 10 or 50.

It will be remembered that the coordination number ($\bar{n}$) of heme iron was about 1.0 in the polymer heme complexes, except for PLL-heme (cf. Table 4). This means that the sixth position of heme iron may be weakly coordinated by a solvent molecule as represented in Fig. 4(a). Therefore, it is reasonable to consider that a ligand-exchange reaction with an oxygen molecule at the sixth position occurs easily and that the oxygen absorption behavior is of the Langmuir type. In the PLL system on the contrary, the deoxyheme complex has a chelate structure and its formation constant is large ($K = 3.5 \times 10^9$ l^2/mol^2 in Table 4). The ability of the PLL-heme complex to bind molecular oxygen or carbon monoxide is suppressed at lower pressures of oxygen or carbon monoxide, and the $P_{1/2}$ value of the PLL system is larger than for other systems, as also listed in Table 11. The binding ability of the PLL complex increases at higher oxygen or carbon monoxide pressures for some reason, which may be related to the structure that has have molecules situated on an α-helical PLL chain.

As mentioned in Section IIB, the helical structure of PLL is believed to be partially destroyed by complex formation with bulky heme, because the helix content of PLL gradually falls from 1.00 to 0.91 while the heme content [heme]/[residual group of amino acid of PLL] increases from 1/100 to 1/5. We also found that the helix content increased again to 1.00 after the reaction of the PLL complex with oxygen or carbon monoxide[107]. These results indicate that the strain in the

helical structure of the PLL complex is reduced by reaction with oxygen or carbon
monoxide and suggest that the reduction of strain initiates a cooperative interaction
through a conformational change in the polymer chain. The cooperative interaction
between the heme complexes in the PLL system is hypothetically illustrated in
Scheme 11.

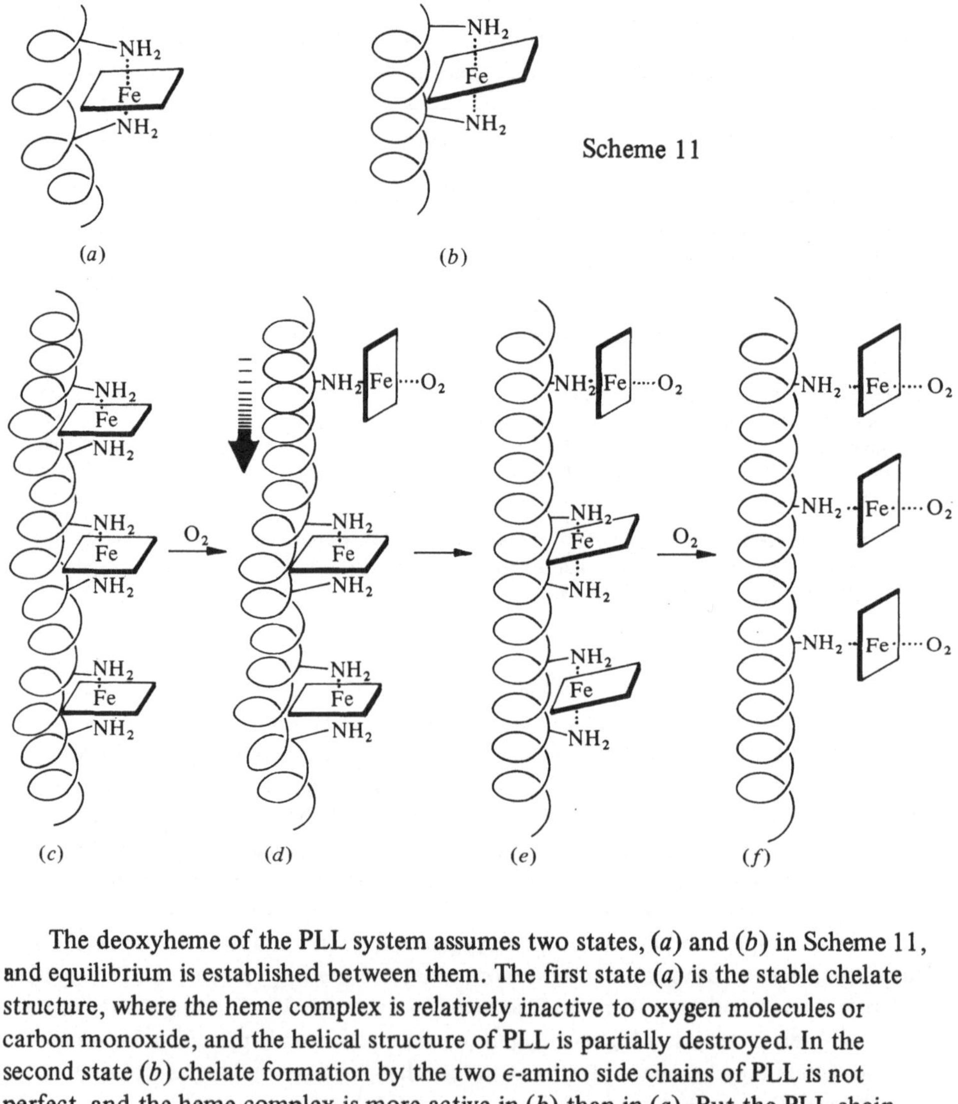

Scheme 11

The deoxyheme of the PLL system assumes two states, (a) and (b) in Scheme 11,
and equilibrium is established between them. The first state (a) is the stable chelate
structure, where the heme complex is relatively inactive to oxygen molecules or
carbon monoxide, and the helical structure of PLL is partially destroyed. In the
second state (b) chelate formation by the two ϵ-amino side chains of PLL is not
perfect, and the heme complex is more active in (b) than in (a). But the PLL chain
is coiled up in an α-helix in (b). As illustrated in (c), a PLL molecule contains many
heme complexes (a) (in our PLL system, [heme]/[residual group of amino acid of
PLL] = 1/7.5 and [heme]/[PLL molecule] = 47). When one of the heme complexes
combines with molecular oxygen, the chelate structure of heme changes to that of
the mixed complex, $-NH_2-Fe-O_2$, according to Eqs. (12) or (13). The formation
of the mixed complex reduces the strain in the PLL chain and the helical structure

of PLL is much stabilized. This conformational change of PLL may induce the deformation of the coordination structure of the nearest neighbors of the deoxy-hemes from (d) to (e), as illustrated. That is, the helical structure becomes more complete, while the lower deoxyhemes compensate by changing from state (a) to the more active state (b). The PLL-heme complexes in (e) react rapidly with second and third oxygen molecules. In the same way, the overall reaction with oxygen proceeds at a stroke.

The helical structure of PLL plays an important role in transferring information on oxygenation to the neighboring heme groups. We call this cooperative phenomenon the "pseudo-allosteric effect" of the PLL-heme complex.

This hypothesis is further supported by the effect on the absorption behavior of molecular oxygen of additives that destroy the α-helix of PLL. When we added to the PLL system non-coordinative polymers, which do not act as ligands but are able to interact with PLL to form polymer aggregates, the cooperative parameter decreased to unity, as shown in Table 11. In these cases, it was found that the α-heli-cal conformation was made unstable by the formation of polymer aggregates[111]. The fact that no change was observed in the visible spectrum after the addition of the polymers showed that the structure of the deoxy-heme complex was not affected thereby. Therefore, the added polymers must cause the decrease in the cooperative interaction through instabilization of the helical structure of PLL. The cooperative parameter was also reduced by the addition of sodium chloride, which destroys the α-helical structure.

In general, the reactivity of a polymer-metal complex can be controlled by the higher structure of the polymer chain and the dynamic change in its conformation. In other words, the structural change within the coordination sphere of a polymer-metal complex, which is accompanied by a reaction, brings about the conformational change in the polymer chain, or conversely, the conformational change of the poly-mer ligand induces the rearrangement of the coordination structure, i.e. the change in reactivity.

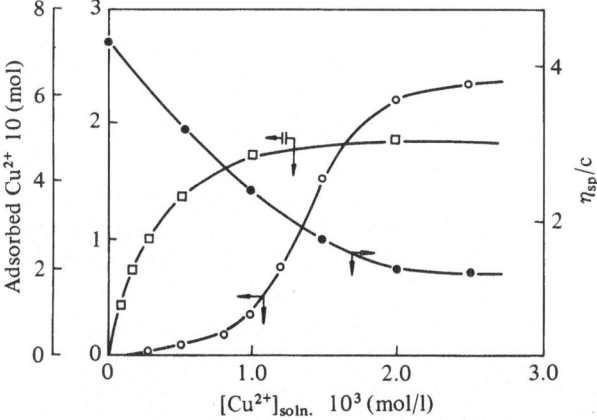

Fig. 25. Adsorption of Cu ions of polymer ligand[112]
○, □: adsorption of Cu(II) on partially quaternized poly(4-vinylpyridine), Cu-tempelate DBQP resin (crosslinking 42%), ●: viscosity of polymer solution

On this occasion, we would like to emphasize that so-called "cooperative phenomena" or "allosteric phenomena" are not characteristic of biological systems alone but are often observed also in synthetic polymer systems. For example, we cite Fig. 25, which shows the adsorption of metal ions on a synthetic polymer ligand[112]. The adsorption of Cu ions on the polymer ligand is sigmoidal. This cooperative binding of metal ions is easily understood by following Scheme 12.

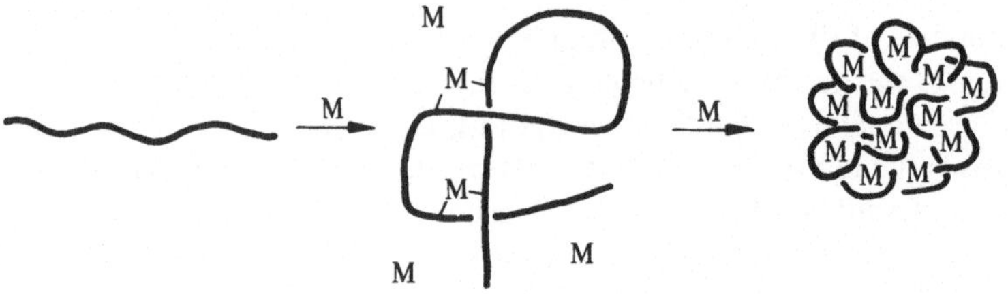

Scheme 12

Cu ions form an intra-polymer chelate with a polymer ligand. Due to chelation the polymer chain is markedly contracted, although the polymer ligand itself is extended. At a low concentration of Cu ions, the conformation of the polymer is not sufficiently contracted to form stable Cu chelates, but at higher Cu concentrations, the tight packing of the polymer ligand, caused by intra-polymer chelation with a large number of Cu ions, progressively facilitates chelate formation. The conformational change in the polymer ligand from an extended shape to a contracted one, which occurs with the binding reaction, enhances the Cu ion-binding ability (refer to Section IIIC).

VI. Reactions Catalyzed by Polymer-Metal Complexes

Organic reactions are extensively catalyzed by metal complexes. The catalytic cycle of a metal complex-catalyzed reaction is illustrated in Scheme 13.

$$ML_n + S \rightleftharpoons L_{n-1}M{-}S$$
$$\uparrow \qquad\qquad \downarrow$$
$$M'L + S^* \rightleftharpoons L_{n-1}M'{-}S^*$$

M = metal ion, L = ligand, S = substrate   Scheme 13

In the first step, a substrate coordinates to a metal catalyst and forms an intermediate mixed complex (LMS in Scheme 13). The substrate is then activated by metal ions and dissociates from the catalyst. The complex catalyst, having accomplished its purpose, is regenerated to the original complex. The catalytic action of a metal ion depends substantially on the nature of the ligands in the intermediate mixed complex. Certain ligands induce an increase in catalytic activity, while others, *e.g.* multidentate ligands such as ethylenediaminetetracetic acid, inhibit the catalytic action of a metal ion. Therefore, if a polymer ligand is used as one component of a metal-complex catalyst, its properties may affect the catalytic action of the metal ion.

A. Oxidative Reactions

Oxidative reactions of organic compounds with molecular oxygen take place with high efficiency and selectivity in the presence of Werner-type metal complexes used as catalysts. The catalytic effects of metal complexes have received attention also as models for metalloenzymes, which are catalysts that possess special high efficiency and high selectivity for oxidative reactions *in vivo*.

Metal complexes catalyze oxidation of compounds having mobile hydrogens, such as ascorbic acid, hydroquinone, phenols, and amines, in the presence of molecular oxygen [Eq. (16)]. In this reaction, a substrate coordinates to the metal catalyst,

$$2\ AH_2 + O_2 \xrightarrow{\ M^+\ } 2\ A + 2\ H_2O \tag{16}$$

and then the substrate is one-electron oxidized by the metal ion with higher valence.

The first example of catalysis by a polymer-metal complex was presented by Lautsch et al.[113]. Metalloporphyrin was linked to a poly(phenylalanine) chain by a peptide bond. The catalytic properties of this polymer-Fe(III)porphyrin complex were compared with Fe(III)porphyrin in the oxidative reaction of phenylenediamine. The catalytic activity of the polymer complex was twice as large as that for the corresponding analog.

Pecht et al.[114] reported some detailed studies on the catalysis of the Cu(II) complex of poly(histidine). The kinetic pattern of poly(histidine)-Cu catalyzed oxidation was a Michaelis-Menten one, whereas the monomeric Cu complex induced first-order oxidation. It was found that the poly(histidine) ligand enhanced catalytic efficiency in the oxidation of negatively charged or neutral substrates, *e.g.* ascorbate, homogentisate anion, and hydroquinone, but retarded the oxidation of positively charged substrates, *e.g.* tetramethyl-p-phenylenediamine ammonium and 1(2,5-dihydroxyphenyl)isopropylammonium cations. The specific effect of the poly-(histidine)-Cu catalyst was cancelled by the addition of a neutral salt, so that enhancement and retardation were interpreted as due to an electrostatic effect of the poly(electrolyte). The polycationic poly(histidine)-Cu catalyst was thought to concentrate the negatively charged substrate within the domain of the polymer catalyst and to facilitate coordination of the substrate to Cu.

The electrostatic effect of the poly(4-vinylpyridine)-Cu(II) catalyst was also reported by Dadze et al.[115]. Oxidation of ascorbic acid, salicylic acid, and tri-t-butylphenol was accelerated and that of p-phenylenediamine retarded by the poly(4-vinylpyridine) ligand at lower pH. As described in Section IIIA, a polymer-metal complex behaves as a polycation in aqueous solution, and the reaction is sensitive to the charge of low-molecular-weight species. The electrostatic effect due to the polycationic domain of a polymer-metal catalyst is also predominant in the oxidation of charged substrates.

Kabanov et al.[116] studied the oxidation of ascorbic acid by the Cu(II) complexes of poly(4-vinylpyridine) partially alkylated by bromoacetic acid. It was considered from kinetic and thermodynamic data that the higher catalytic activity of the polymer-Cu complex was caused by binding of the substrate to the catalytic site, represented as *48*.

~CH$_2$–CH–CH$_2$–CH–CH$_2$–CH~

Br$^-$ ... Cu^{2+} ... Br$^-$ *48*

CH$_2$ ... CH$_2$
COOH ... COOH

Hatano *et al.*[117] studied the poly(L-lysine)(PLL)- and poly(DL-lysine)-Cu(II) complexes as catalysts in the oxidation of 3,4-dihydroxyphenylalanine. The catalytic activity of PLL-Cu was found to be greater than that of ethylenediamine-Cu. The oxidation was asymmetrically catalyzed by PLL-Cu; the D-isomer of the substrate coordinated to PLL-Cu more strongly and its oxidation rate was greater than that of the L-enantiomer. The asymmetrical oxidation was ascribed to the helical structure of PLL-Cu in an aqueous solution of pH 10.5[118].

The decomposition of hydrogen peroxide is effected by using various metal complexes as catalysts. This catalysis has its origin in the redox action of a metal complex. A transient intermediate is formed between the metal catalyst and the substrate, at least in the initial step of the reaction, but the reaction also proceeds through chain decomposition. Although the catalytic mechanism has not been elucidated, the decomposition of hydrogen peroxide is often employed as a standard reaction to determine the catalytic activity of a polymer-metal complex, because the catalytic action is always observed regardless of metal ion species and ligand species, because the reaction takes place even if the catalyst is insoluble in the reaction solvent, and because the experimental technique to follow the reaction is not difficult.

Polymer-metal complexes often exhibit high efficiency in the catalytic decomposition of hydrogen peroxide. The following reasons for this activity have been advanced. (i) Some polymer-metal complexes contain incomplete complexes due to steric hindrance, and this contributes to their catalytic activity[121, 122]. (ii) In other polymer complexes, the coordinate bond between polymer ligand and metal ion is relatively weak and the substrate coordinates with high frequency[124]. (iii) Chain decomposition of the substrate proceeds rapidly because the concentration of the catalytic site is locally enhanced in polymer-metal systems[119, 120, 123]. Many polymer-metal complexes have been reported to catalyze the decomposition of hydrogen peroxide; poly(β-diketone)-Cu[119], poly(β-ketoester)-Cu[120], poly-(methacroylacetone)-Cu[121], poly(acrylhydroxamic acid)-Cu[122], poly(ethylene-imine)-, poly(acrylic acid)-, and poly(methacrylic acid)-Fe[123], poly(4-vinylpyridine)-Co[124], various poly(amino acid)-Cu[125], and RNA- and DNA-Cu[126].

The oxidation of various alcohols by the poly(ϵ-carbobenzoxy-L-lysine)-Cu complex was studied by Welch *et al.*[127]. The polymer catalyst showed selectivity in oxidation by virtually excluding alcohols of bulky structure such as diisopropyl and diisobutyl carbinol while admitting simple alcohols such as n-butyl, iso-butyl, and sec-butyl. It was suggested from structural studies that the selectivity of the polymer catalyst resulted from the highly complex geometry of the molecular cage formed by the helix and the amino acid side chain around the coordinated Cu. The oxidation of toluene by poly(methylvinylketone)-Cu, -Mn and -Ni complexes was also reported[128].

As mentioned above, in metal ion-catalyzed oxidations many polymer-metal complexes have been found to exhibit high catalytic efficiency in comparison with their low-molecular-weight analogs. Table 12 summarizes the catalytic activity of the Cu complexes of synthetic polymers. It is noteworthy that high efficiency is observed for complexes composed of "simple" synthetic polymers such as poly-(ethyleneimine), poly(vinylpyridine) and poly(acrylic acid).

Table 12. Catalytic activity of polymer-Cu complexes for oxidation of various substrates

Substrate	log [(Activity of polymer-Cu catalyst)/(Activity of monomeric Cu catalyst)]			
	-1	0	1	2
Ascorbic acid		◯ ◯◯ ◯		
		a a a c		
p-Hydroquinone				◯
				c
2,6-Dimethylphenol	◯	◯	◯◯	
	d	a	a a	
Homogentisic acid				◯
Dioxycinnamic acid				c
			◯	
			c	
p-Phenylenediamine	◯	◯◯ ◯		
	c	c a b		
Hydrogen peroxide		◯◯◯	◯◯ ◯	
		d e e	b a d	

Polymer ligand; a) poly(4-vinylpyridine), b) poly(L-lysine), c) poly(L-hystidine), d) poly-(ethyleneimine), e) poly(acrylic acid).

B. Reductive Reactions

The instances of reductive reactions catalyzed by polymer-metal complexes of the Werner type are few compared with the oxidative reactions.

Hirai et al.[129] studied the hydrogenation of olefins catalyzed by poly(acrylic acid)-Rh(II) complexes in homogeneous solutions. The catalytic activity of the polymer-Rh complex was about 10^3 times that of the acetato-Rh complex. When olefins having another functional group, such as diallylether, allylaldehyde, and cyclohexene-1-one, were used as the substrates, the olefinic bond was preferentially hydrogenized by the polymer-Rh complex. The polymer ligand was presumed to exercise a steric effect.

Asymmetric hydrogenation catalyzed by poly(L-methylethyleneimine)-Ru(III) complexes was also reported[130]. The L-polymer-Ru complex homogeneously catalyzed hydrogenation of methylacetoacetate and of mesityloxide and produced more D-isomer than L-isomer. But the Ru complex of poly(DL-methylethylene-imine) exhibited no selectivity. It was considered that the optically active polymer ligand coordinated to Ru as a bidentate ligand to form the active site for asymmetric hydrogenation, as shown in 49. The relationship between the secondary structure

49

of the polymer ligand and asymmetric selectivity was discussed in the hydrogenation of methylacetoacetate catalyzed by poly(L-glutamic acid)-Ru(III) complexes[131].
The yield of L-isomer increased with the helix content of the poly(L-glutamic acid)
ligand. The use of metal complexes of polymer ligands having coordinating groups
with optically active atoms has often accomplished asymmetric syntheses and
stereoselective reactions[132, 133].

In these polymer-metal complexes of the Werner type, however, organometallic
compounds are formed as reaction intermediates and/or activated complexes. As a
result, the properties of polymer-metal catalysts in reductive reactions are different
from those of polymer-metal catalysts in oxidative reactions. In the former, the
catalytic reactions are very sensitive to moisture and air, and the complex catalysts
often decompose in the presence of water and oxygen. Thus, reductive catalytic
reactions are carried out under artificial conditions such as organic solvent, high
pressure, and high temperature. Oxidative catalytic reactions, on the other hand,
proceed under mild conditions: aqueous solution, oxygen atmosphere, and room
temperature. Therefore, it is to be expected that the catalytic effects of a polymer
ligand will differ from the latter to the former.

Many polymer-metal complexes of non-Werner type (organometallic compounds)
have been successfully applied to catalytic processes: hydrogenation, isomerization,
disproportionation, oligomerization, and hydroformylation. Styrene-divinyl-benzene
copolymers containing phosphine ligands have been most often used as polymer
ligands. A few short reviews on the catalysis of polymer-supported organometallic
compounds have appeared[134–136]. The effects of polymer ligands on the catalysis
of non-Werner type complexes have been mentioned. (i) They allow the metal-
complex catalyst to be easily separated from the reaction mixture, possibly regenerated, and recycled. The advantage of homogeneous catalysis over heterogeneous
catalysis is its greater selectivity. By immobilizing a homogeneous complex catalyst on a polymer matrix, the catalytic activity of the resulting polymer-metal complex can be made to partake of both homogeneous and heterogeneous character.
(ii) Polymer-metal catalysts are not susceptible to poisoning by impurities, since
the catalytic site is somehow protected by the polymer matrix. (iii) As the amount
of catalyst employed increases, the efficiency of a homogeneous catalyst decreases.
This is due to aggregation and insolubilization of a metal complex of non-Werner
type. With a polymer complex, aggregation is physically prevented by the rigidity
of the polymer matrix. A polymer catalyst has the advantage of maintaining its

catalytic activity over a wide range of concentrations. (iv) As regards selectivity in the catalysis, additional selectivity may arise from the steric hindrance and/or chemical environment of the polymer matrix.

Metal complexes immobilized on crosslinked polymer matrices are hard to characterize by the customary physicochemical techniques, which is why no quantitative studies have been made on the catalytic activity of polymer complexes.

C. Hydrolysis

The catalytic hydrolysis of oligophosphates by poly(L-lysine)-Cu(II) complexes has been reported[137]. The PLL-Cu complex showed strong catalytic activity and attacked pentaphosphate exclusively, thus producing orthophosphate as the main product. This result was explained by the chelate structure of PLL-Cu.

The hydrolysis of sodium pyrophosphate was effected by using some metal complexes of poly(methacrylacetone) as catalysts[138]. The catalytic activity of the polymer complexes declined in the following order: $Zr(IV)O > U(VI)O_2 > Cr(III) \simeq Ce(III) \simeq Cu(II)$. The enhancement of activity was ascribed not only to the electrostatic effect of the polymer complexes but also to a possible multiplying effect by $ZrOL^+$ and $ZrOL(OH)$ along the polymer-ligand chain.

Hatano et al.[139] found that the poly(L-lysine)-Cu(II) complex exerted asymmetrically selective catalysis on the hydrolysis of phenylalanine ester, whereas Cu ions and bis(bipyridyl)Cu had no catalytic activity. The great stability of the intermediate PLL-Cu complex with the D-ester was considered responsible for the catalytic activity.

Breslow et al.[140] attached the Ni(II) complex to cyclohexaamylose, which formed a hydrophobic cavity, as drawn in 50, and studied the hydrolysis of p-nitro-

50

phenyl acetate. The reaction was accelerated by a factor of about 10^3 over the uncatalyzed system, the increased reactivity being the result of binding of the substrate in the hydrophobic cavity.

In addition to the above-mentioned reactions, metal complexes catalyze decarboxylation of keto acids, hydrolysis of esters of amino acids, hydrolysis of peptides, hydrolysis of Schiff bases, formation of porphyrins, oxidation of thiols, and so on. However, polymer-metal complexes have not yet been applied to these reactions.

In the next chapter, we describe the catalytic mechanisms of polymer-metal complexes, using as an example the oxidative polymerization of phenols catalyzed by polymer-Cu complexes.

VII. Phenol Oxidation Catalyzed by Polymer-Cu Complexes

We have selected the Cu-complex-catalyzed oxidative polymerization of phenols as a model reaction in which a polymer-metal complex acts as a catalyst. This reaction has some merits for the study of the catalytic mechanism of a polymer ligand. (i) The complex homogeneously catalyzes the reaction. (ii) The reaction proceeds via the intermediate substrate-coordinated complex (substrate-metal ion-ligand mixed complex), so that the property of the ligand is clearly reflected in the catalysis. (iii) The complex catalyst affects not only the reaction rate but also the characteristics of the resulting polymer, *e.g.* molecular weight and composition. Since the oxidative polymerization of phenols is the industrial process used to produce poly(phenyleneoxide) s, the application of polymer catalysts may well be of interest.

Organic compounds having mobile hydrogens, such as phenols[141], phenylenediamine[142], and acetylene[143], are oxidatively coupled in the presence of a metal complex to form polymeric compounds, as represented by Eqs. (17)–(19). These

$$ n \underset{}{\bigcirc}\!\!-\!OH + (n/2)\, O_2 \longrightarrow \left[\!\underset{}{\bigcirc}\!\!-\!O\!\right]_n + n\, H_2O \qquad (17) $$

$$ n\, H_2N\!-\!\underset{}{\bigcirc}\!\!-\!NH_2 + n\, O_2 \longrightarrow \left[\!N\!=\!N\!-\!\underset{}{\bigcirc}\right]_n + 2n\, H_2O \qquad (18) $$

$$ n\, HC\!\equiv\!CH + (n/2)\, O_2 \longrightarrow \left[C\!\equiv\!C\right]_n + n\, H_2O \qquad (19) $$

reactions are called oxidative polymerization. It is well known that 2,6-disubstituted phenols with an amine-Cu complex catalyst produce poly(2,6-disubstituted phenyleneoxide)s by C–O coupling[144]. Poly(2,6-dimethylphenyleneoxide)(PPO) *51* and poly(2,6-diphenylphenyleneoxide)(PPPO) *52* are produced from 2,6-dimethylphenol and 2,6-diphenylphenol, respectively.

In the Cu-complex-catalyzed oxidative polymerization of phenols, the substrate (phenol) coordinates to the Cu(II) complex, is then activated, and the

51   *52*

selective coupling of the activated phenol occurs. Oxidation proceeds rapidly at room temperature under an air atmosphere. Moreover, the polymerization follows the kinetics of Michaelis-Menten type[145]. This catalytic behavior is similar to that of the copper oxidases. Indeed, enzymic oxidation of phenols is an important pathway in the biosynthesis of plants[146], *e.g.* "lignin" is a natural polymeric product of phenols, such as coniferyl alcohol and sinapyl alcohol, catalyzed by a metalloenzyme (Fig. 26).

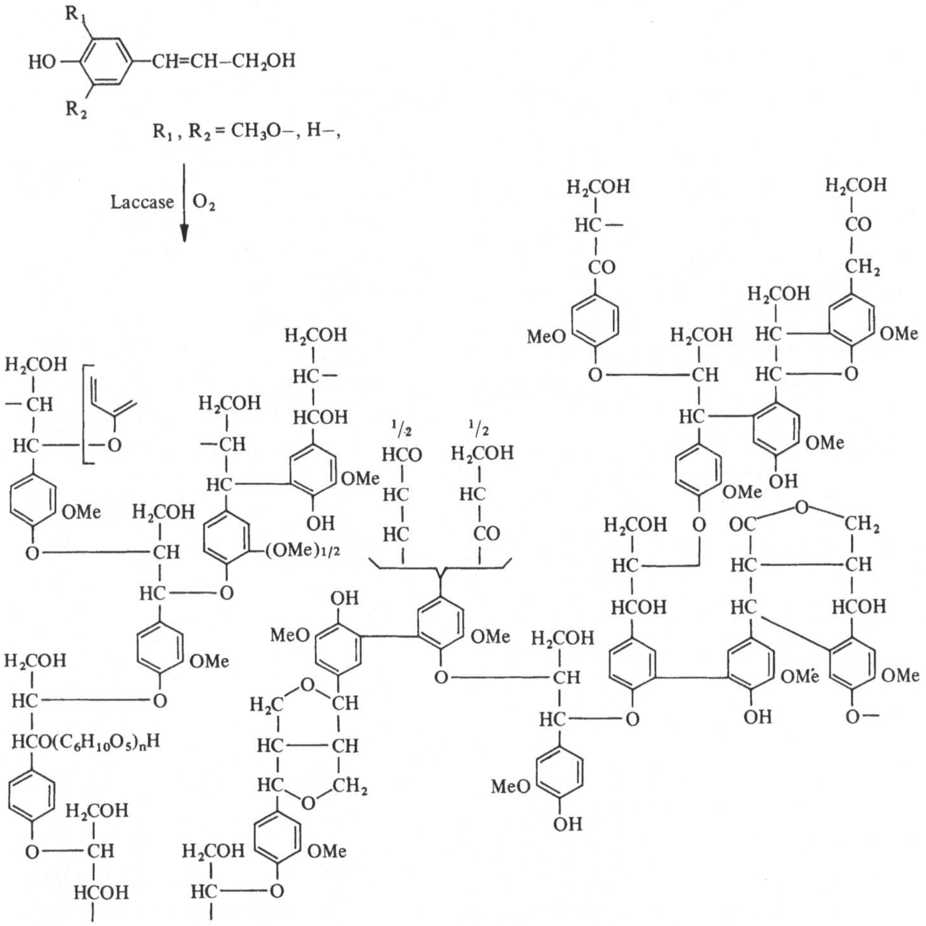

$R_1, R_2 = CH_3O-, H-,$

Laccase | O_2

$HCO(C_6H_{10}O_5)_nH$

Fig. 26. Polymeric structure of lignin

A. Oxidative Polymerization Catalyzed by Polymer-Cu Complexes

The mechanism of the oxidative polymerization of 2,6-dimethylphenol (XOH) with an amine-Cu complex is represented by Eq. (20)[145, 147].

$$\underset{X}{\overset{L}{\diagdown}}\underset{L}{\overset{X}{\diagup}}Cu^{II} + {}^{-}O{-}\bigcirc \underset{}{\overset{-X^-}{\rightleftharpoons}} \underset{X}{\overset{L}{\diagdown}}\underset{L}{\overset{-O-\bigcirc}{Cu^{II}}} \longrightarrow \underset{X}{\overset{L}{\diagdown}}\underset{L}{\overset{*O-\bigcirc}{Cu^{I}}}$$

$$\tag{20}$$

$$\underset{X}{\overset{L}{\diagdown}}\underset{L}{\overset{}{Cu^{I}}} + {}^{*}O{-}\bigcirc$$

L = amine ligand, X = anionic ligand such as chloro and hydroxyl ions

The activated phenols are C—O coupled each other. The dimer thus formed is activated by a similar mechanism, and polymerization occurs. The effects of the amine ligand (L) are to improve the solubility and stability of the Cu ion, to affect the stability of the substrate-coordinated complex, and to control the redox potential of the Cu ion. The Cu-complex catalyst not only enhances the rate of polymerization, but it also has an important effect on the coupling reaction.

We describe below the change in catalytic behavior when an amine ligand (L) is substituted by a polymer ligand. The Cu(II) complexes of poly(4-vinylpyridine) (PVP), partially quaternized PVP (QPVP) 18, partially diethylaminomethylated poly(styrene)(PDA) 53, and poly(ethyleneimine) proved effective as catalysts in the

53

oxidative polymerization of phenols (Table 14)[148−150]. The oxidation rate and the kinetic and thermodynamic parameters for the polymerization of XOH catalyzed by Cu complexes in dimethylsulfoxide are listed in Table 13[149]. The polymerization rates are higher for the polymer-complex-catalyzed polymerization than for that catalyzed by the pyridine-Cu complex; the QPVP-Cu catalyst shows especially high activity. For the QPVP-Cu-catalyzed system, the polymerization rate is not only much larger, but the kinetic and thermodynamic parameters differ considerably from those of the monomeric catalyst. A difference in the catalytic mechanism is suggested.

The polymerization of XOH with an insoluble polymer-Cu complex was provoked by vigorous stirring of the reaction mixture (Table 14)[151]. The polymer

Table 13. Kinetic and activation parameters for poylmerization of 2,6-dimethylphenol catalyzed by Cu complexes

Catalyst	Polymn. rate · 10^4 mol/l · min	1/K_m[a] 1/mol	k_2[b] 1/min	$\Delta H^{\ddagger}$ kcal/mol	$\Delta S^{\ddagger}$ e. u.
QPVP-Cu	1.4	190	0.10	15	−4.9
QPVP-Cu[c]	0.68	66	0.059	25	29
PVP-Cu	0.93	68	0.062	26	33
pyridine-Cu	0.10	22	0.038	30	42

[a] Michaelis constant.
[b] Rate constant based on Michaelis-Menten equation.
[c] A large amount of neutral salt was added (NaClO$_4$ 0.1 mol/l). DMSO solvent, 30 °C, in air atmosphere. QPVP = partially quaternized PVP, PVP = poly(4-vinylpyridine).

catalyst was readily filtered in high yield and was reused repeatedly. The following results confirm that the insoluble polymer complex heterogeneously catalyzed the reaction. (i) The amount of Cu ions eluted in the solution phase did not exceed 0.1% when the reaction was completed. (ii) The polymer-Cu(I) complex was quantitatively formed by treatment of the insoluble polymer-Cu(II) complex with a large excess of XOH under a nitrogen atmosphere, hence all of the Cu(II) ions in the insoluble polymer complex were considered to be effective as catalytic sites. (iii) Coordination of XOH to the insoluble polymer-Cu complex and a change in the oxidation number of Cu were observed to accompany the catalysis. The first requisite of the catalyst which produces the PPO polymer of higher molecular weight was presumed to be that it must be composed of polymer ligands having an affinity with the solvent used, e.g. poly(styrene) derivatives in benzene, further that its effective Cu concentration should be high and its steric hindrance small.

Polymerization proceeded about 5 times faster in an alkaline solution, and the side reaction that forms biphenoquinone was suppressed[151]. The acceleration effect is due to the acid dissociation of XOH by alkali, but the monomeric pyridine-Cu complex was hydrolyzed at the same time, therefore the acceleration was not observed in the pyridine-Cu system. The PVP-Cu complex is relatively stable toward alkali due to its chelate structure; thus, the PVP-Cu catalyst was active during the polymerization even in an alkaline solution.

The QPVP-Cu complex, which is an emulsifier, catalyzed the emulsion polymerization of XOH[152]. An aqueous solution of the QPVP-Cu complex and a benzene solution of XOH were mixed and stirred, then the system was emulsified and polymerization occurred. After polymerization, the reaction mixture could be separated into two layers, the aqueous catalyst solution, and the benzene solution containing the PPO polymer. The recovered catalyst solution could be used repeatedly. Polymerization was influenced by the pH value of the aqueous catalyst phase. Below pH 7 the main product was biphenoquinone and at pH 8–10 it was PPO. The polymerization rate catalyzed by the Cu complex decreased in the following order: QPVP-Cu > pyridine-Cu with benzylpyridinium chloride added as emul-

Table 14. Oxidative polymerization of 2,6-dimethylphenol catalyzed by polymer-Cu complexes

Catalyst	Solvent	System	[Cu]	React. time h	Yield of PPO wt%	M.W. of PPO $\times 10^{-3}$	Recovery of catalysts wt%	Ref.
QPVP-Cu	DMSO	Homogeneous	0.01	1	96	1.5	–	149)
Py-Cu	DMSO	Homogeneous	0.01	5.5	64	1.6	–	149)
PVP-Cu	Benzene	Homogeneous	0.005	1.5	99	7.1	–	148)
Py-Cu	Benzene	Homogeneous	0.005	1.5	88	5.8	–	148)
PDA-Cu	CCl$_4$	Homogeneous	0.005	5	90	6.6	–	158)
Et$_3$N-Cu	CCl$_4$	Homogeneous	0.005	5	89	13.2	–	158)
PVP-Cu	Benzene	Heterogeneous	0.05	2	86	3.0	86	151)
DBQP-Cu	Methanol	Heterogeneous	0.05	5	42	1.2	96	155)
IDA-Cu	Benzene	Heterogeneous	0.05	10	22	–	100	155)
QPVP-Cu	Water/benzene	Emulsion	0.01	1.5	88	4.8	87	152)

PPO = poly(2,6-dimethylphenyleneoxide), QPVP = partially quaternized PVP, Py = pyridine, PVP = poly(4-vinylpyridine), PDA = partially diethylaminomethylated poly(styrene), DBQP-partially crosslinked PVP with 1,4-dibromobutane, IDA = chelating resin containing iminodiacetic acid group.

sifier > pyridine-Cu. Polymerization was enhanced by addition of the low-molec-ular-weight emulsifier because it increased the interface area between the aqueous catalyst solution and the benzene solution of the monomer. The rate of QPVP-Cu catalyzed polymerization was much larger than those of the other systems. It was considered that QPVP not only acted as an emulsifier but also concentrated the Cu catalyst at the interface where polymerization proceeded (*54*).

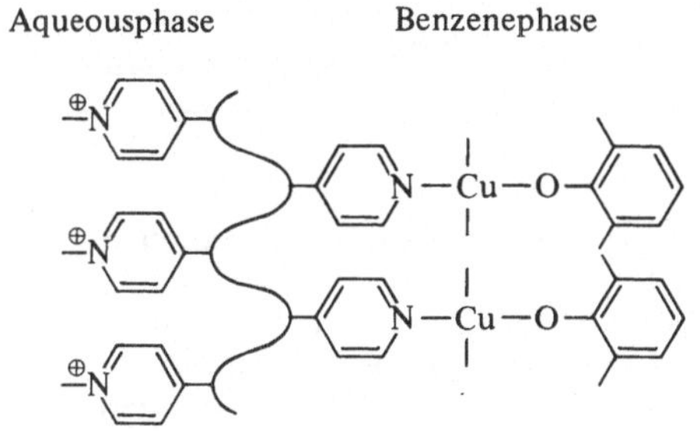

Aqueousphase   Benzenephase

54

B. Elementary Reactions of Phenol Oxidation

The catalytic cycle of Cu-complex-catalyzed oxidation is illustrated in Scheme 14, the example used being the dimerization of 2,6-dimethylphenol $(XOH)^{153)}$. In the first step, the substrate coordinates to the Cu(II) complex and one electron trans-fers from the substrate to the Cu(II) ion. Then the activated substrate dissociates

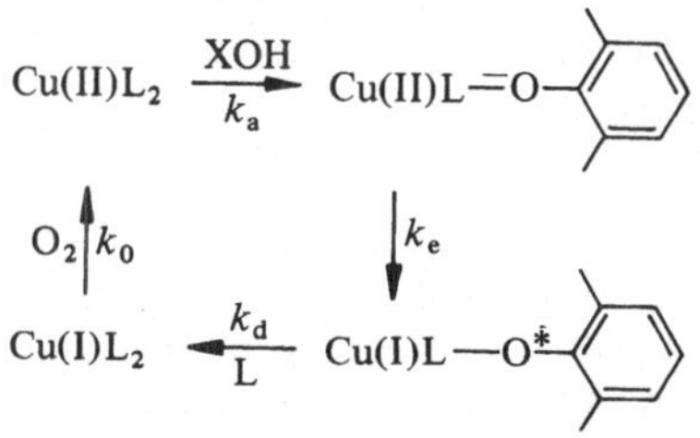

Scheme 14

from the catalyst and the reduced catalyst is reoxidized to the original Cu(II) com-plex by oxygen. The rate constants of each step are represented by k_a, k_e, k_d, and k_0, respectively. It is important to study these elementary reactions in order to determine the catalytic mechanism, especially the catalytic effects of a polymer ligand, which are really complicated. Many investigations of metal-complex-cata-lyzed reactions have been reported, but most of these results were obtained by observing the overall reaction. The details of the catalysis are still undetermined because of its complexity and speed.

    With reference to Scheme 14, the *d-d* spectrum of the Cu complex is expected to change at each step of the catalysis, because the composition of the Cu complex and the oxidation number of the Cu ion are not the same at each step. The change in the visible spectrum of the reaction solution when XOH was added to the Cu

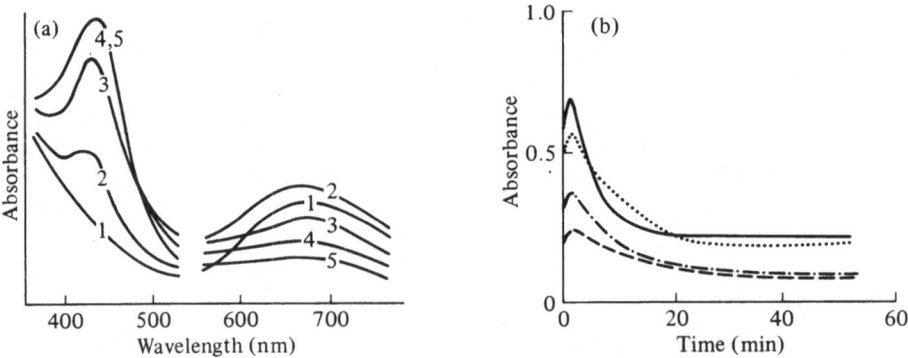

Fig. 27. Change in visible spectra of the reaction system with time (a) and $d-d$ absorption change of polymer-Cu catalyst with time (b)[154]
(a) 1: without substrate(2,6-dimethylphenol), 2: 1 min, 3: 3 min, 4: 10 min, 5: 30 min after addition of substrate, (b) ——: partially quaternized poly(4-vinylpyridine)-Cu catalyst, ······: pyridine-Cu catalyst, —·—: partially diethylaminomethylated poly(styrene)-Cu catalyst, ———: triethylamine-Cu catalyst

catalyst solution was observed by means of a rapid-scanning spectrophotometer [(Fig. 27(a)][154]. The absorbance of the d-d spectrum changed with time, as illustrated in Figure 27(b), so that the rate constants for each elementary step were determined by the following methods[153].

Step 1. Coordination of the substrate: immediately after mixing the Cu(II) catalyst with the substrate, a rapid change in the absorption is observed within several decaseconds [Fig. 27(b)]; this is believed to be caused by the coordination of the substrate to the Cu(II) complex. We measured this rapid change spectroscopically by the stopped-flow method, and calculated the apparent rate constant k_a. When an insoluble polymer complex is used as a catalyst, a decrease in the XOH concentration in the liquid phase corresponds to the coordination of XOH to the Cu catalyst in the solid phase[155].

Step 2. Electron-transfer: the decrease in the d-d absorption subsequent to the steep rise of the absorbance in Fig. 27(b) corresponds to the formation of Cu(I) ions, $i.e.$ electron transfer from the substrate to Cu(II) ions. The d orbitals of Cu(I) are perfectly occupied, so that d-d absorption spectrum does not appear in the Cu(I) complex. The rate constant k_e is obtained from the falling curve of the d-d absorption. The fraction of Cu(II) ions is also determined from the magnetic susceptibility of the reaction system. The Cu(II) ion contains an unpaired electron and is paramagnetic, while the Cu(I) ion is diamagnetic. Magnetic measurement is applicable particularly to an insoluble catalyst system[155].

Step 3. Dissociation of the activated substrate: both before and after the dissociation step of the activated substrate the complex is composed of Cu(I), so that neither spectral nor magnetic measurements on the Cu catalyst are effective. However, this step can be monitored by ESR spectrometry on the substrate. The ESR signals due to the activated phenol (phenoxy radical) appear at about $g = 2.003$.

Step 4. Reoxidation of the catalyst: the catalyst is regenerated by oxidizing the Cu(I) complex with oxygen. The rate constant k_0 of the catalyst-recycle step can therefore be determined by measuring the increase in d-d absorption or in magnetic susceptibility when the Cu(I) complex solution is mixed with a solvent saturated with oxygen.

Fraction of complex species in the steady state: after several decaminutes the steady state appears to be reached because the d-d absorption becomes constant, as shown in Fig. 27(b), and the rate of oxygen consumption which accompanies the reaction is also held constant. The amount of Cu(II) complex (upper half of the catalytic cycle illustrated in Scheme 14) relative to Cu(I) complex (lower half of the cycle) in the steady state is determined from the ratio (absorbance of the d-d spectrum in the steady state)/(maximum absorbance immediately after substrate addition). The fraction of Cu(II) complex is also determined by magnetic measurement. The amount of substrate-coordinated Cu complex (right half of the catalytic cycle) relative to the Cu catalyst (left half) in the steady state is also obtained by spectroscopic measurement.

Table 15. Rate constants of elementary reactions in Cu-catalyzed oxidation of 2,6-dimethylphenol

Catalyst	Solvent	System	k_a l/mol·min	k_e l/min	k_o l/min	$\dfrac{[Cu-XOH]}{\Sigma\,[Cu]}$	$\dfrac{[Cu(II)]}{\Sigma\,[Cu]}$
QPVP-Cu	DMSO	Homogeneous	19	0.027	0.95	–	0.24
pyridine-Cu	DMSO	Homogeneous	71	0.011	11	–	0.28
PVP-Cu	Benzene	Heterogeneous	2.5	0.0015	0.075	0.81	0.62

k_a = rate constant of coordination step of substrate, k_e = rate constant of electron-transfer step, k_o = rate constant of reoxidation step of Cu catalyst, $[Cu-XOH]$ or $[Cu(II)]/\Sigma[Cu]$ = fraction of the substrate-coordinated Cu catalyst or of Cu(II) catalyst in the steady state of the reaction.

Table 15 shows the rate constants of each elementary step as determined by the above-mentioned methods[153, 155]. A comparison of the approximate magnitudes of each rate constant indicates that the electron-transfer step is the slowest one in the catalytic cycle. From the fractions of the catalyst species in the steady state of the reaction, it is possible to construct a profile of the Cu complex on a QPVP ligand, which is acting as a catalyst for the oxidative polymerization of XOH, as illustrated in Fig. 28.

C. Coordination Step of Substrate to Polymer-Cu Catalysts

As Table 15 shows, the rate constant k_a of the polymer catalyst system is smaller than that of the monomeric pyridine-Cu catalyst, because of the bulky polymer ligand obstructing the coordination of the substrate to the catalyst[153].

The formation constant (K_a) of the substrate-coordinated complex, polymer-ligand-Cu(II)-substrate, was kinetically determined for the dimerization of XOH

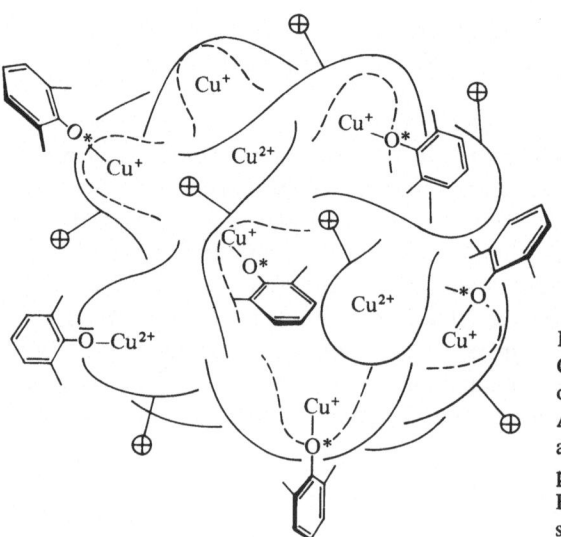

Fig. 28. Schematic profile of polymer-Cu catalyst in steady state of oxidation of 2,6-dimethylphenol
About 72 Cu ions are coordinated on a partially quaternized poly(4-vinyl-pyridine) ligand (DP = 200, Q% = 28). Fraction of Cu(II) = 0.24, fraction of substrate-coordinated Cu = 0.80

$$Cu(II)L + S \underset{K_a}{\overset{}{\rightleftharpoons}} LCu(II)S \overset{k_2}{\longrightarrow} Cu(I) + S* \tag{21}$$

L = ligand, S = substrate

and the polymerization of oligo(2,6-dimethylphenyleneoxide); the results are listed in Table 16[150]. K_a of the PDA-Cu catalyst is small, especially in the polymerization of the oligomer. From this result plus the fact that the molecular weight of the polymer obtained by the PDA-Cu catalyst is smaller (Table 14), it is concluded that the substrate is difficult to coordinate to the Cu(II) ion in the PDA-Cu complex, which has a chelate structure, because a chelate structure causes large steric hindrance and is itself very stable. Consequently, the oligomer of XOH is even more difficult to coordinate and the polymer-Cu-catalyzed polymerization of the oligomer proceeds much more slowly than the dimerization and that catalyzed by the monomeric Cu complex.

Table 16. Kinetic data for polymerization of 2,6-dimethylphenol catalyzed by polymer-Cu complex

Substrate	Catalyst	Rate · 10^4 mol/l · min	K_a 1/mol	k_2 1/min
XOH	PDA-Cu	7.3	9.7	0.19
XOH	Et$_3$N-Cu	3.1	36	0.067
oligo-PO	PDA-Cu	0.45	4.0	0.16
oligo-PO	Et$_3$N-Cu	1.0	24	0.085

XOH = 2,6-dimethylphenol, oligo-PO = oligo(2,6-dimethylphenyleneoxide)(DP 24), PDA = partially diethylaminomethylated poly(styrene), benzene solvent, 30 °C. K_a and k_2 are formation constant and rate constant in Eq. (21).

On the other hand, an attempt to accelerate the step of coordination of the substrate to the Cu catalyst was successful because it used the hydrophobic domain of the polymer ligand[156]. That was the oxidation catalyzed by polymer-Cu complexes in a dilute aqueous solution of phenol, which occurred slowly. The substrate was concentrated in the domain of the polymer catalyst and was effectively catalyzed by Cu in the domain. A relationship was found to exist between the equilibrium constant (K_a') for the adsorption of phenol on the polymer ligand and the catalytic activity (V) of the polymer-ligand-Cu complex for various polymer ligands: $K_a' = 0.21$ l/mol and $V = 10^{-6}$ mol/l min for QPVP, $K_a' = 26$ and $V = 10^{-4}$ for PVP, $K_a' = 52$ and $V = 10^{-4}$ for the copolymer of styrene and 4-vinylpyridine (PSP) (styrene content 20%), and $K_a' = 109$ and $V = 10^{-3}$ for PSP (styrene content 40%). The V value was proportional to the K_a' value, and both K_a' and V increased with the hydrophobicity of the polymer ligand. The oxidation rate catalyzed by the polymer-Cu complex in aqueous solutions depended on the adsorption capacity of the polymer domain.

This hydrophobic polymer-Cu catalyst was utilized to remove phenol from an aqueous solution[157]. A dilute aqueous solution of phenol was made to flow through the column packed with the PSP-Cu resin. Phenol was adsorbed on the polymer and furthermore oxidized to insoluble polymeric compounds in the column.

D. Effects of Polymer Ligands on the Electron-Transfer Step

A phenoxy anion coordinates to a Cu(II) catalyst and an electron is transferred from the coordinated phenoxy anion to the Cu(II) ion. This electron-transfer step [Eq. (22)] is an intracomplex process, and it is expected that the electronic and steric factors of the amine ligands(L) in the Cu complex as well as the electronic

$$(22)$$

and steric factors of the substituent groups (R) of the coordinated phenol directly affect the rate of the electron-transfer step. As mentioned previously with reference to Table 15, the electron-transfer step is the slowest elementary reaction in the catalytic cycle of phenol oxidation. Therefore, it is presumed that the rate constant of the electron-transfer step (k_e) is one of the important parameters to represent the reactivity of Cu-catalyzed phenol oxidation.

The reactivity of phenols in the Cu-complex-catalyzed oxidation was studied by measuring the oxidation rate, the rate constant k_e, and the redox potentials of the Cu complex and of the phenol (Fig. 29)[158]. The logarithm of the oxidation rate is proportional to log k_e, which supports the assumption that the electron-transfer step is rate-determining. A linear relationship is observed between log k_e and Hammett's σ value of the phenol, which is proportional to the oxidation potential of the phenol.

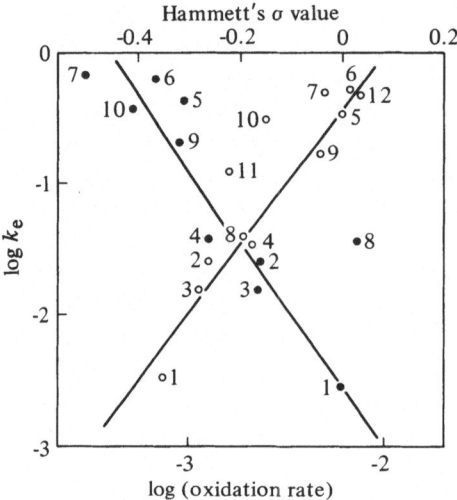

Fig. 29. Relationship between rate constant of electron-transfer step (k_e), oxidation rate, and Hammett's σ value of phenol for the pyridine-Cu-catalyzed oxidation of phenols[158]

Phenol	R^1	R^2	R^3
1	H	H	H
2	CH_3	H	H
3	H	H	CH_3
4	OCH_3	H	H
5	CH_3	CH_3	H
6	$C(CH_3)_3$	$C(CH_3)_3$	H
7	OCH_3	OCH_3	H
8	C_6H_5	C_6H_5	H
9	CH_3	H	CH_3
10	OCH_3	H	CH_3
11	OCH_3	H	$CH_2-CH=CH_2$
12	OCH_3	H	$CH=CH-CH_3$

A linear relationship is also observed between log k_e and the reduction potential of the Cu-complex catalyst[158]. The plots for Cu complexes involving polymer ligands, however, deviate from the linear relationship in that the k_e values for polymer-Cu complexes are often larger than those for Cu complexes involving low-molecular-weight ligands[153, 159]. This deviation is presumed to be due to the effects of polymer ligands.

The overall reaction rate and the rate constant of the electron-transfer step are summarized in Table 17 for the polymer-Cu-catalyzed oxidation of substrates such as 2,6-dimethylphenol (XOH) and ascorbic acid[159]. The k_e values for polymer-Cu-catalyzed oxidation are larger than those for monomeric-Cu-catalyzed oxidation. Particularly in the oxidative polymerization of XOH, it is obvious that the electron-transfer step is accelerated by polymer ligands, and the large value of k_e is in agreement with the higher rate of polymer-Cu-catalyzed polymerization. Therefore, the high catalytic activity of the polymer-Cu complex is considered to be due to the larger value of k_e, which is the rate constant of the slowest step in the catalytic cycle of the polymerization.

From Table 17, one notices that the k_e value of the particularly quaternized poly(4-vinylpyridine)(QPVP)-Cu complex is reduced and is nearly equal to the k_e value of the poly(4-vinylpyridine) catalyst when a large amount of neutral salt is added to the system. It is also found by comparing the activation parameters of the electron-transfer step that the acceleration by the QPVP-Cu catalyst is due to the remarkable lowering of the activation enthalpy.

In order to discuss the reasons why the k_e value of the QPVP-Cu catalyst is large, the effect of the quaternized part of the QPVP-ligand on k_e was studied under conditions where the concentration of Cu and the unquaternized pyridine unit in QPVP were held constant[159]. Figure 30 shows that k_e rises with the increase in the percentage of the quaternized part of the QPVP ligand (Q%) and k_e is reduced by the addition of neutral salt, which shields the charges on the QPVP ligand. In contrast to k_e, the rate constant of the reoxidation step of the catalyst (k_0) decreases with Q%. The electron-transfer step is the reduction of the Cu(II) complex, and the reoxidation step is the oxidation of the Cu(I) complex. Thus, a factor which is favorable for k_e is unfavorable for k_0. An increase in Q% also brings an increase in the viscosity of the QPVP-Cu-complex solution. This profile suggests that the shape of the QPVP catalyst is an important factor.

QPVP is a positively charged polymer ligand. When Cu(II) ion was added to the QPVP solution, the viscosity of the solution fell to one eighth that of the free QPVP ligand [$\eta_{sp}/c = 6.23$ dl/g for QPVP, 0.84 for QPVP-Cu(II)][160]. This indicates that the QPVP chain is contracted into a compact structure due to intra-polymer

Table 17. Rate constant and activation parameters of electron-transfer step in Cu-catalyzed oxidation (30 °C)

Substrate	Catalyst	Solvent	Rate · 10^3 mol/l · min	$k_e \cdot 10^2$ l/min	$\Delta H^{\ddagger}$ kcal/mol	$\Delta S^{\ddagger}$ e.u.
XOH	QPVP-Cu	DMSO	1.4	2.7	5.3	−48
XOH	QPVP-Cu[a]	DMSO	0.68	1.4		
XOH	PVP-Cu	DMSO	0.93	1.8		
XOH	Pyridine-Cu	DMSO	0.10	1.1	19	−4.1
XOH	PDA-Cu	Benzene	0.73	1.1		
XOH	Et$_3$N-Cu	Benzene	0.31	0.12		
DOH	QPVP-Cu	DMSO	2.7	3.3		
DOH	Pyridine-Cu	DMSO	2.0	2.6		
AA	QPVP-Cu	DMSO	0.46	210	7.0	−34
AA	Pyridine-Cu	DMSO	0.55	170	12	−19
AA	QPVP-Cu	Methanol	0.23	3400		
AA	Cu	Methanol	0.08	1400		

XOH = 2,6-dimethylphenol, DOH = the dimer of XOH, AA = ascorbic acid, QPVP = partially quaternized PVP, PVP = poly(4-vinylpyridine), PDA = partially diethylaminomethylated poly-(styrene).

[a] NaClO$_4$(0.1 mol/l) was added as a neutral salt.

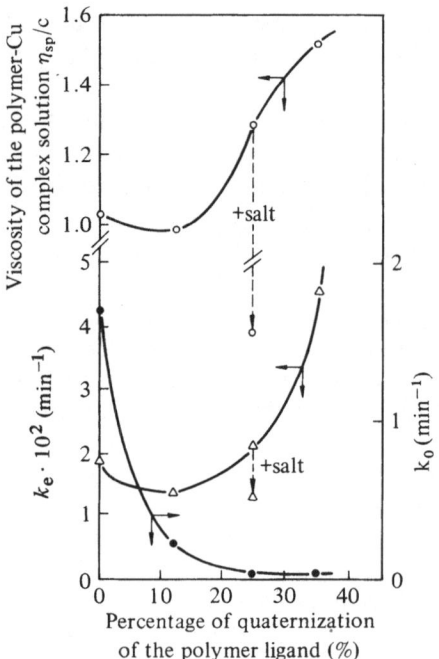

Fig. 30. Effect of quaternization of poly-(4-vinylpyridine) ligand on rate constant of electron-transfer step (k_e), rate constant of reoxidation step of catalyst (k_o), and viscosity of polymer-Cu-catalyst solution[159]

chelation, as illustrated in Fig. 8 (b). When the Cu(II) complex was reduced to the Cu(I) complex, the viscosity showed an increase [$\eta_{sp}/c = 1.27$ for QPVP-Cu(I)]. This result indicates that the contracted shape of the polymer complex expands somewhat during the electron-transfer step.

The expansion and contraction of the polymer chain which accompanies the redox of Cu ions can also be visually confirmed by means of the mechanochemical system proposed by Kuhn[161], as illustrated in Fig. 31. A film is prepared with a poly(vinylalcohol)-Cu(II) complex and is suspended with a sinker in water. The film is extended by about 20% on the reduction of Cu(II) to Cu(I) and shrinks back to its original length on the oxidation of Cu(I). The poly(vinylalcohol) chain is densely crosslinked by the extremely stable Cu(II) chelate, but is loosened when Cu ion forms the unstable Cu(I) chelate. This change is reversible as may be observed.

On the basis of these results, the acceleration mechanism of the electron-transfer step in the polymer catalyst is illustrated schematically in Fig. 32 and explained

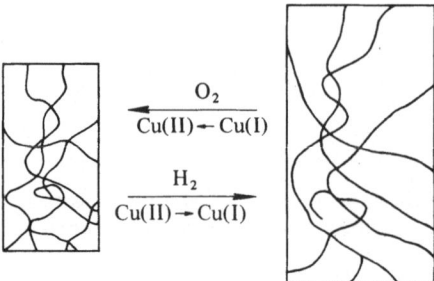

Fig. 31. Mechanochemical system poly-(vinylalcohol)-Cu film[161]

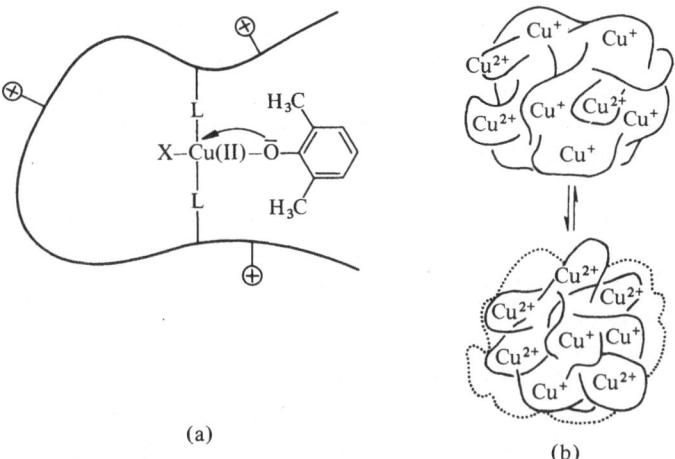

(a)   (b)

Fig. 32. Electron transfer from substrate to Cu(II) ion in polymer-Cu catalyst (a) and fluctuating profile of polymer catalyst (b)

as follows[150, 160]. The activation energy of an electron-transfer reaction is the energy required to rearrange the complex structure. Thus, before electron transfer the coordinate bond between the Cu(II) ion and the ligands must be stretched and the Cu(II) complex must be rearranged to a structure which can accept one electron. In the polymer complex, the coordinate bonds between the Cu(II) ion and the polymer ligand are probably weakened by the strain produced in the polymer-ligand chain by electrostatic repulsion between the cations on the polymer chain or by the steric bulkiness of the polymer ligand. Thus, less energy is required to stretch the coordinate bonds and electron transfer occurs more easily than in the monomeric analogue.

In the catalyst reoxidation step, contrary to the electron-transfer step, the polymer ligand should shrink because of the formation of the Cu(II) complex. Therefore, the polymer chain may partially repeat are expansion and contraction occurring during the catalytic cycle. When one has a view of the polymer-Cu catalyst as a whole, each part of the polymer catalyst domain, which is drifted in solution, is seen to be fluctuating during the catalytic process [Fig. 32(b)]. The fluctuating shape of biopolymers in enzymic reactions has been pointed out, and the dynamically conformational change of a flexible polymer chain is considered to be one of the effects of the polymer catalyst.

Other attempts have been reported to control the electron-transfer step by using a polymer ligand.

The catalytic activity of the metal complex on the oxidative reaction in solution is much influenced not only by the species and the structure of the complexes but also by the chemical environment around them. For instance, in the oxidative polymerization of phenols catalyzed by a Cu complex, the reaction rate varied about 10^2 times with changes in the composition of the solvent, and the highest rate was observed for polymerization in a benzene solvent[162]. Thus, we used the copolymer of styrene and 4-vinylpyridine(PSP) as the polymer ligand and studied the effect on the catalysis of the non-polar field formed by the polymer ligand[163].

The catalytic activity of the PSP-Cu complex was greater than that of the other polymer-Cu complexes, and increased with the styrene content of the PSP ligand. The local chemical environment of the polymer-metal complex can be considered to be fairly different from that of the monomeric analog due to the effect of the polymer chain and the neighboring groups, even if part of the complex is the same in both.

It was stated above that the metal complex is reduced to a lower valence state on the electron-transfer step, and then is oxidized again by molecular oxygen. But it is more correct to say that oxygen not only acts on the reoxidation step but also participates in and influences the electron-transfer step[147]. When molecular oxygen acts as an oxidizing agent, 1 mol of oxygen needs one electron and 4 protons to produce 2 mol of H_2O. The following three processes are known for the course of this reaction[164]. Equation (23) is thermodynamically unfavorable: in the first step, which mainly determines the velocity of the reaction, a high energy barrier

(i) Four one-electron-transfer steps (slow) (redox potential in V)
$$O_2 + H^+ + e^- \rightarrow HO_2\cdot \ (-0.32 \ V); HO_2^- + H^+ + e^- \rightarrow H_2O_2 \ (1.68 \ V);$$
$$H_2O_2 + e^- \rightarrow HO\cdot + OH^- \ (0.80 \ V); HO\cdot + H^+ + e^- \rightarrow H_2O \ (2.74). \qquad (23)$$

(ii) Two two-electron-transfer steps (faster)
$$O_2 + 2H^+ + 2e^- \rightarrow H_2O_2 \ (0.68 \ V); H_2O_2 + 2H^+ + 2e^- \rightarrow 2H_2O \ (1.77 \ V). \quad (24)$$

(iii) One four-electron-transfer step (the fastest step)
$$O_2 + 4H^+ + 4e^- \rightarrow 2H_2O \ (1.23 \ V). \qquad (25)$$

must be overcome to get the starting radical. If the oxidation proceeds in two-electron-transfer steps (24) the redox potential is higher. The highest redox potential can be realized in a four-electron-transfer step (25), and oxygen should be a powerful oxidant.

Manecke et al.[165] synthesized a semiconducting polymeric complex which possessed both bis(ethylene-1,2-dithiolato)Cu(II) and a phthalocyanine-Cu(II)-type structure 54. This Cu complex exhibited high catalytic activity in the oxidative polymerization of XOH, about 50 times higher than that of pyridine-Cu. A synchronous four-electron-transfer mechanism was proposed for the catalysis of 54. The phthalocyanine-Cu(II) type structure of 54 is presumed to form a complex with molecular

Dithiolato-
Cu(II) structure
(coordination site)

Cu(II)phthalocyanine-
-type structure
(electron-accepting site)

54   55

oxygen, and the four end groups of *54*, namely the (ethylene-1,2-dithiolato)Cu(II) structure, probably act as coordination sites for the phenol. Therefore, it should be possible that the complex *54* catalyzes a four-electron-transfer step *55*.

Such a mechanism might play an important role in a metal-ion-catalyzed enzymic oxidation *in vivo*, in which metal ions work cooperatively[166]. A synchronous four-electron-transfer requires a specific spatial arrangement which should be possible in a macromolecular environment.

Control of the electron-transfer step was also attempted by combining two metal species on a polymer ligand[167]. We prepared polymer-metal complexes involving both the Cu(II) and Mn(III) ions. The oxidative polymerization of XOH catalyzed by the PVP-Cu, Mn mixed complex or the diethylaminomethylated poly(styrene)(PDA)-Cu Mn mixed complex proceeded 10 times faster than the polymerization catalyzed by either PVP- or PDA-metal complex. The maxima of the activity observed at [Cu]/[Mn] $\simeq$ 1 and [polymer]/[Cu,Mn] $\simeq$ moderately small where Cu and Mn ions were crowded within the contracted polymer chain. Cooperative interaction between Cu and Mn was inferred. The rate constant of the electron-transfer step (k_e in Scheme 14) for Cu(II) $\rightarrow$ Cu(I) was much larger than that for Mn(III) $\rightarrow$ Mn(II). The rate constants of the reoxidation step (k_o) were polymer-Mn $\simeq$ polymer-Cu,Mn $>$ polymer-Cu, so the rapid redox reaction

$$Cu(I) + Mn(III) \xrightarrow{\text{polymer}} Cu(II) + Mn(II)$$ was recognized. The schematic redox cycles were proposed where two catalytic cycles geared each other: Cu ion and Mn ion, which existed on the polymer chain in high concentration, interacted each other and accelerated the catalytic reaction (Scheme 15).

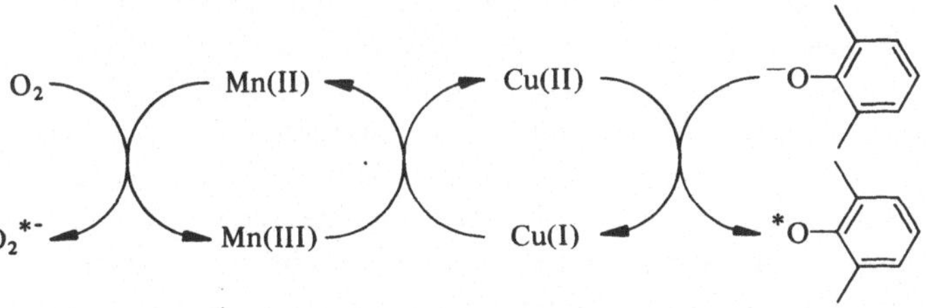

Scheme 15

Even the basic properties of mixed metal or heteronuclear complexes and mixed valence metal complexes have not been studied, so that their catalytic effects are far from clearly understood.

At present, it is not yet possible to predict whether a polymer ligand influences a reaction as catalyst or as inhibitor, because the catalytic mechanism of one reaction differs from another and the rate-determining step varies with the reaction conditions. Furthermore, as seen in Fig. 30, the effect of the polymer ligand on the electron-transfer step was positive, but was negative on the reoxidation step. It is also presumed that the steric hindrance of a polymer ligand has a negative effect on the coordination step of the substrate but a positive one on the dissociation step of the activated substrate. Thus, the effects of the polymer ligand would cancel each other out during the catalytic cycle. There still remain many difficulties in designing a polymer catalyst which will exhibit high efficiency and selectivity.

VIII. Conclusion and Further Studies

The characteristics of synthetic polymer-metal complexes having uniform structure were illustrated. The chemical reactivity of a metal complex is often affected by the addition of a polymer ligand that exists outside the coordination sphere and surrounds the metal complex. The effects of polymer ligands have been summarized under two heads: steric effects, and environmental effects.

1. The steric effects are determined by the conformation of the polymer-ligand chain. A complex attached to a rigid and extended polymer chain at low concentration behaves virtually as in a solution at infinite dilution. This dilution effect has been well established in the oxygenation of polymer-Fe(II) porphyrin complexes (cf. Section VB). The irreversible oxidation of Fe(II)porphyrin via dimerization was inhibited by the polymer ligand. When, on the contrary, metal complexes were attached to a polymer ligand at high concentration as in the polymer-Co(III) complex (see Fig. 2) and in the intrapolymer Cu(II) chelate (Fig. 28), the interaction between the crowded metal ions was observed in the catalytic action of the latter Cu complex. A locally high concentration of ligand was produced by the densely coiled conformation of the polymer ligand (Section IIIA). The stability of the Cu chelate is closely associated with the contracted conformation of the polymer ligand, so that its metal ion-binding ability is determined by the primary structure and the crosslinking structure of the polymer ligand (Section IIIB).

The steric effect of a flexible polymer ligand is not static. The coordinating ability of a polymer ligand varies with the conformation of the polymer chain; the coordinate bond was weakened by the addition of a poor solvent of the polymer ligand because of the steric hindrance of the contracted polymer ligand chain (Fig. 6). The dynamic change in the higher structure of a polymer ligand was clearly demonstrated in the pseudo-allosteric phenomenon of the poly(L-lysine)-heme complex. The helical structure of poly(L-lysine) and its conformational change are considered to play an important role in transferring the information of oxygenation to the neighboring heme groups (Section VC). The effect of the dynamic conformational change was also pointed out in the electron-transfer step of the polymer-Cu catalyst (Fig. 31). The reaction of the metal complex brought about a structural change within its coordination sphere, which itself induced a conformational change in the polymer ligand.

2. The environmental effects are caused by the micro-environments constituted by the domain of a polymer ligand. The electrostatic domain of a polymer-metal complex was demonstrated in the reaction of the polymer-Co(III) complex with ionic species (Section IVA), and was shown to be utilized in the catalytic activity of the polymer-Cu complex (Section VIA). In another case, the hydrophobic domain was predominant, i.e. in the reaction with hydrophobic substrates (Sections IVB and VIIC). The environmental effects of a polymer ligand also include dynamic effects, which vary with the solution conditions (Section IIIC).

As summarized above, by using polymer-metal complexes of uniform structure, much progress has already been achieved in the quantitative study of the relationship between the chemical function and the effects of a polymer chain, in comparison with previous studies in the fields of other polymeric chemical reagents; organic polymer catalysts such as hydrolysis catalysts, organometallic polymer catalysts, polymeric carriers as seen in peptide syntheses, polymer-supported enzymes, ion-exchange resins, and redox polymers. This progress is mainly attributed to the fact that much information about the structure and properties of a metal complex, *i.e.* its functional site, can be obtained by physicochemical measurements, *e.g.* of visible spectra, ESR, NMR, CD, magnetic effects, and redox potentials. However, more detailed study is required for further development. The aims should be (i) to determine the microstructure of metal complexes; *e.g.* configuration within the coordination sphere, strain of the coordinate bond, and spin state of the metal ion in a polymer-metal complex; (ii) to study the higher structure, the microconfiguration, and the dynamic conformational change of a polymer ligand; (iii) to develop new methods in order to follow both the reaction of the complex and the conformation of the chain at the same time; (iv) to carry out a statistical treatment of the stability and structure of a polymer complex involving a labile metal ion; (v) to evaluate the strength of the environments constructed by polymer ligands, etc.

It seems to us that the potential applications of polymer-metal complexes outweight the limitations of the study. The combination of a metal complex with an organic polymer often offers possibilities not attainable in other ways. For example, polymer-metal complexes that bind molecular oxygen reversibly will be developed for the purpose of isolating pure oxygen from air. Other applications of polymeric oxygen carriers include their use as catalysts for reactions of molecular oxygen. It is even possible that they may eventually find a use in artifical blood. The catalytically active polymer-metal complexes are not only produced in the form of resins and films easily separated from the reaction mixture, they are also catalysts of high efficiency and selectivity. Although previous studies have been concerned with the catalysis of Cu, Co and Fe, attention must be paid to the activity of other metal ion species such as Zn, Mo, and Mn. A series of catalytic cycles geared to each other is finally expected in which various metal ions act cooperatively. In another application, polymer-metal complexes will be utilized as photoresponsive materials and developed into electron-transferring agents. The polymer ligands which selectively adsorb a metal ion involve interesting applications for the removal of heavy metal ions from effluents and the recovery of rare metal ions.

Study of the polymer-metal complex is still a new field, and not all possible applications of metal complexes have been explored yet. Further demonstrations of new, unique applications of the polymer-metal complexes offering advantages over the conventional properties of both polymers and metal complexes are to be expected in the near future.

IX. References

1) Kurimura, Y., Tsuchida, E., Kaneko, M.: J. Polymer Sci. Al, 9, 3511 (1971).
2) Tsuchida, E., Nishide, H., Takeshita, M.: Makromol Chem. 175, 2293 (1974).
3) Tsuchida, E., Karino, Y., Nishide, H., Kurimura, Y.: Makromol. Chem. 175, 171 (1974).
4) Tsuchida, E., Nishide, H., Ohkawa, T.: Bull. Sci. Res. Lab., Waseda Univ. 69, 31 (1975).
5) Nishide, H., Tsuchida, E.: Makromol. Chem., to be published.
6) Biedermann, H., Wickmann, K.: Z. Naturforsch. 27, 1332 (1972); 28, 182 (1973).
7) Bedetli, R., Carunchio, V., Cernia, E.: J. Polymer Sci., Letter 13, 329 (1974).
8) Tsuchida, E., Nishide, H., Kurimura, Y.: Functional polymers, Tokyo: Kyoritsu Shuppan 1974, p. 92.
9) Blauer, G.: Acta. Chem. Scand. 17, 8 (1963); Biochim. Biophys. Acta 133, 206 (1976); Arch. Biochem. Biophys. 121, 587 (1967).
10) Yamamoto, S., Nozawa, T., Hatano, M.: Polymer 15, 330 (1974).
11) Tohjo, M., Shibata, K.: Arch. Biochem. Biophys. 103, 401 (1963).
12) Hatano, M.: Kagaku to Kogyo 18, 926 (1965).
13) Tsuchida, E., Honda, K., Sata, H.: Biopolymers 13, 2147 (1974).
14) Tsuchida, E., Honda, K., Hasegawa, E.: Biochim. Biophys. Acta 393, 483 (1975).
15) Nishide, H., Hata, S., Tsuchida, E.: Biopolymers, to be published.
16) Tsuchida, E., Shigehara, K., Miyamoto, K.: J. Polymer Sci., Chem. Ed. 14, 911 (1976).
17) Miller, J. R., Drough, C. D.: J. Am. Chem. Soc. 74, 3977 (1952).
18) Tsuchida, E., Hasegawa, E., Honda, K.: Biochim. Biophys. Acta 427, 520 (1976).
19) Tsuchida, E. Hasegawa, E., Honda, K.: J. Polymer Sci., Chem. Ed. 13, 1747 (1975).
20) Wang, J. H., Bringar, W. S.: Proc. Natl. Acad. Sci. U.S. 45, 958 (1960).
21) Gallagher, W. A., Elliott, W. B.: Biochem. J. 97, 187 (1965); 108, 131 (1968).
22) Scheler, W. A.,Rahmel, G.: Acta. Biol. Med. Germ. 10, 218 (1963); Simplicio J. Schwenzer, K., Maepa, F.: J. Am. Chem. Soc. 97, 7319 (1975).
23) Nishide H Mihayashi, K., Tsuchida, E.: Biochim. Biophys. Acta, (1977) in press.
24) Abe, K., Koide, M., Tsuchida, E.: J. Polymer Sci., Chem. Ed., (1977) in press.
25) Beetlestone, J., George, P.: Biochemistry 3, 707 (1961); George, P., Beetlestone, J., Griffith, J. S.: Revs. Modern Phys. 36, 441 (1964).
26) Tsuchida, E., Honda, K., Hata, S., Suwa, H., Kohn, K.: Chem. Letters 1975, 761.
27) Tsuchida, E., Honda, K., Hata, S.: Bull. Chem. Soc., Japan 49, 868 (1976).
28) Yonetani, T., Yamamoto, H., Erman, J. E., Jeigh, J. S., Reed, G. H.: J. Biol. Chem. 247, 2447 (1972).
29) Hata, S., Nishide, H., Tsuchida, E.: J. Chem. Soc., to be published.
30) Kobayashi, K., Tamura, M., Hayashi, K.: Polymer Preprints, Japan 25 (1), 48 (1976).
31) Cassidy, H. G., Kun, K. A.: Oxidation-reduction polymers. New York: Interscience 1965; Pittman, Jr., C. U., Voges, R. L., Jones, W. R.: Macromolecules 4, 291, 298, 302 (1971).
32) Collison, E., Dainton, F. S., Mile, F. R. S. B., Tazuke, S., Smith, P. R.: Nature 193, 26 (1963).
33) Tazuke, S., Okamura, S.: J. Polymer Sci., Al, 4, 141, 2461 (1966).
34) Tomono, T., Honda, K., Tsuchida, E.: J. Polymer Sci., Chem. Ed. 12, 1243 (1974).
35) Osada, Y.: Nippon Kagaku Kaishi 1975, 1089, 1893 (1975).
36) Osada, Y.: Makromol. Chem. 176, 1893 (1975).
37) Nishikawa, H., Terada, E., Tsuchida, E.: Eur. Polymer J., to be published.
38) Gregor, H. P., Luttinger, L. B., Label, E. M.: J. Phys. Chem. 59, 34 (1955).
39) Marinsky, J. A.: Ion exchange and solvent extraction, Vol. 4, Chapter 5. New York: Marcel Dekker 1973.
40) Nishikawa, H., Tsuchida, E.: Bull. Chem. Soc., Japan 49, 1545 (1976).
41) Nishikawa, H., Tsuchida, E.: J. Polymer Sci., Chem. Ed. 14, 1557 (1976).
42) Kirsh, Y. E., Kovner, U. Y., Kokorin, A. I, Zamaraev, K. I., Chernyak, V. Y., Kabanov, V. A.: Eur. Polymer J. 10, 671 (1974).
43) Nishide, H., Deguchi, J., Tsuchida, E.: Bull. Chem. Soc., Japan 49, 3498 (1976).

44) Kokorin, A. I., Zamaraev, K. I., Kovner, V. Y., Kirsh, Y. E., Kabanov, V. A.: Eur. Polymer J. *11*, 719 (1975).
45) Nishide, H., Deguchi, J., Tsuchida, E.: J. Polymer Sci., Chem. Ed., to be published.
46) Nishide, H., Tsuchida, E.: Makromol. Chem. *177*, 2295 (1976).
47) Mandel, M., Leyte, J. C.: J. Polymer Sci., A, *2*, 2883, 3771 (1964).
48) Teyssie, P. H., Decoere, C., Teyssie, M. T.: Makromol. Chem. *84*, 51 (1965).
49) Hojo, K., Shirai, H.: Kogyo Kagaku Zasshi *73*, 1962 (1970).
50) Marinsky, J. A.: Ion exchange, Vol. 1. New York: Marcel Dekker 1966.
51) Gregor, H. P., Luttinger, L. B.: J. Phys. Chem. *59*, 34 (1955).
52) Anspach, W. M., Marinsky, J. A.: J. Phys. Chem. *79*, 433, 439 (1975).
53) Hojo, K., Shirai, H., Hayashi, S.: J. Polymer Sci., Symp. *47*, 299 (1974).
54) Nozawa, T., Hatano, M., Kambara, S.: Kogyo Kagaku Zasshi *72*, 369 (1969).
55) Nishikawa, H., Tsuchida, E.: J. Phys. Chem. *79*, 2072 (1975).
56) Gold, D. H., Gregor, H. P.: J. Phys. Chem. *64*, 1461 (1960).
57) Kimura, K., Inaki, Y., Takemoto, K.: Makromol. Chem. *175*, 83 (1974).
58) Liu, K. J., Gregor, H. P.: J. Phys. Chem. *69*, 1252 (1965).
59) Lobel, E. M., Luttinger, L. B., Gregor, H. P.: J. Phys. Chem. *59*, 559 (1955).
60) Fay, D. P., Purdie, N.: J. Phys. Chem. *73*, 3462 (1969); *75*, 1136 (1971).
61) Hojo, K., Shirai, H.: Nippon Kagaku Kaishi *1972*, 1316, 1518, 1954.
62) Hojo, K., Shirai, H.: Polymer Preprints, Japan *23* (1), 77 (1974).
63) O'neill, J. J., Loeble, E., Kandanian, A. Y., Morawetz, H.: J. Polymer Sci A, *3*, 4201 (1965).
64) Sunahara, M., Muto, N., Komatsu, T., Nakagawa, T.: Nippon Kagaku Kaishi *1974*, 2414.
65) Hojo, K., Shirai, H.: Nippon Kagaku Kaishi *90*, 823, 827 (1969); *91*, 833 (1970).
66) Gregor, H. P., Luttinger, L. B., Loebl, E. M.: J. Phys. Chem. *59*, 366 (1955).
67) Gustafson, R. L., Lirio, J. A.: J. Phys. Chem. *69*, 2849 (1965); *72*, 1502 (1968).
68) Nonogaki, S., Makishima, S., Yoneda, Y.: J. Phys. Chem. *62*, 601 (1958).
69) Dingman, J., Siggia, S., Barton, C., Hiscock, K. B.: Anal. Chem. *44*, 1351 (1972).
70) Nishide, H., Deguchi, J., Tsuchida, E.: Chem. Letters *1976*, 169.
71) Nishide, H., Deguchi, J., Tsuchida, E.: J. Polymer Sci., Chem. Ed., to be published.
72) Hering, R.: Chelatbildende Ionenaustauscher. Berlin: Akademie Verlag 1967.
73) Harada, K.: Nature, *205*, 590 (1965);
    Marcus, Y., Eliezer, I.: Coord. Chem. Rev. *4*, 273 (1969).
74) Davankov, V. A., Rogozhin, S. V.: J. Chromatogr. *60*, 280 (1971); *91*, 493 (1974).
75) Snyder, R. V., Angelici, R. J., Meck, R. B.: J. Am. Chem. Soc. *94*, 2660 (1972).
76) Tsuchida, E., Nishikawa, H., Terada, E.: Eur. Polymer J. *12*, 611 (1976).
77) Fujii, Y.: Bull. Chem. Soc., Japan *45*, 2459 (1972); *47*, 2856 (1974).
78) Kurimura, Y., Yamada, K., Kaneko, M., Tsuchida, E.: J. Polymer Sci., Al, *9*, 3521 (1971).
79) Tsuchida, E., Karino, Y., Nishide, H., Kurimura, Y.: Makromol. Chem. *175*, 161 (1974).
80) Kurimura, Y., Sekine, I, Tsuchida, E., Karino, Y.: Bull. Chem. Soc., Japan *47*, 1823 (1974).
81) Kurimura, Y., Tsuchida, E.: Polymer Preprints, Japan *21* (2), 338 (1972).
82) Bruckner, S., Crescenzi, V., Quadrifoglio, F.: J. Chem. Soc., A, *1970*, 1168.
83) Gold, E. S.: J. Am. Chem. Soc. *92*, 6797 (1970).
84) Morawetz, H., Gordimer, G.: J. Am. Chem. Soc. *92*, 7532 (1970).
85) Tsuchida, E., Nishide, H., Ohkawa, T., Kurimura, Y.: Nippon Kagaku Kaishi *1974*, 1768.
86) Tsuchida, E., Shigehara, K., Kurimura, Y.: J. Polymer Sci., Chem. Ed. *13*, 1457 (1975).
87) Tsuchida, E., Shigehara, K., Kurimura, Y.: J. Polymer Sci., Chem. Ed. *12*, 2207 (1974).
88) Tsuchida, E., Honda, K., Sata, H.: Makromol. Chem. *176*, 2251 (1975).
89) Tsuchida, E., Honda, K., Sata, H.: Inorg. Chem. *15*, 352 (1976).
90) Tsuchida, E., Nishide, H., Hata, S.: J. Polymer Sci., Chem. Ed., to be published.
91) Nishide, H., Hata, S., Mihayashi, K., Tsuchida, E.: Biopolymers, to be published.
92) Stynes, D. V., Stynes, H. C., James, B. R., Ibers, J. A.: J. Am. Chem. Soc. *94*, 1559 (1972); *95*, 1142, 1796 (1973).
93) Walker, F. A.: J. Am. Chem. Soc. *92*, 4235 (1970); *95*, 1154 (1973).
94) Stynes, H. C., Ibers, J. A.: J. Am. Chem. Soc. *94*, 1559 (1972).
95) Spilburg, C. A., Hoffman, B. M., Petering, D. H.: J. Biol. Chem. *247*, 4219 (1972).

96) Keyes, M. H., Falley, M., Lumry, R.: J. Am. Chem. Soc. 93, 2038 (1971).
97) Wang, J. H., Nakahara, A., Fleischer, E. B.: J. Am. Chem. Soc. 80, 1109, 6526 (1958);
Wang, J. H.: Accounts Chem. Res. 3, 3168 (1970).
98) Tsuchida, E., Honda, K., Hata, S.: Bull. Chem. Soc., Japan 49, 868 (1976).
99) Collman, J. P., Reed, C. A.: J. Am. Chem. Soc. 95, 2048 (1973).
100) Leal, O., Anderson, D. L., Bowman, R. G., Basolo, F., Burwell, R. L.: J. Am. Chem. Soc.
97, 5125 (1975).
101) Collman, J. P., Gagne, R. G., Halbert, T. R., Harchon, J. C., Reed, C. A.: J. Am. Chem.
Soc. 95, 7868 (1973);
Collman, J. P., Gagne, R. G., Reed, C. A.: J. Am. Chem. Soc. 96, 2699, 6522 (1974);
Collman, J. P., Gagne, R. G., Reed, C. A., Halbert, T. R., Lang, G., Robinson, W. T.:
J. Am. Chem. Soc. 97, 1427 (1975).
102) Baldwin, J. E., Huff, J.: J. Am. Chem. Soc. 95, 5757 (1973);
Almog, J. Baldwin, J. E., Dyer, R. L., Huff, J., Wilkerson, C. J.: J. Am. Chem. Soc. 96,
5600 (1974);
Almog, J., Baldwin, J. E., Dyer, R. L., Peters, M.: J. Am. Chem. Soc. 97, 226 (1975).
103) Chang, C. K., Traylor, T. C.: J. Am. Chem. Soc. 95, 8475 (1973);
Briniger, W. S., Chang, C. K., Geibel, J., Traylor, T. C.: J. Am. Chem. Soc. 96, 5597 (1974).
104) Shirakawa, H., Kagami, M., Ikeda, S.: Polymer Preprints, Japan 25 (1), 47 (1976).
105) Hasegawa, E., Kanayama, T., Tsuchida, E.: J. Polymer Sci, Chem. Ed. 15, (1977) in press.
106) Nichol, L. W., Lackson, J. H., Winzor, D. J.: Biochemistry 6, 2449 (1967).
107) Tsuchida, E., Hasegawa, E., Honda, K.: Biochim. Biophys. Acta 427, 520 (1976).
108) Tsuchida, E., Hasegawa, E., Honda, K.: Biochim. Biophys. Res. Commun. 67, 864 (1975).
109) Hill, A. V.: J. Physiol, 40, 190 (1910); Wyman, J.: Advances in protein chemistry, Vol. 19.
New York: Academic Press 1964, p. 223.
110) Wyman, J.: Adv. Protein Chem. 4, 407 (1948).
111) Tsuchida, E., Osada, Y.: Makromol. Chem. 175, 583, 593, 603 (1974).
112) Nishide, H., Deguchi, J., Tsuchida, E.: J. Polymer Sci., Chem. Ed, to be published.
113) Lautsch, W., Broser, W., Rothkegel, W., Biedermann, W., Doering, V., Zoschke, H.:
J. Polymer Sci. 8, 191 (1952); 17, 479 (1955).
114) Pecht, I., Levitzki, A., Amber, M.: J. Am. Chem. Soc. 89, 1587 (1967).
115) Dadze, T. P., Khidekel, M. L.: Izv. Akad. Nauk SSSR, Ser. Khim. 1970, 2722.
116) Vengerova, N. A., Kirsh, Y. E., Kabanov, V. A.: Vysokomol. Soed., A, 13, 2509 (1971).
117) Hatano, M., Nozawa, T., Ikeda, S., Yamamoto, T.: Makromol. Chem. 141, 11 (1971).
118) Nozawa, T., Hatano, M.: Makromol. Chem. 141, 31 (1971).
119) Nose, Y., Hatano, M., Kanbara, S.: Makromol. Chem. 98, 136 (1966).
120) Davydova, S. L., Barabanov, V. A., Plate, N. A., Kargin, V. A.: Vysokomol. Soed., A, 10,
1004 (1968).
121) Nozawa, T., Hatano, M., Kanbara, S.: Kogyo Kagaku Zasshi 72, 369 (1969).
122) Nozawa, T., Nose, Y., Hatano, M., Kanbara, S.: Makromol. Chem. 112, 73 (1968); 115. 10
(1968).
123) Kapanchan, A. T., Pschezhetskii, V. S., Kabanov, V. A.: Vysokomol. Soed, B, 11, 5 (1969).
124) Sasaki, T., Matsunaga, F.: Bull. Chem. Soc., Japan 41, 2440 (1968).
125) Sigel, H., Blamer, G.: Hev. Chem. Acta 51, 1246 (1968).
126) Siegel, H., Prijs, B., Erlenmeyer, H.: Experimentia 23, 170 (1967).
127) Welch, R. C. W., Rase, H. F.: Ind. Eng. Chem. Fundamentals 8, 389 (1969).
128) Slinkin, A. A., Dulov, A. A., Rubinstein, A. M.: Izv. Akad. Nauk. SSSR, Otd. Khim. Nauk.
1963, 1140.
129) Nakamura, Y., Hirai, H.: Chemistry Lett. 1974, 645, 809.
130) Hirai, H., Furuta, T.: J. Polymer Sci., Letters 9, 459, 729 (1971).
131) Hirai, H.: Kagaku Zokan 51, 195 (1971).
132) Beamer, R. L., Fickling, C. S.: J. Pharm. Sci. 56, 1029 (1967); 58, 1142, 1419 (1969);
Harada, K., Yoshida, T.: Naturwiss. 57, 131, 306 (1970).
133) Plate, N. A., Pavydova, S. L., Alieva, E. D., Kargin, V. A.: Eur. Polymer J. 6, 1371 (1970).
134) Pittmann, Jr., C. U., Evans, G. O.: Chemtechn. 1973, 560.

135) Cernia, E. M., Graziani, M.: J. Appl. Polymer Sci. *18*, 2725 (1974).
136) Michalska, Z. M., Webster, D. E.: Chemtech. *1975*, 117.
137) Moriguchi, Y.: Bull. Chem. Soc., Japan *37*, 2656 (1966).
138) Nozawa, T., Hatano, M., Kanbara, S.: Kogyo Kagaku Zasshi *72*, 373 (1969).
139) Nozawa, T., Akimoto, Y., Hatano, M.: Makromol. Chem. *158*, 21 (1972).
140) Breslow, R., Overman, L. E.: J. Am. Chem. Soc. *92*, 1075 (1970).
141) Hay, A. S., Blanchard, H. S., Endres, G. F., Eustance, J. W.: J. Am. Chem. Soc. *81*, 6335 (1959);
   Hay, A. S.: Adv. Polymer Sci. *4*, 496 (1967).
142) Tsuchida, E., Kaneko, M., Kurimura, Y.: Makromol. Chem. *132*, 209, 215 (1970).
143) Hay, A. S.: J. Org. Chem. *25*, 1275, *27*, 3320 (1962).
144) Hay, A. S.: Encycl. Polymer Sci., Technol. *10*, 92 (1969).
145) Tsuchida, E., Kaneko, M., Nishide, H.: Makromol. Chem. *151*, 221 (1972); *155*, 45 (1972).
146) Taylor, W. I., Battersby, A. R.: Oxidative coupling of phenols. New York: Marcel Dekker 1968.
147) Tsuchida, E., Kaneko, M., Nishide, H.: Makromol. Chem. *151*, 235 (1972).
148) Tsuchida, E., Kaneko, M., Nishide, H.: Makromol. Chem. *164*, 203 (1973).
149) Tsuchida, E., Nishide, H., Nishiyama, T.: J. Polymer Sci., Symp. *47*, 35 (1974).
150) Tsuchida, E., Nishide, H., Nishiyama, T.: Makromol. Chem. *175*, 3047 (1974).
151) Tsuchida, E., Nishide, H., Kamata, K.: Nippon Kagaku Kaishi *1972*, 1313.
152) Tsuchida, E., Nishide, H., Nishikawa, H.: Nippon Kagaku Kaishi *1972*, 2416.
153) Tsuchida, E., Nishide, H., Nishikawa, H.: J. Polymer Sci., Symp. *47*, 47 (1974).
154) Nishide, H., Nishikawa, H., Tsuchida, E.: Nippon Kagaku Kaishi *1974*, 771.
155) Nishide, H., Deguchi, J., Tsuchida, E.: Makromol. Chem., to be published.
156) Tsuchida, E., Nishikawa, H., Terada, E.: J. Polymer Sci, Chem. Ed. *14*, 825 (1976).
157) Tsuchida, E., Nishide, H.: J. Appl. Polymer Sci., to be published.
158) Tsuchida, E., Nishide, H., Nishiyama, T.: Makromol. Chem. *176*, 1349 (1975).
159) Tsuchida, E., Nishide, H., Nishikawa, H.: Macromolecules, to be published.
160) Nishide, H., Nishiyama, T., Tsuchida, E.: Nippon Kagaku Kaishi *1974*, 1565.
161) Kuhn, W., Ramel, A., Walter, D. H.: Chimica *12*, 123 (1958).
162) Nishide, H., Nishikawa, H., Tsuchida, E.: Nippon Kagaku Kaishi *1974*, 809.
163) Tsuchida, E., Nishikawa, H., Terada, E.: J. Polymer Sci., Chem. Ed. *14*, 825 (1976).
164) Fallab, S.: Angew. Chem. *79*, 500 (1967).
165) Kaneko, M., Manecke, G.: Makromol. Chem. *175*, 2795, 2811 (1974).
166) Frieden, E., Osaki, S., Kobayashi, H.: J. Gen. Physiol. *49*, 213 (1965).
167) Nishide, H., Deguchi, J., Tsuchida, E.: Makromol. Chem., to be published.

Received October 15, 1976

Strain Energy Density Functions of Rubber Vulcanizates from Biaxial Extension

Sueo Kawabata and Hiromichi Kawai

Department of Polymer Chemistry, Kyoto University, Kyoto 606, Japan

Table of Contents

I. Introduction

Since Meyer[1] introduced the concept of kinetic molecular chain into the physics of polymers in 1932, remarkable progress has been made in the molecular-theoretical interpretation of elastic behavior of rubber vulcanizates and polymer solids in general[2-6], and one can appreciate the present status of knowledge on this subject by a number of review articles and reference books. On the other hand, the phenomenologic approach to rubber elasticity has not aroused much interest in the field of polymer research. This is understandable because polymer scientists are primarily concerned with affairs of the molecular world.

Probably, Rivlin[7] in 1947 was the first to introduce the mathematical treatment into deformation analysis of hyperelastic materials. His procedure was applied to rubber elasticity by Rivlin and Saunders[11] and by Rivlin and Thomas[12], both in 1951. This was followed by investigations of several groups of authors who aimed at determining the strain energy density function (see Section II. A for its definition) of typical rubber vulcanizates. Though a significant amount of information about this function and related quantities has been accumulated, there still remains much to be investigated.

In the present article, we summarize typical approaches to the evaluation of the strain energy density function from biaxial extension experiments and illustrate some important data. This article is not a review in the ordinary sense, as it deals to a large extent with a series of experiments carried out in our laboratory. By this we do not mean to bias or ignore any of the many important contributions by other authors. Our principal data are reported here because they are not only detailed and systematic enough to illustrate the characteristic features of elastic behavior of rubbers, but also because there have been few opportunities to have them published in the well-established journals.

II. The Strain Energy Density Function and the Phenomenologic Equations for Elasticity

A. Strain Energy Density Function

The ensuing discussion is restricted to continua that are homogeneous, isotropic, and elastic. Actually, we idealize vulcanized rubbers as such. If such a body is strained or deformed by application of external forces the work done is stored in it as free energy. The stored energy depends only on the state of deformation of the body. It is not always distributed uniformly throughout the deformed body but its "intensity", conveniently defined as its amount per unit volume of the undeformed body, may vary from point to point. This energy intensity is called the strain energy density. If the strain and the corresponding strain energy density are given at every point in the body, the stress at any given point can be calculated by using the known relations in the theory of elasticity, regardless of whether the strain is infinitesimal or finite.

A number of so-called constitutive equations, which relate strain and stress in the general coordinate system, have been published[13-16]. Details can be found, for

example, in a book by Eringen[17]. With the help of such equations the distributions of the strain and the strain energy density in a body can be found. Between
these two quantities there is a definite relation characteristic of the material under
consideration. This relation defines the strain energy density function, which is
designated by W in the following. W is a scalar quantity, while the strain is represented
by a tensor of second order. However, the latter can be expressed in terms of three
quantities called principal strains if the cartesian coordinate system, to which it refers,
is taken so that its orthogonal axes coincide with the directions of the principal
strains at the same point. Thus we may write for homogeneous, isotropic, and elastic
material

$$W = W(\epsilon_1, \epsilon_2, \epsilon_3)$$
$$= W(\lambda_1, \lambda_2, \lambda_3) \tag{1}$$

where ϵ_i ($i = 1, 2, 3$) and λ_i denote the principal strains and the principal stretch ratios,
respectively. Note that we have the relation $\epsilon_i = \lambda_i - 1$.

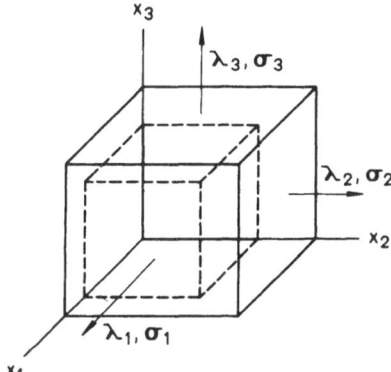

Fig. 1. Extension in the directions of
principal axes

As shown in Fig. 1, a cubic body of material under consideration is deformed in
the directions of orthogonal axes X_i. If this mode of deformation, the coordinate
axes coincide with the principal strain axes. In the principal stresses σ_i corresponding
to the principal strains are measured as functions of stretch ratios λ_i in the directions
of X_i, W can be calculated from

$$W(\lambda_1, \lambda_2, \lambda_3) = \sum_{i=1}^{3} \int_1^{\lambda_i} \sigma_i d\lambda_i \tag{2}$$

Note that here σ_i are referred to unit area of the undeformed body. Such stresses are
called engineering stresses. Eq. (2) is equivalent to a set of relations:

$$\sigma_i = \frac{\partial W}{\partial \lambda_i} \quad (i = 1, 2, 3) \tag{3}$$

This is the fundamental for phenomenologic analyses of elastic bodies.

The form of W as a function of the set of λ_i must be derived from molecular theory or from experimental measurement. It cannot be deduced from the phenomenologic theory of continua, just as the free energy cannot be deduced from thermodynamics. However, the phenomenologic theory imposes the following restrictions on the form of W if the material is isotropic[18]. First, W must be an even power function of λ_i (restriction A). Second, W must be invariant for permutations of X_i (restriction B). Third, W must be invariant for the transformation of coordinate axes (restriction C).

Customarily, the following three invariants of Cauchy-Green deformation tensor, I_1, I_2, and I_3, are used in place of λ_1, λ_2, and λ_3 as the independent variables for the function of W.

$$I_1 = \lambda_1^2 + \lambda_2^2 + \lambda_3^2$$
$$I_2 = (\lambda_1\lambda_2)^2 + (\lambda_2\lambda_3)^2 + (\lambda_3\lambda_1)^2 \tag{4}$$
$$I_3 = (\lambda_1\lambda_2\lambda_3)^2$$

Thus the task is reduced to either theoretical or experimental determination of W as a function of the set of I_i:

$$W = W(I_1, I_2, I_3) \tag{5}$$

In the undeformed state, $\lambda_i = 1$ (i = 1, 2, 3), and hence I_1 = 3, I_2 = 3, and I_3 = 1. Accordingly, we have

$$W(3, 3, 1) = 0 \tag{6}$$

This condition is all that can be obtained from the phenomenologic theory about

$$W(I_1, I_2, I_3)$$

With this condition, there are a great many possible choices for the form of W as a function of I_i. Our ultimate purpose in the phenomenologic study of rubber elasticity is to find out its form applicable for an accurate and coherent description of the elastic behavior of rubber-like materials under various modes of deformation. We may use $J_1, \mathring{J}_2$, and J_3 for the set of I_i, which are defined by

$$\left.\begin{array}{l} J_1 = I_1 \\ J_2 = I_2/I_3 \\ J_3 = (I_3)^{1/2} \end{array}\right\} \tag{7}$$

This new set of invariants was introduced by Blatz and Ko[19] in order to discuss compressible materials under relatively small deformations.

Valanis and Landel[20] put forward the following form

$$W = w(\lambda_1) + w(\lambda_2) + w(\lambda_3) \tag{8}$$

Thus, in this case, one may determine a single function $w(\lambda)$ by experiment. Eq. (8) satisfies the symmetry condition imposed by isotropy (restriction B). If its use is limited to the coordinate system whose axes are taken in the directions of the principal strains, restriction A mentioned above does not matter. Valanis and Landel deduced this form of W from the kinetic theory of network, in which the entropy change Δs upon deformation is represented by the sum of three components, each corresponding to the deformation in one of the X_1, X_2, and X_3 directions and having the same functional dependence on the argument. Thus

$$\Delta s = s(X_1) + s(X_2) + s(X_3) \tag{9}$$

B. Stress-Strain Relation Under Finite Deformation

Eq. (3) may be rewritten

$$\sigma_i = \frac{\partial W}{\partial I_1} \frac{\partial I_1}{\partial \lambda_i} + \frac{\partial W}{\partial I_2} \frac{\partial I_2}{\partial \lambda_i} + \frac{\partial W}{\partial I_3} \frac{\partial I_3}{\partial \lambda_i} \tag{10}$$

$$(i = 1, 2, 3)$$

Substitution of $\partial I_k / \partial \lambda_i$ $(i, k = 1, 2, 3)$ calculated from Eq. (4) yields

$$\sigma_k = 2\lambda_k \left[\frac{\partial W}{\partial I_1} + (\lambda_l^2 + \lambda_m^2) \frac{\partial W}{\partial I_2} + \lambda_l^2 \lambda_m^2 \frac{\partial W}{\partial I_3} \right] \tag{11}$$

$$(k, l, m = 1, 2, 3; \; k \neq l \neq m)$$

which gives the general stress-strain relations expressed in terms of W. If the engineering stresses in these equations are to be converted to the true stresses t_i, which refer to unit area of the deformed body, we may use the relations

$$t_k = \frac{\sigma_k}{\lambda_l \lambda_m} = \lambda_k \sigma_k \quad (k = 1, 2, 3) \tag{12}$$

where the last equality is valid only under the condition of incompressibility $\lambda_1 \lambda_2 \lambda_3 = 1$.

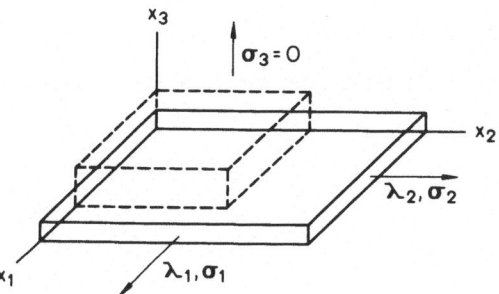

Fig. 2. Biaxial extension

In the case of biaxial deformation of incompressible material (Fig. 2), $I_3 = 1$ and $\sigma_3 = 0$. Hence, Eq. (11) yields

$$\sigma_1 = \frac{2}{\lambda_1}\left(\lambda_1^2 - \frac{1}{\lambda_1^2\lambda_2^2}\right)\left(\frac{\partial W}{\partial I_1} + \lambda_2^2 \frac{\partial W}{\partial I_2}\right) \tag{13}$$

$$\sigma_2 = \frac{2}{\lambda_2}\left(\lambda_2^2 - \frac{1}{\lambda_1^2\lambda_2^2}\right)\left(\frac{\partial W}{\partial I_1} + \lambda_1^2 \frac{\partial W}{\partial I_2}\right) \tag{14}$$

$$\sigma_3 = 0 \tag{15}$$

Furthermore, the incompressibility condition allows I_1 and I_2 to be expressed only in terms of λ_1 and λ_2 as follows:

$$I_1 = \lambda_1^2 + \lambda_2^2 + (\lambda_1\lambda_2)^{-2}$$

$$I_2 = (\lambda_1\lambda_2)^2 + \lambda_1^{-2} + \lambda_2^{-2} \tag{16}$$

The stress-strain relations for some special cases of biaxial deformation are derived from Eqs. (13) to (15) in the following way.

"Strip biaxial extension" of incompressible material is defined as the mode of deformation in which one of the λ_i, say λ_2, is kept at unity, while the other, λ_1, varies. This deformation is also called "pure shear". We have for it

$$\sigma_1 = \frac{2}{\lambda_1}\left(\lambda_1^2 - \frac{1}{\lambda_1^2}\right)\left(\frac{\partial W}{\partial I_1} + \frac{\partial W}{\partial I_2}\right) \tag{17}$$

$$\sigma_2 = 2\left(1 - \frac{1}{\lambda_1^2}\right)\left(\frac{\partial W}{\partial I_1} + \lambda_1^2 \frac{\partial W}{\partial I_2}\right) \tag{18}$$

$$\sigma_3 = 0 \tag{19}$$

and the invariants I_i are given by

$$I_1 = I_2 = \lambda_1^2 + \lambda_1^{-2} + 1 \tag{20}$$

In the case of $\lambda_1 = \lambda_2$, which is called equal biaxial deformation, Eqs. (13) to (15) reduce to

$$\sigma_1 = \sigma_2 = 2\left(\lambda - \frac{1}{\lambda^5}\right)\left(\frac{\partial W}{\partial I_1} + \lambda^2 \frac{\partial W}{\partial I_2}\right) \tag{21}$$

$$\sigma_3 = 0 \tag{22}$$

where $\lambda_1 = \lambda_2 \equiv \lambda$, and the invariants are

$$\left.\begin{array}{l} I_1 = 2\lambda^2 + \lambda^{-4} \\[2mm] I_2 = \lambda^4 + 2\lambda^{-2} \end{array}\right\} \tag{23}$$

Finally, for simple uniaxial extension of incompressible material we obtain with $\sigma_2 = \sigma_3 = 0$

$$\sigma_1 = 2\left(\lambda_1 - \frac{1}{\lambda_1^2}\right)\left(\frac{\partial W}{\partial I_1} + \frac{1}{\lambda_1}\frac{\partial W}{\partial I_2}\right) \tag{24}$$

where

$$\left.\begin{array}{l} I_1 = \lambda_1^2 + 2(\lambda_1)^{-1} \\[2mm] I_2 = (\lambda_1)^{-2} + 2\lambda_1 \end{array}\right| \tag{25}$$

Valanis and Landel [20] derived the stress-strain relations from their assumed form of W, Eq. (8), as follows:

$$t_1 - t_2 = \lambda_1 w'(\lambda_1) - \lambda_2 w'(\lambda_2)$$

$$t_2 - t_3 = \lambda_2 w'(\lambda_2) - \lambda_3 w'(\lambda_3) \tag{26}$$

$$t_3 - t_1 = \lambda_3 w'(\lambda_3) - \lambda_1 w'(\lambda_1)$$

where $w'(\lambda) = \partial W/\partial \lambda$ and t_i are true stresses which are related to engineering stresses by Eq. (12). These relations allow the function $w'(\lambda)$ to be determined from measurements of the principal stresses and strains. One may perform only strip-biaxial extension experiment for this determination.

III. Search for the Strain Energy Density Function of Vulcanized Rubbers

A. Kinetic Theory

The search for the form of W of vulcanized rubbers was initiated by polymer physicists. In 1934, Guth and Mark [2] and Kuhn [3] considered an idealized single chain which consists of a number of links jointed linearly and freely, and derived the probability P that the end-to-end distance of the chain assumes a given value. The resulting probability function of Gaussian type was then substituted into the Boltzmann equation for entropy s, which reads,

$$s = k \ln P \qquad (27)$$

and the force required to stretch the chain was computed by the standard thermo-dynamic method. During the period 1942–43 an attempt was made to extend this type of calculation to the tensile force of rubber network[21–23, 37], and it was shown that the W for the network model used is represented by

$$W = (1/2)NkT(I_1-3) \qquad (28)$$

Here N is the number of chains in unit volume of the rubber and T is the absolute temperature. Note that the chain is defined here as part of a long polymer molecule between neighboring cross-linking points. The fundamental assumptions made in deriving Eq. (28) are that all chain ends are displaced to new positions by affine trans-formation when the rubber is deformed, that there are no intermolecular interactions, and that the end-to-end distance of each chain is much smaller than the contour length of the chain.

From Eq. (28) it follows that

$$\left. \begin{array}{l} \partial W/\partial I_1 = (1/2)NkT \\[2mm] \partial W/\partial I_2 = 0 \end{array} \right\} \qquad (29)$$

Therefore, the early kinetic theory of network predicts that $\partial W/\partial I_1$ is independent of deformation and $\partial W/\partial I_2$ is identically zero.

Attempts were made to remove the third assumption above, and it was shown that correct considerations of the limited extensibility of the chain adequately explain the S-shaped feature of stress-strain curves observed in uniaxial extension of vulcanized rubbers. However, the improved theory still gave zero for $\partial W/\partial I_2$. Up to now there is no molecular theory available which predicts $\partial W/\partial I_2$ that varies with I_i.

B. Phenomenologic Approach

This consists of experimental measurements of stress-strain relations and analysis of the data in terms of the mathematical theory of elastic continua. Rivlin[7–10] was the first to apply the finite (or large) deformation theory to the phenomenologic analysis of rubber elasticity. He correctly pointed out the above-mentioned restrictions on W, and proposed an empirical form

$$W = \sum_{i=0} \sum_{j=0} \sum_{k=0} C_{ijk}(I_1 - 3)^i(I_2 - 3)^j(I_3 - 1)^k \qquad (30)$$

where C_{ijk} are constants, with C_{000} being zero. For essentially incompressible materials such as rubber vulcanizates this form may be replaced by

$$W = \sum_{i=0} \sum_{j=0} C_{ij}(I_1 - 3)^i(I_2 - 3)^j \qquad (31)$$

where C_{ij} are constants with $C_{00} = 0$. The first-order approximation to Eq. (31) is

$$W = C_1(I_1 - 3) + C_2(I_2 - 3) \tag{32}$$

where C_1 and C_2 denote C_{10} and C_{01}, respectively. This was the form given by Mooney[24] in 1940. It is instructive to note here that the kinetic theory of network in the early days gives

$$W = C_1(I_1 - 3) \tag{33}$$

which lacks the second term, the so-called C_2 term, in the Mooney expression. Rivlin referred to material as neo-Hookian when its W obeys Eq. (33).

C. Determination of $\partial W/\partial I_1$ and $\partial W/\partial I_2$ from Biaxial Extension Experiments

In principle, W can be determined from Eq. (2) if principal stresses σ_i are measured as functions of applied principal stretch ratios λ_i. However, since $\partial W/\partial I_i$ rather than W itself are more directly connected with the stress-strain relations [see Eq. (11)], their determination from the measurements of σ_i and λ_i is more feasible than that of W.

For incompressible material, the stress-strain relations for biaxial extension are given by Eqs. (13) and (14), which may be solved for $\partial W/\partial I_1$ and $\partial W/\partial I_2$ to give

$$\frac{\partial W(I_1, I_2)}{\partial I_1} = \frac{1}{2(\lambda_1^2 - \lambda_2^2)} \left[\frac{\lambda_1^3 \sigma_1}{\lambda_1^2 - (\lambda_1 \lambda_2)^{-2}} - \frac{\lambda_2^3 \sigma_2}{\lambda_2^2 - (\lambda_1 \lambda_2)^{-2}} \right] \tag{34}$$

$$\frac{\partial W(I_1, I_2)}{\partial I_2} = \frac{1}{2(\lambda_2^2 - \lambda_1^2)} \left[\frac{\lambda_1 \sigma_1}{\lambda_1^2 - (\lambda_1 \lambda_2)^{-2}} - \frac{\lambda_2 \sigma_2}{\lambda_2^2 - (\lambda_1 \lambda_2)^{-2}} \right] \tag{35}$$

These allow determination of $\partial W/\partial I_1$ and $\partial W/\partial I_2$ from observed values of σ_i and λ_i $(i = 1, 2)$.

The first attempt at measuring $\partial W/\partial I_i$ was reported by Treloar[29], who used a square sheet of natural rubber vulcanizate for biaxial extension shown in Fig. 3. After Treloar's experiment, Rivlin and Saunders[11] carried out a similar experiment by using the device illustrated in Fig. 4. The sheet is equipped with wires at its four edges, and homogeneous stretching is applied biaxially by pulling them carefully. Tensions in the stretching directions are measured by springs connected to the wires. This simple device has often been adapted in the studies of subsequent investigators, e.g., Zapas[26] used it to measure biaxial creep of rubbers.

Blatz and Ko[19] added a special attachment to their uniaxial tensile testing equipment (Fig. 5). In this, a sheet specimen of rubber is stretched by chucks attached to its four edges, but, differing from the apparatus of Rivlin and Saunders, the chucks can be moved smoothly on the rigid tracks so that it is possible to strech

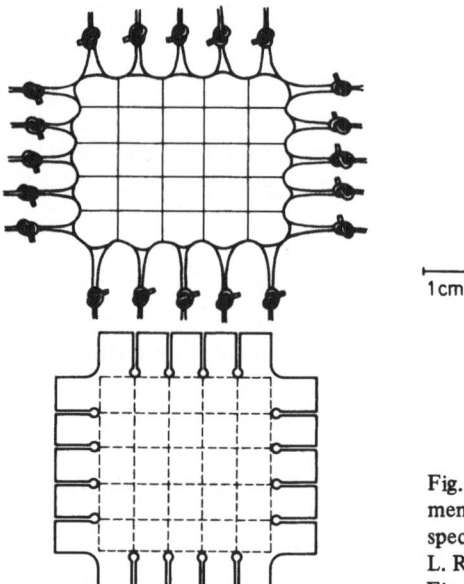

1cm

Fig. 3. Treloar's biaxial extension experiment; *lower figure* shows undeformed specimen. [Reproduced from Treloar, L. R. G.: Proc. Phys. Soc. *60*, 135 (1948), Fig. 1.]

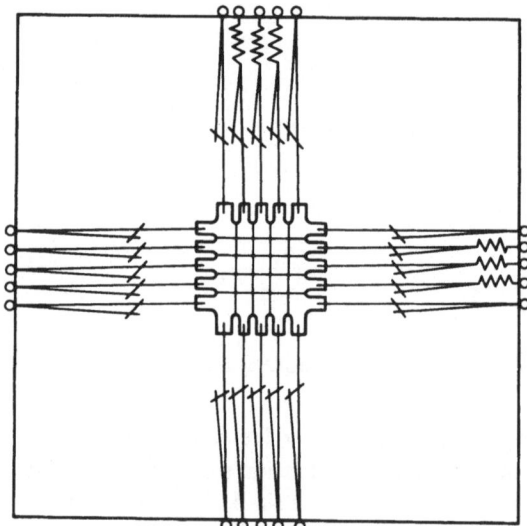

Fig. 4. Rivlin-Saunders' apparatus for biaxial extension. Specimen is 8 cm x 8 cm and 0.7 mm thick. [Reproduced from Rivlin, R. S., Saunders: Phil. Trans. Roy. Soc. *A243*, 251 (1951), Fig. 3.]

the sheet uniformly without manual adjustment of the positions of the chucks. However, this apparatus is of use only for equal biaxial extension, and the experimental data from this type of deformation do not allow separate evaluation of $\partial W/\partial I_1$ and $\partial W/\partial I_2$, because with $\lambda_1 = \lambda_2 = \lambda$ and $\sigma_1 = \sigma_2 = \sigma$ we find from Eq. (21)

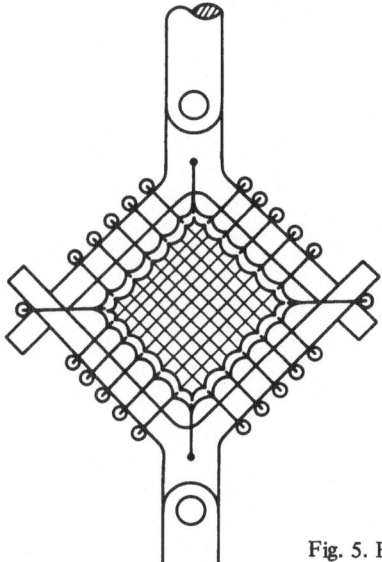

Fig. 5. Blatz-Ko's apparatus for equal biaxial extension.
Specimen in undeformed state is 3 in × 3 in and 0.0187, in thick

$$\frac{\sigma}{2\,(\lambda - \lambda^{-5})} = \frac{\partial W}{\partial I_1} + \lambda^2\,\frac{\partial W}{\partial I_2} \tag{36}$$

Only in the special case in which both $\partial W/\partial I_1$ and $\partial W/\partial I_2$ are independent of λ, a plot of the left-hand quantity of Eq. (36) against λ^2 gives a straight line, and the values of these two derivatives are found from its intercept and slope.

Blatz and Ko[19] also devised an apparatus which allows strip-biaxial extension testing to be performed (Fig. 6). Here one of the stretch ratios, say λ_2, is held at unity while the sample sheet is extended in the direction of the stretch ratio λ_1. These authors measured not only σ_1 but also σ_2, with a small transducer being mounted on the rod connected to a chuck. In this way, they determined $\partial W/\partial J_1$ and $\partial W/\partial J_2$ for a foam rubber, where J_1 and J_2 are the invariants defined in Eq. (7).

Becker[25] improved the biaxial extension apparatus of Blatz and Ko in such a way that pairs of different values of λ_1 and λ_2, *i.e.*, general biaxial extensions, can be obtained. Here, pairs of two rectangular tracks are connected to the pulling rods in asymmetric fashion, as shown in Fig. 7. With the use of this apparatus he obtained valuable information about the behavior of $\partial W/\partial I_1$ and $\partial W/\partial I_2$ for natural rubber vulcanizate. His principal results are presented in Section V.

Another apparatus for general biaxial extension testing, illustrated in Fig. 8, was built at our laboratory[27]. A square sheet of rubber, 11.5 cm long and 1 mm thick, is clamped at the edges by several pieces of movable chucks and stretched biaxially by means of the two bars that can be displaced separately back and forth by a servo-meachnism. Tensile forces acting on the sheet in the stretching directions are transmitted to the bars and measured by two transducers mounted at the ends of each bar. After many modifications and improvements on the mechanical and electric parts,

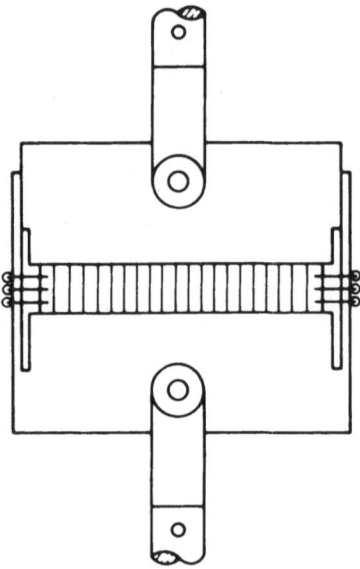

Fig. 6. Blatz-Ko's apparatus for strip biaxial extension. Specimen is rectangular in shape, 7 in × 1 in and 0.0187 in thick

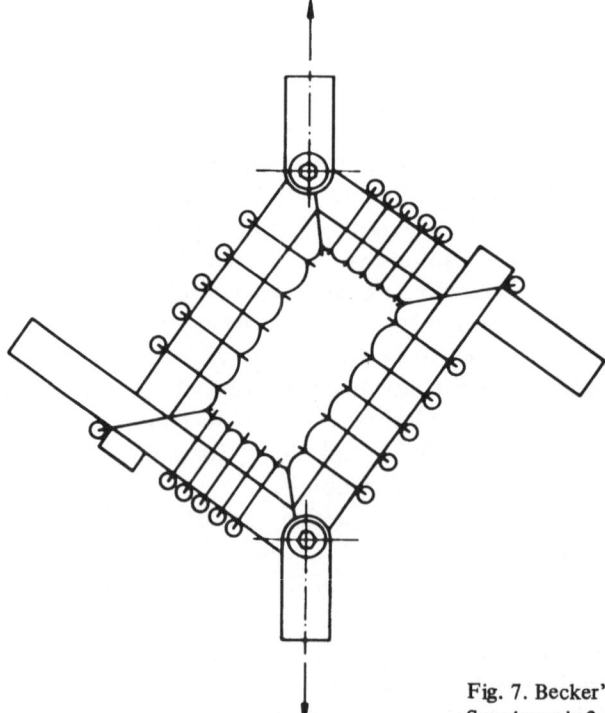

Fig. 7. Becker's apparatus for biaxial extension. Specimen is 2.5 in × 2.5 in and 1 mm thick

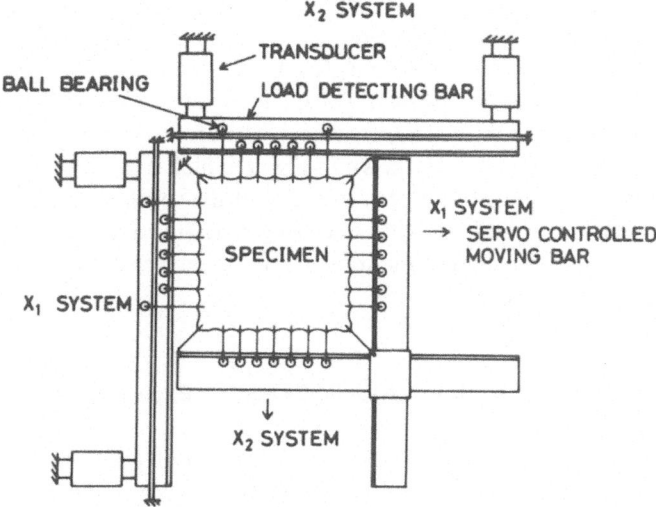

Fig. 8A. Our apparatus for biaxial extension. Specimen is 11.5 cm × 11.5 cm and 1 mm thick

Fig. 8B. View of the apparatus in Fig. 8A

we were able to determine with this apparatus both $\partial W/\partial I_1$ and $\partial W/\partial I_2$ for several rubber vulcanizates.

Measurements of stresses σ_i and stretch ratios λ_i must be made with extreme care and precision, for small errors in them lead to a gross uncertainty in the calculation of $\partial W/\partial I_i$. Naturally, the specimen must be deformed as homogeneously as possible.

In actuality, the corners of a square sheet of specimen undergo inhomogeneous deformations. To minimize this effect on the final results we arranged the cramping chucks in the manner shown in Fig. 8 A, and picked up the tensile force on each edge from the inner five chucks. It was found that in the region of small deformations, the determination of $\partial W/\partial I_i$ by our apparatus was subject to gross errors. Therefore, we undertook the construction of still another apparatus [28], shown in Fig. 9, which permits accurate measurement of small stresses and strains. In this, the strains of the specimen were measured by a travelling microscope, and the force-detecting mechanism was greatly improved. Moreover, for especially small deformations the specimen was dipped in water to eliminate the effect of gravitational force.

D. Determination of $\partial W/\partial I_1$ and $\partial W/\partial I_2$ from Uniaxial Extension Experiments

From the theoretical point of view, it is impossible to determine $\partial W/\partial I_1$ and $\partial W/\partial I_2$ separately from uniaxial extension measurements. Eq. (24) for this type of deformation indicates that only the sum $(\partial W/\partial I_1) + (\lambda_1)^{-1}(\partial W/\partial I_2)$ can be estimated. If and only if the derivatives $\partial W/\partial I_i$ are constant, $i.e.$, W is of the Mooney type, Eq. (32), it follows from Eq. (24) that

$$\frac{\sigma_1}{\lambda_1 - \lambda_1^{-2}} = 2\,C_1 + \frac{2\,C_2}{\lambda_1} \tag{37}$$

so that plots of $\sigma_1/(\lambda_1 - \lambda_1^{-2})$ versus λ_1^{-1} are linear and the constants C_1 and C_2, and hence $\partial W/\partial I_1$ and $\partial W/\partial I_2$, respectively, can be evaluated from the intercept and slope of the line. This type of plot for uniaxial extension data is customarily referred to as the Mooney-Rivlin plot.

If material is neo-Hookean, its Mooney-Rivlin plot ought to give a horizontal line and hence yield $C_2 = 0$. Thus one is tempted to consider that nonzero C_2 must be associated in one way or another with the deviation of a given material from the idealized network model, and it is understandable why so many rubber scientists have concerned themselves with evaluating the C_2 term from the Mooney-Rivlin plot of uniaxial extension data. However, the point is that a linear Mooney-Rivlin plot, if found experimentally, does not always warrant that its intercept and slope may be equated to $2(\partial W/\partial I_1)$ and $2(\partial W/\partial I_2)$, respectively. This fact is illustrated below with actual data on natural rubber (NR) and styrene-butadiene copolymer rubber (SBR).

IV. Some Reported Forms of W for Actual Rubbers

Rivlin and Saunders [11] suggested from their biaxial extension experiments on vulcanized natural rubber that the form of W for rubber-like material would be

$$W = C_{10}(I_1 - 3) + f(I_2 - 3) \tag{38}$$

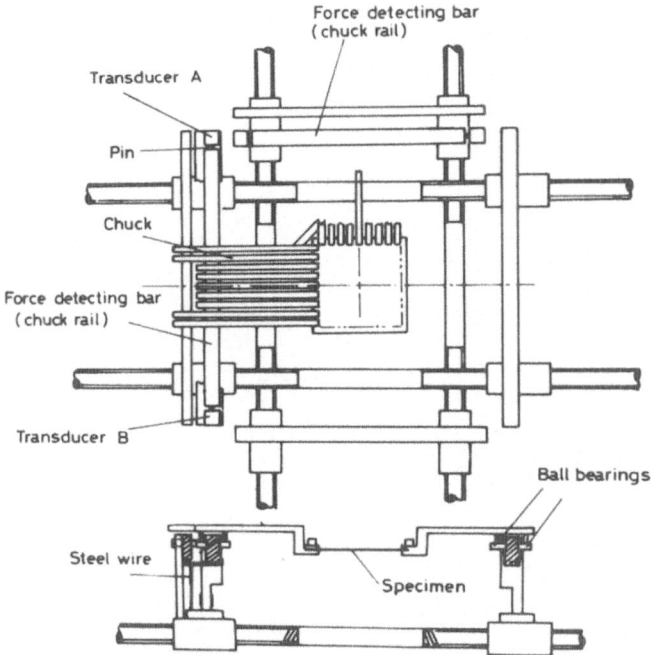

Fig. 9A. Our new apparatus for small deformation testing. Specimen is 11.5 cm × 11.5 cm and 1 mm thick

Fig. 9B. Photo of the new apparatus

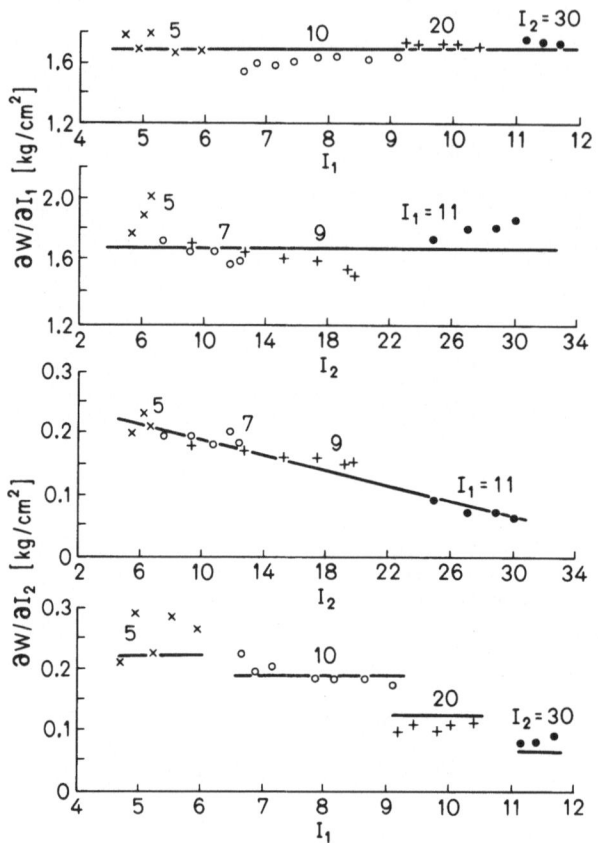

Fig. 10. $\partial W/\partial I_i$ measured by Rivlin and Saunders for natural rubber vulcanizate. [Reproduced from Rivlin, R. S., Saunders: Phil. Trans. Roy. Soc. *A243*, 251 (1951), Fig. 3.]

where C_{10} is a constant and f is a function only of $I_2 - 3$. This equation indicates that $\partial W/\partial I_1$ is independent of deformation and that $\partial W/\partial I_2$ varies only with I_2. Actually, the experimental results of Rivlin and Saunders illustrated in Fig. 10 indicate that f up to $I_2 = 30$ decreases linearly with increasing I_2.

Thomas[30] derived theoretically

$$W = C_1(I_1 - 3) + C_2 f(\lambda_1, \lambda_2) \tag{39}$$

where f is a complicated function of the principal stretch ratios. Gent and Thomas[31] showed that $f(\lambda_1, \lambda_2)$ is actually close to $\ln (I_2/3)$. Thus

$$W = C_1(I_1 - 3) + C_2 \ln (I_2/3) \tag{40}$$

Tschoegl[32] tried to evaluate the coefficients in Eq. (31) by fitting this general expression for W to experimental data from uniaxial extension experiments and found for pure SBR

$$W = C_{10}(I_1 - 3) + C_{01}(I_2 - 3) + C_{22}(I_1 - 3)^2(I_2 - 3)^2 \tag{41}$$

and for carbon-reinforced natural rubber

$$W = C_{10}(I_1 - 3) + C_{01}(I_2 - 3) + C_{11}(I_1 - 3)(I_2 - 3) \tag{42}$$

From a theory of non-Gaussian network Hert-Smith[34, 35] derived

$$\partial W/\partial I_1 = G \exp[k_1(I_1 - 3)^2]$$
$$\tag{43}$$
$$\partial W/\partial I_2 = Gk_2/I_2$$

where G, k_1, and k_2 are constants. It should be noted that this expression for $\partial W/\partial I_2$ gives rise to a logarithmic term in W similar to that found in the Gent-Thomas equation (40). Alexander[33] improved the theory of Hert-Smith and obtained an expression of W which contains five adjustable parameters. He confirmed his theory with data from uniaxial and equal biaxial extension measurements.

The work of Becker[25] quoted above marked a very important step in this field, for it probably provided the first detailed information about $\partial W/\partial I_i$ evaluated from general biaxial extension experiments. From this work the behavior of these derivatives of W at small strains was found to be very complex. Our data displayed in the next section conform in many points to his important findings.

Valanis and Landel[20] showed that if an empirical form is assumed for $w'(\lambda)$

$$w'(\lambda) = 2\mu \ln \lambda \tag{44}$$

where μ is a constant taken to be the initial shear modulus, Eq. (8) leads to good agreements with the uniaxial and biaxial extension data of Becker[25], Rivlin and Saunders[11], and Treloar[37]. In this connection, we note that the Mooney equation for W, Eq. (32), can be written in the form of Eq. (8) if $w(\lambda)$ is taken to be

$$w(\lambda) = C_1\lambda^2 + C_2\lambda^{-2} \tag{45}$$

Valanis and Landel[20] suggested that a possible form of W for compressible material would be

$$W = \frac{1}{2}\hat{\lambda}[\ln(\lambda_1\lambda_2\lambda_3)]^2 + 2\mu \sum_{i=1}^{3} \lambda_i(\ln \lambda_i - 1) \tag{46}$$

where $\hat{\lambda}$ and μ are the Lamé constants. This equation was constructed so that it reduces to the W predicted from the classic theory for infinitesimal deformations at the limit of small strains. It has not as yet been subjected to experimental test.

As is illustrated in the next section, we have recently tested the applicability of Eq. (44) to our biaxial extension data, but have seen only poor agreement between predicted and observed results. Recently, Jones and Treloar[36] derived from biaxial extension measurements, with an apparatus similar to the original one of Rivlin and Saunders[11], an Ogden-type[38] of expression for $w'(\lambda)$:

$$\lambda w'(\lambda) - c = 0.69\,(\lambda^{1.3} - 1) + 0.01\,(\lambda^{4.0} - 1)$$

$$- 0.0122\,(\lambda^{-2.0} - 1) \tag{47}$$

where c is a constant actually equal to $w'(1)$, and the numerical coefficients are in $N\,mm^{-2}$.

Still another equation for $w(\lambda)$ has been proposed by Blatz et al.[39] which reads

$$w(\lambda) = 2\,G(\lambda^n - 1)/n \tag{48}$$

so that

$$W = (2\,G/n)I_E + BI_E{}^m \tag{49}$$

where G, n, B, and m are material constants and I_E is defined by

$$I_E = \sum_{i=1}^{3} (\lambda_i^n - 1)/n \tag{50}$$

These authors have shown that this four-parameter equation for W fits the early data of Treloar[29] on natural rubber.

Our own experience, as well as that of other authors, has shown that very precise measurement for the stress-strain relationship under general biaxial deformation is required to investigate the behavior of the strain energy density function of rubber vulcanizates. Unfortunately, available biaxial extension data are still too meager to deduce the general form of the strain energy density function of rubber-like substances. We wish to take this opportunity to summarize the principal results from our recent efforts, in the hope that they may serve to illustrate the interesting and complex nature of the derivatives $\partial W/\partial I_i$ of such substances.

V. Summary of Results from the Authors' Biaxial Extension Experiments

The samples we used were vulcanizates of natural rubber (NR) and styrene-butadiene copolymer rubbers (SBR), carbon-filled and unfilled. Table 1 summarizes their preparative data. Incompressibility of these vulcanizates and some other vulcanizates were checked by dipping, stretching uniaxially, and weighing a specimen in water. For example, the volume change of an acrylonitrile-butadiene rubber (NBR)[40] sample at $\lambda = 2$ relative to the volume of its undeformed state was about 5×10^{-4}, and the values for the other vulcanizates were less than this. We therefore assumed that the use of Eqs. (34) and (35) is warranted for the computation of $\partial W/\partial I_i$ for our rubber samples, except at very small deformations for which $I_i < 3.02$. In most cases, stress relaxation was allowed to proceed at given stretch ratios and 1- and 10-min isochronal stress values were taken for the calculations.

Table 1. Samples[a]

	NR	SBR 1	SBR 2	SBR 3	SBR 4	SBR Carbon-filled
NR RSS 1	100					
JSR SBR 1500		100	100	100	100	100
ZnO	3	5	5	5	5	5
St. acid	1	1	1	1	1	1
S		1	1.5	2.0	2.5	1
Dicup	4					
MgO	1					
M.B.T.		1.5	1.5	1.5	1.5	1.5
Carbon black						40

[a] 145 °C, 30-min cure.

A. Deformation Dependence of $\partial W/\partial I_i$

Figures 11 and 12 illustrate the contour plots of 1-min values of $\partial W/\partial I_1$ and $\partial W/\partial I_2$, respectively, for SBR 1 [41]. Each thin line represents a locus of the paired values of I_1 and I_2 for which either of these derivatives assumes a given value. The equal biaxial line is defined by the condition that $\lambda_1 = \lambda_2$ and $\lambda_1 \lambda_2 \lambda_3 = 1$, while the uniaxial line is defined by the condition that $\lambda_2 = \lambda_3 = (\lambda_1)^{1/2}$ with $\lambda_1 \lambda_2 \lambda_3 = 1$. Another representation of the data is given in Fig. 13, in which the values of $\partial W/\partial I_i$ for fixed I_1 are plotted against I_2. The data obtained for other samples showed similar features[42, 43], but they are not illustrated here. These experimental results demonstrate that $\partial W/\partial I_i$ varies appreciably with I_1 and I_2 in the region of small strains, while, in the region of relatively large strains, $\partial W/\partial I_1$ remains almost constant and $\partial W/\partial I_2$ decreases only gradually as I_2 increases.

Becker[25] measured 10-min isochronal values of $\partial W/\partial I_i$ for natural rubber vulcanizate; the results are shown in Figs. 14 and 15 for $\partial W/\partial I_1$ and $\partial W/\partial I_2$, respectively. His data are almost similar in the form of $\partial W/\partial I_1$ and $\partial W/\partial I_2$ to those in Fig. 12 and 13, but less sensitive to deformation except in the region of very small deformation.

The recent data reported by Jones and Treloar[36] for natural rubber vulcanizate are depicted in Fig. 16. They first determined $w'(\lambda)$ in the Valanis-Landel form of W, represented it in terms of Eq. (47), and then used it to evaluate $\partial W/\partial I_i$.

All the data obtained by these three groups suggest that both $\partial W/\partial I_1$ and $\partial W/\partial I_2$ are markedly dependent on deformation, especially in the region of small deformations. The data of Becker and the present authors indicate a rapid decrease in $\partial W/\partial I_2$ as I_2 approaches 3, while Jones and Treloars data do not. This discrepancy must have arisen from the difference in the methods employed for evaluating $\partial W/\partial I_i$.

B. Temperature Dependence of $\partial W/\partial I_i$

It is important to examine the temperature dependence of $\partial W/\partial I_i$ for development of a more exact molecular theory of rubber elasticity. Figure 17[44, 45] illustrates this

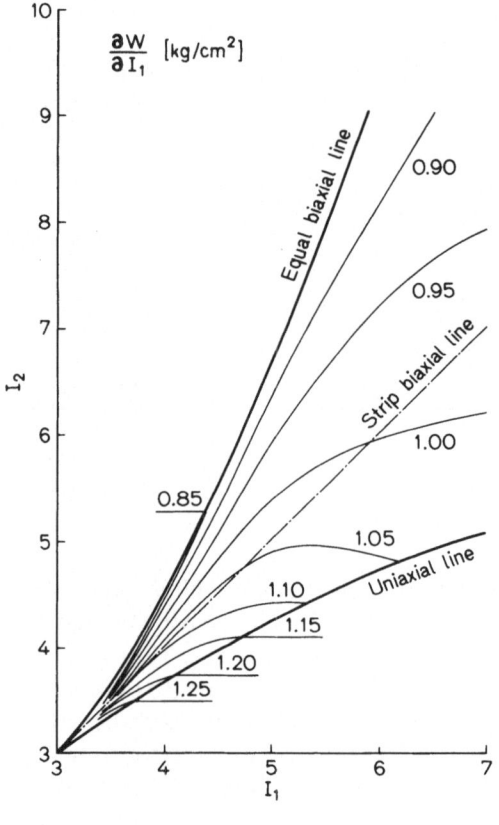

Fig. 11. Contour map of $\partial W/\partial I_1$ for SBR 1 at 1 min and 22 °C

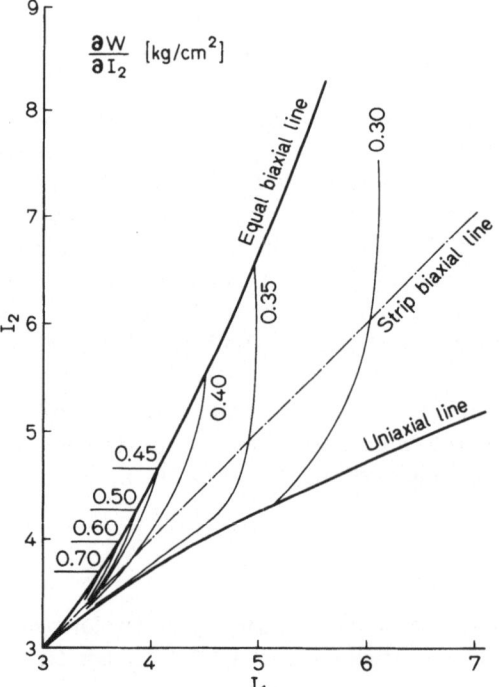

Fig. 12. Contour map of $\partial W/\partial I_2$ for SBR 1 at 1 min and 22 °C

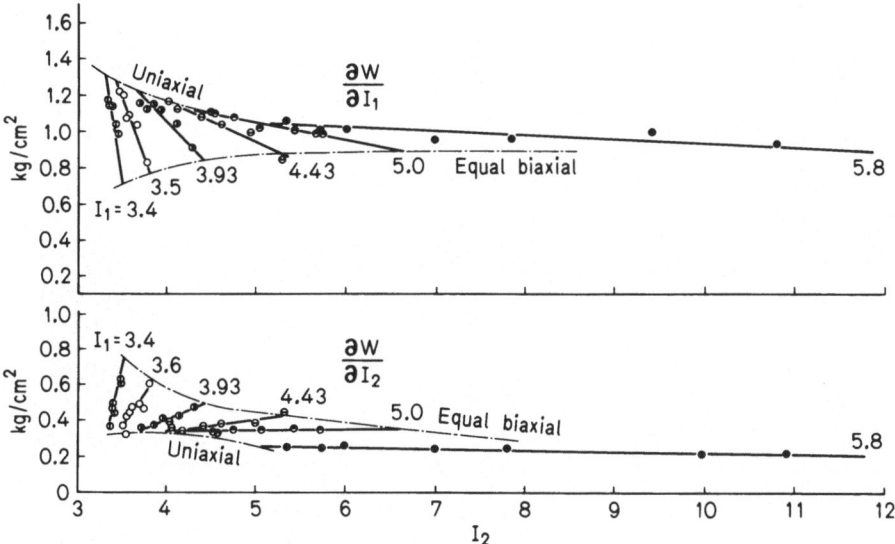

Fig. 13. $\partial W/\partial I_1$ and $\partial W/\partial I_2$ at fixed I_1 as a function of I_2 for SBR 1 at 1 min and 22 °C. [Reproduced from Kawabata, S.: J. Macromol. Sci. Phys. *B8* (3−4), 605 (1973), Figs. 3, 6, 7 and 8.]

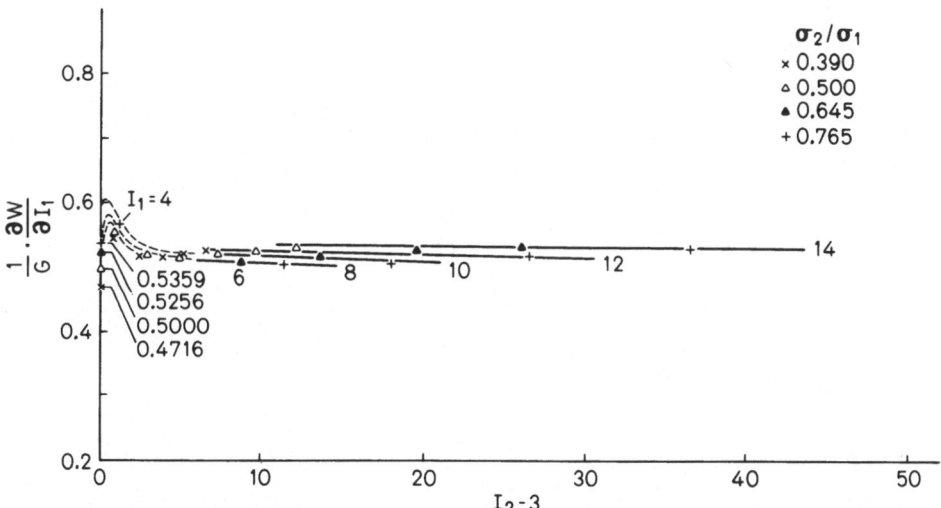

Fig. 14. $\partial W/\partial I_1$ measured by Becker at 10-min relaxation for natural rubber vulcanizate. These values are normalized by shear modulus, G. [Reproduced from Becker, C. W.: J. Polymer Sci., Part C, *16*, 2893, a part of Fig. 4 and a part of Fig. 5.]

dependence for the NR sample, with three isochronal values of $\partial W/\partial I_i$ at $I_1 = I_2 = 6.0$. The corresponding data for SBR 1 is presented in Fig. 18. In both figures, the derivative $\partial W/\partial I_1$ depends markedly on temperature, while the temperature dependence of $\partial W/\partial I_2$ is essentially negligible. It should be noted that the data of $\partial W/\partial I_1$ for NR above 0 °C follow the prediction from the kinetic theory (broken line). Their upswing

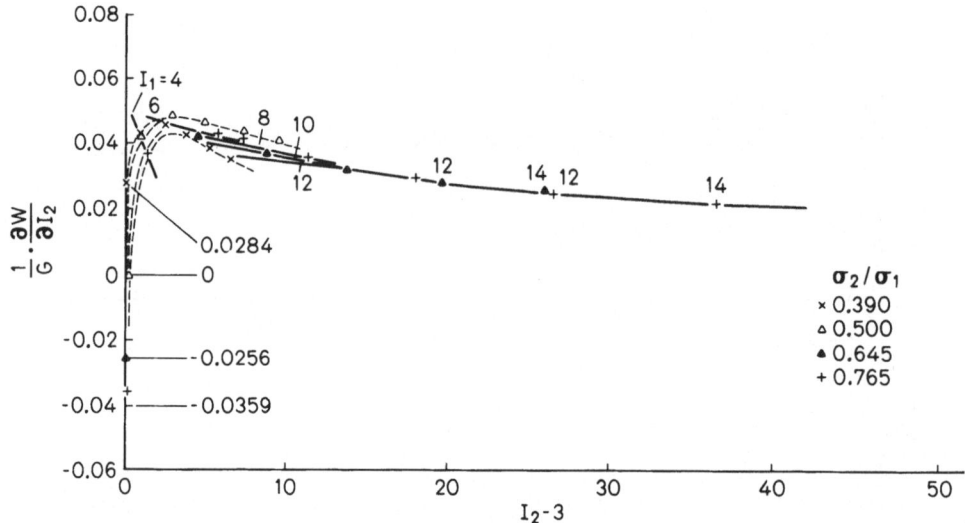

Fig. 15. $\partial W/\partial I_2$ measured by Becker at 10-min relaxation for natural rubber vulcanizate. These values are normalized by shear modulus, G. [Reproduced from Becker, C. W.: J. Polymer Sci., Part C, 16, 2893 (1967), a part of Fig. 4 and a part of Fig. 5.]

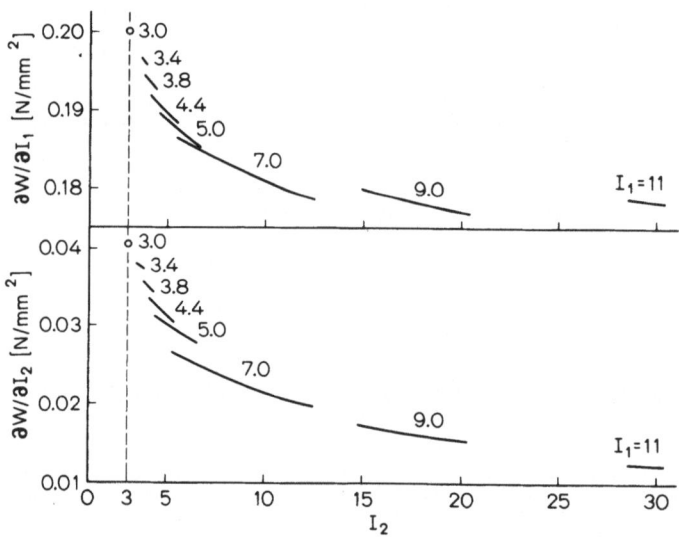

Fig. 16. $\partial W/\partial I_1$ and $\partial W/\partial I_2$ derived from $w'(\lambda)$ by Jones and Treloar for natural rubber vulcanizate. Temperature is between 19 and 24 °C. [Reproduced from Jones, D. F., Treloar, L. R. G.: J. Phys. D. Appl. Phys. 8, 1285 (1975), Fig. 7.]

below 0 °C appears to be a manifestation of the transition of the specimen from rubbery to glassy state. With SBR 1 only the upswing behavior, much more appreciable than with NR, can be seen in the temperature range studied. These apparent transitions occur at temperatures much higher than the ordinary glass transition temperatures observed with these rubbers. At present, no adequate mechanism for them is

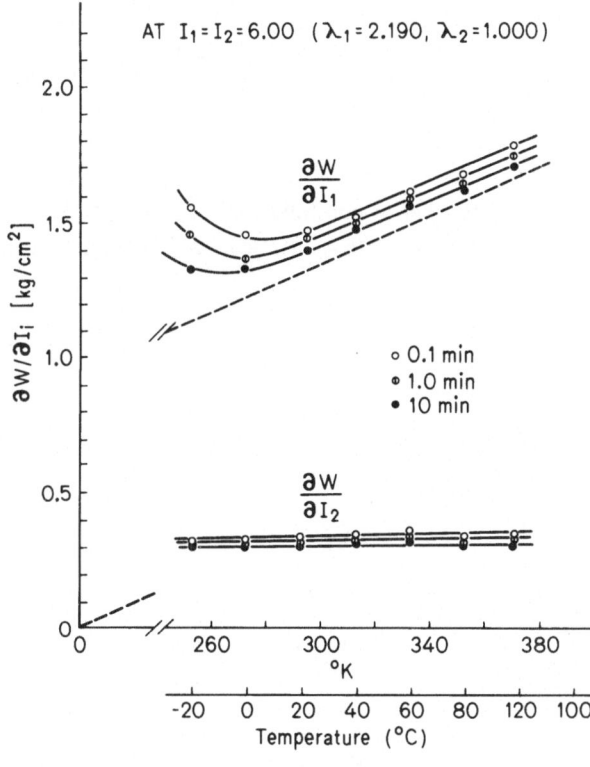

Fig. 17. Temperature dependence of $\partial W/\partial I_1$ and $\partial W/\partial I_2$ for NR. [Reproduced from Kawabata, S.: J. Macromol. Sci. Phys. B8 (3–4), 605 (1973), Figs. 3, 6, 7 and 8.]

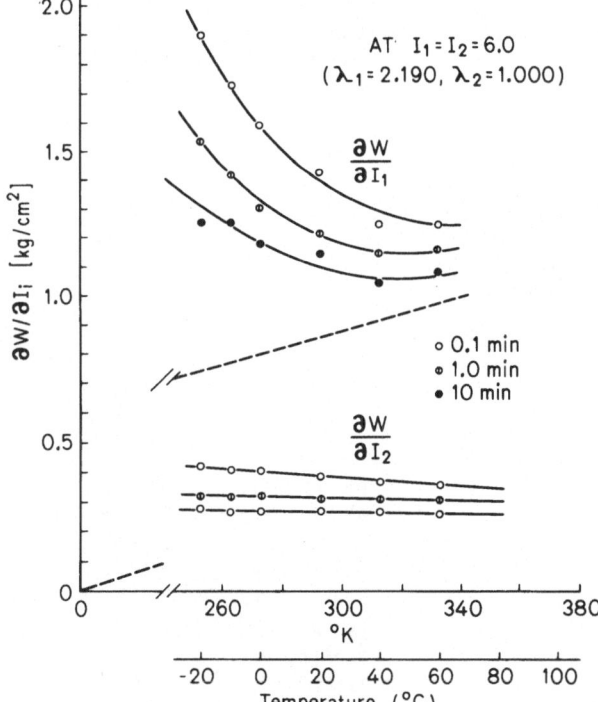

Fig. 18. Temperature dependence of $\partial W/\partial I_1$ and $\partial W/\partial I_2$ for SBR 1. [Reproduced from Kawabata, S.: J. Macromol. Sci. Phys. B8 (3–4), 605 (1973), Figs. 3, 6, 7 and 8.]

conceivable, but we hope that attention of those who are concerned with molecular theoretical aspects of rubber elasticity will be drawn to this behavior.

C. Time Dependence of $\partial W/\partial I_i$

This property examined with SBR 1 at a number of combinations of I_1 and I_2 is displayed in Figs. 19A and 19B [41, 42]. The data include both general biaxial and strip

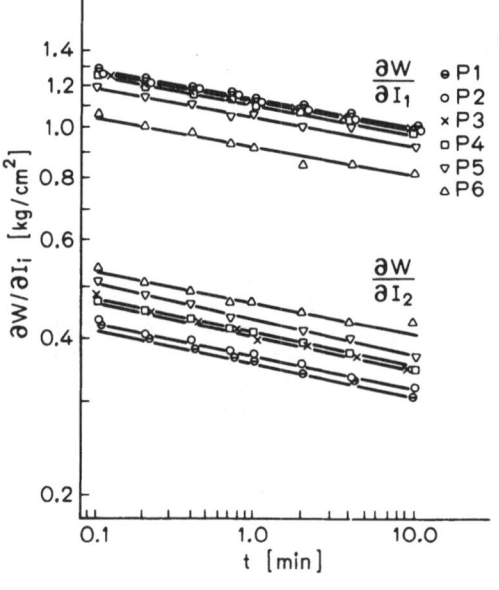

Fig. 19A. Time dependence of $\partial W/\partial I_1$ and $\partial W/\partial I_2$ for SBR 1 at various points on the $I_1 - I_2$ plane:
P_1 ($I_1 = 3.916, I_2 = 3.710$),
P_2 ($I_1 = 3.935, I_2 = 3.777$),
P_3 ($I_1 = 3.934, I_2 = 3.858$),
P_4 ($I_1 = 4.002, I_2 = 3.908$),
P_5 ($I_1 = 3.923, I_2 = 4.135$),
P_6 ($I_1 = 3.919, I_2 = 4.292$)

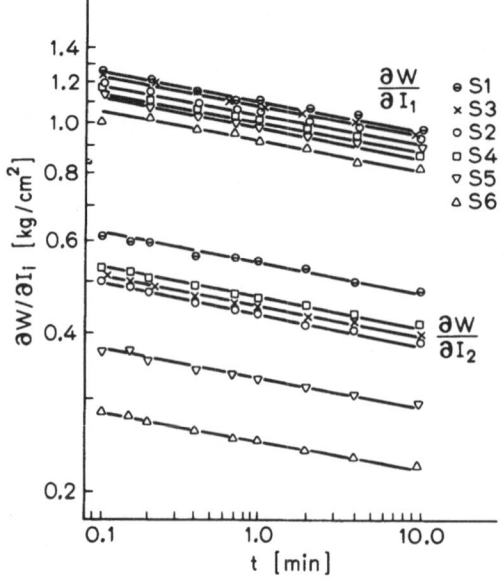

Fig. 19B. Time dependence of $\partial W/\partial I_1$ and $\partial W/\partial I_2$ for SBR 1 at various strip biaxial deformations:
S_1 ($I_1 = 3.402, I_2 = 3.404$),
S_2 ($I_1 = 3.611, I_2 = 3.611$),
S_3 ($I_1 = 3.910, I_2 = 3.926$),
S_4 ($I_1 = 4.427, I_2 = 4.429$),
S_5 ($I_1 = 5.136, I_2 = 5.142$),
S_6 ($I_1 = 6.519, I_2 = 6.606$)

biaxial extension measurements. Figure 20 shows that both $\partial W/\partial I_1$ and $\partial W/\partial I_2$ as functions of time determined from the general biaxial extension experiment follow the forms suggested by Smith and Dickie [45], *i.e.*,

$$
\left.
\begin{aligned}
\frac{\partial W(I_1, I_2, t)}{\partial I_1} &= \phi_1(t) \, \frac{\partial W(I_1, I_2)}{\partial I_1} \\[2ex]
\frac{\partial W(I_1, I_2, t)}{\partial I_2} &= \phi_2(t) \, \frac{\partial W(I_1, I_2)}{\partial I_2}
\end{aligned}
\right\}
\tag{51}
$$

and satisfy the condition $\phi_1(t) = \phi_2(t)$. A similar result was also obtained from strip biaxial extension measurements. However, as illustrated below, the Smith-Dickie form does not always apply for rubber vulcanizates.

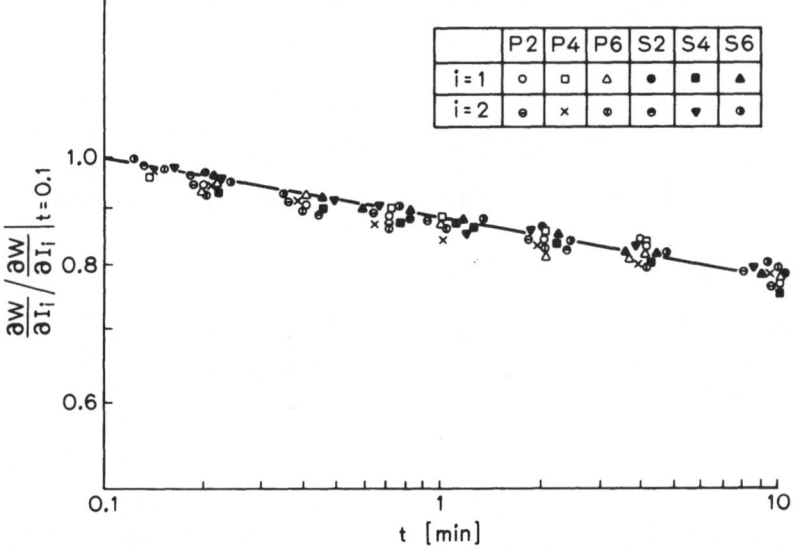

Fig. 20. Reduced time dependence of $\partial W/\partial I_1$ and $\partial W/\partial I_2$ for SBR 1: P_2, P_4, P_6, S_2, S_4 and S_6 correspond to the deformations given in Figs. 19A and B. [Reproduced from Kawabata, S.: J. Macromol. Sci. Phys. B8 (3−4), 605 (1973), Figs. 3, 6, 7 and 8.]

D. Dependence of $\partial W/\partial I_i$ on Degree of Cross-linking

SBR 2, 3, and 4 were used to investigate the effects of the degree of cross-linking on $\partial W/\partial I_i$ at $I_1 = I_2$. In Fig. 21 [46], the values of these derivatives relative to those for the sample with 1.5 part sulfur content are plotted against both M_c^{-1} and sulfur content in parts. Here M_c is the average molecular weight of the chain between neighboring cross-linking points. Actually, its values for the given samples were estimated by the usual swelling method. It is seen from Fig. 21 that the values of $\partial W/\partial I_i$ depend on the degree of cross-linking approximately in the fashion predicted from the classical network theory.

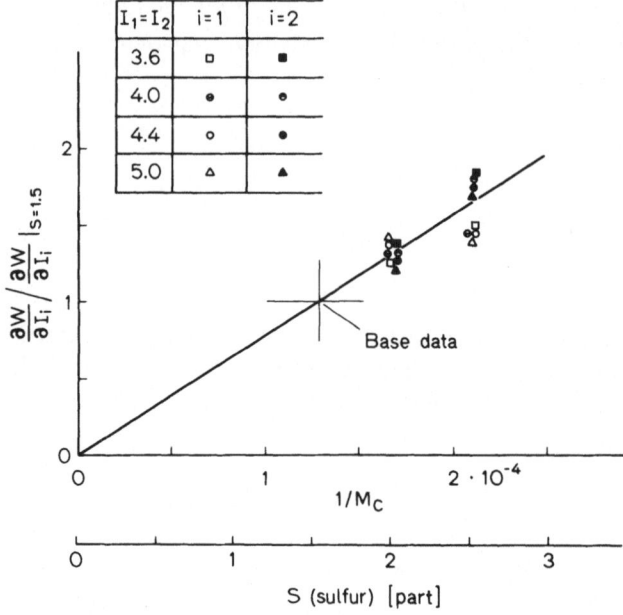

Fig. 21. Dependence of $\partial W/\partial I_1$ and $\partial W/\partial I_2$ for SBR vulcanizates on degree of cross-linking

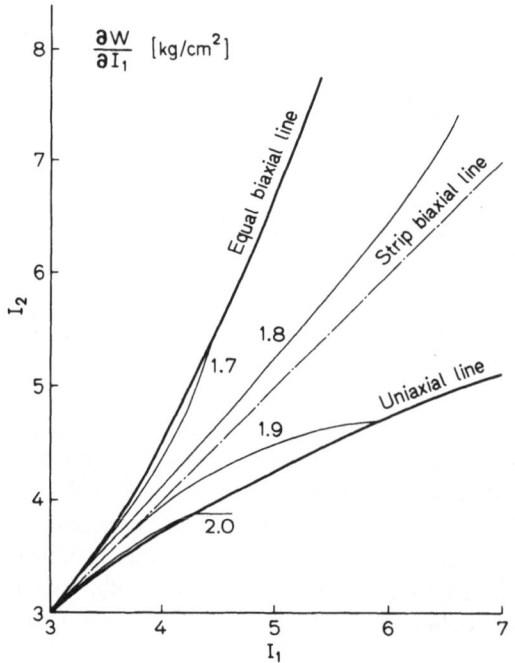

Fig. 22. Contour map of $\partial W/\partial I_1$ for carbon-filled SBR at 1 min and 23 °C

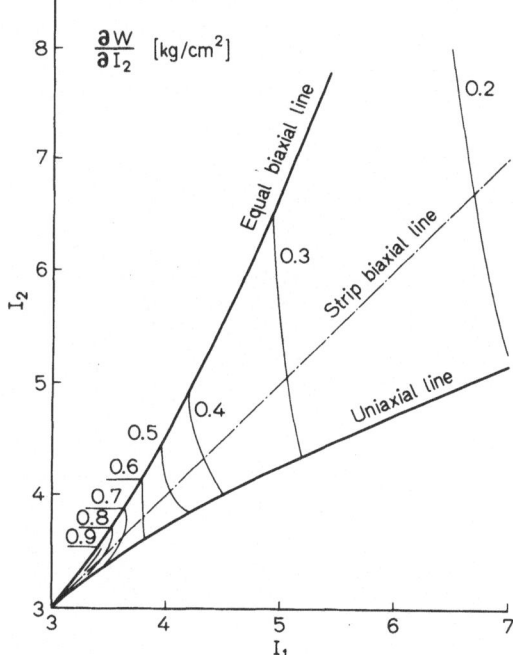

Fig. 23. Contour map of $\partial W/\partial I_2$ for carbon-filled SBR at 1 min and 23 °C

E. Effects of Reinforcement on $\partial W/\partial I_i$

The contour maps of $\partial W/\partial I_1$ and $\partial W/\partial I_2$ obtained from biaxial extension measurements on the carbon-filled SBR sample are depicted in Fig. 22 and 23 [41]. A com-
parison of these graphs with those for unfilled SBR (Figs. 11 and 12) reveals effects
of carbon-filling. Roughly speaking, for both $\partial W/\partial I_1$ and $\partial W/\partial I_2$, the contour lines
of the filled SBR in the region of I_1 and I_2 close to 3 are similar in shape to those of
the unfilled one in the region of larger I_1 and I_2. This behavior agrees with the prediction that carbon-filling tends to stretch the chain molecules in the rubber network.
The time dependence of $\partial W/\partial I_i$ for the filled SBR (Fig. 24) is not as simple as that
for the unfilled SBR (Fig. 19), clearly not obeying the Smith-Dickie form mentioned
above.

F. Behavior at Very Small Strains

As noted in the foregoing lines, the values of $\partial W/\partial I_i$ in the region of small deformations are very sensitive to errors in the experimental determinations of strains and
stresses. We carried out the study in this region with the specially designed apparatus
shown in Fig. 9. Figure 25 [47] shows the resulting data, in which the values of $\partial W/\partial I_i$
for a series of different I_1 are plotted against I_2, with both I_1 and I_2 restricted to the
range 3.03–3.10. A special feature to note is that $\partial W/\partial I_2$ for fixed I_1 decreases
sharply and falls into the region of negative values as I_2 approaches 3. Thus, in the

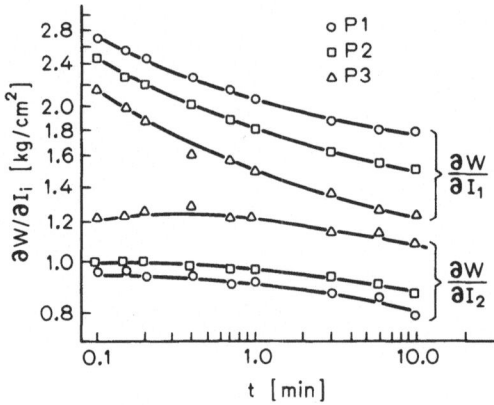

Fig. 24A. Time dependence of $\partial W/\partial I_1$ and $\partial W/\partial I_2$ for carbon-filled SBR: P_1 ($I_1 = 3.395, I_2 = 3.345$), P_2 ($I_1 = 3.399, I_2 = 3.437$), P_3 ($I_1 = 3.390, I_2 = 3.457$)

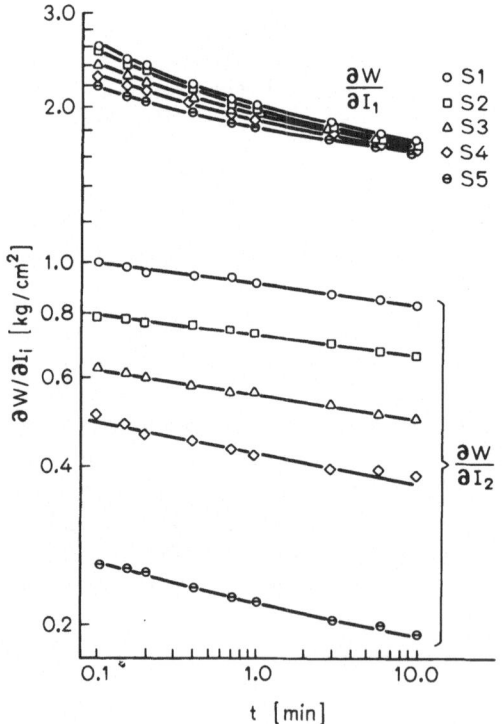

Fig. 24B. Time dependence of $\partial W/\partial I_1$ and $\partial W/\partial I_2$ for carbon-filled SBR at various strip biaxial deformations: S_1 ($I_1 = 3.405$, $I_2 = 3.403$), S_2 ($I_1 = 3.619$, $I_2 = 3.619$), S_3 ($I_1 = 3.932$, $I_2 = 3.948$), S_4 ($I_1 = 4.339$, $I_2 = 4.326$), S_5 ($I_1 = 6.664$, $I_2 = 6.781$)

case of uniaxial extension, the energy ought to be liberated from the specimen as it is deformed slightly. This characteristic behavior was confirmed by repeating the experiment many times. The graph for $\partial W/\partial I_1$ is approximately a mirror image of that for $\partial W/\partial I_2$ about a horizontal line. Therefore, we have an approximate relation

$$\partial W/\partial I_1 + \partial W/\partial I_2 = \text{a positive constant } c \qquad (51)$$

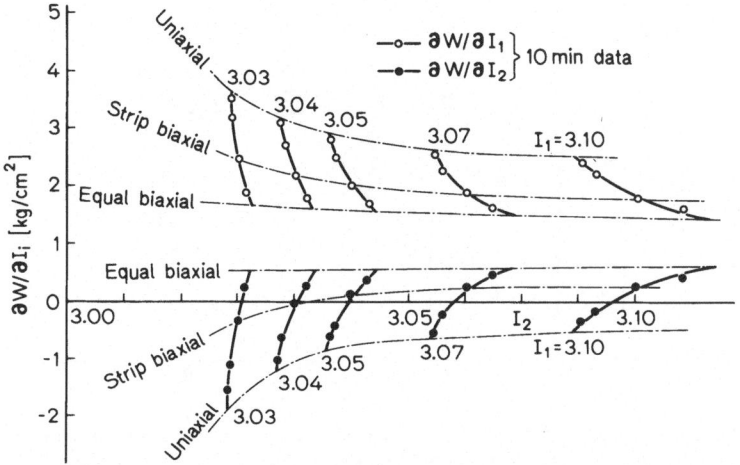

Fig. 25. $\partial W/\partial I_1$ and $\partial W/\partial I_2$ at small deformations and 10 min for NR at 17 °C

for small deformations. Interestingly, this relation agrees with the prediction from the elasticity theory for infinitesimal deformations, in which the constant c equals six times the Young modulus of the material.

The data obtained by Becker are quoted in Figs. 26 A and B, where the details at small deformations are illustrated. His data also indicate $\partial W/\partial I_2$ tends to be negative as the deformation becomes smaller, but no attention was paid to this feature nor was it discussed. The functional forms of $\partial W/\partial I_1$ and $\partial W/\partial I_2$ by Becker and the present authors are considerably different. This is probably due to the difference in the experimental techniques used.

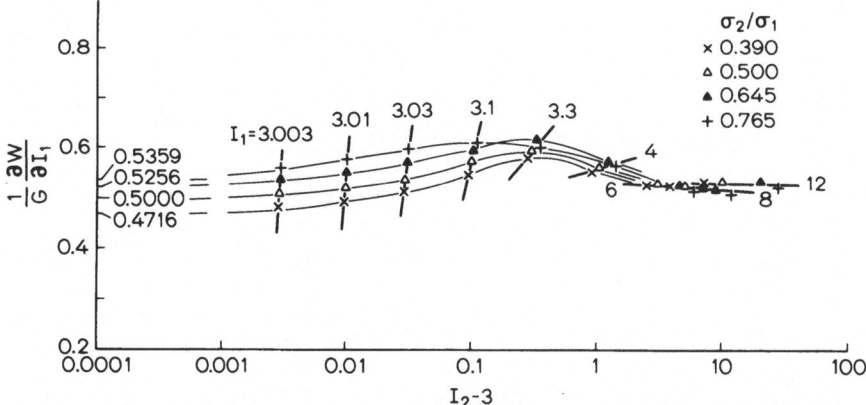

Fig. 26 A. $\partial W/\partial I_1$ measured by Becker for natural rubber vulcanizate at small deformations and 10 min. [Reproduced from Becker, C. W.: J. Polymer Sci., Part C, 16, 2893 (1967), a part of Fig. 4 and a part of Fig. 5.]

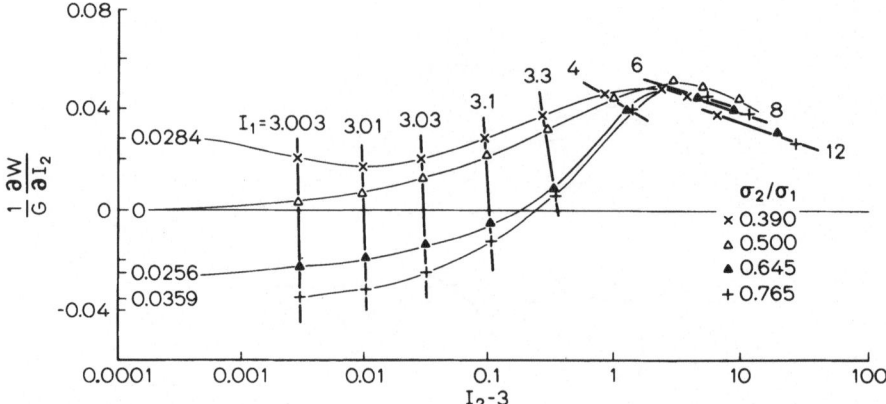

Fig. 26 B. $\partial W/\partial I_2$ measured by Becker for natural rubber vulcanizate at small deformations and 10 min. [Reproduced from Becker, C. W.: J. Polymer Sci., Part C, *16*, 2893 (1967), a part of Fig. 4 and a part of Fig. 5.]

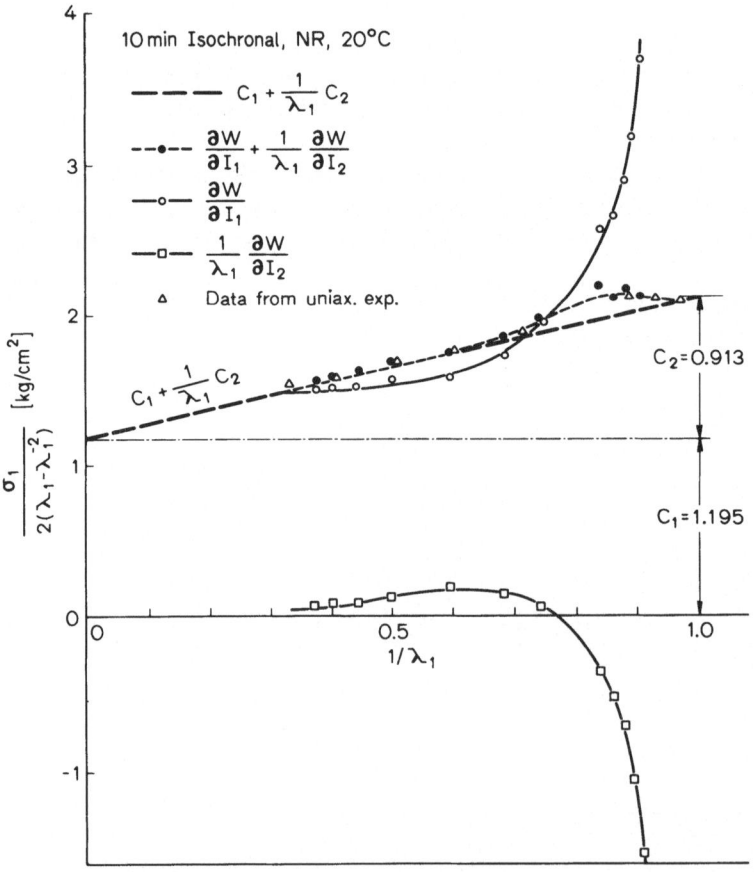

Fig. 27. Mooney-Rivlin plot of uniaxial extension data for NR (Δ) compared with the sum of $\partial W/\partial I_1$ and $\lambda^{-1}\partial W/\partial I_2$, where $\partial W/\partial I_i$ were extrapolated for uniaxial extension from biaxial data. The contributions of $\partial W/\partial I_1$ and $\lambda^{-1}\partial W/\partial I_2$ to their sum are also shown

G. The Mooney-Rivlin Plot

The triangular symbols in Fig. 27 show the 10-min isochronal Mooney-Rivlin plots for the NR specimen [48], obtained from uniaxial extension experiments. The thick broken line fitting them appears to indicate that its intercept and slope may be equated to $\partial W/\partial I_1$ and $\partial W/\partial I_2$, respectively. That this cannot be done is clearly seen from the actual values of $\partial W/\partial I_1$ and $\partial W/\partial I_2$ plotted separately in the figure with open circles and squares, respectively. These values were estimated from the graphs, as in Fig. 13, by extrapolating the 10-min isochronal biaxial data of $\partial W/\partial I_i$ to uniaxial deformation. The sum $\partial W/\partial I_1 + \lambda^{-1}\partial W/\partial I_2$ is also plotted in the figure with filled circles. The good agreement between the uniaxial data (triangles) and the extrapolated ones from the biaxial data (filled circles) demonstrates not only a high accuracy of the experimental results, but also the danger of attributing a linear Mooney-Rivlin plot to the Mooney equation for W, Eq. (32). In fact, the slope of the Mooney-Rivlin plot here is primarily associated with a λ_1-dependence of $\partial W/\partial I_1$. A similar conclusion was reached by Jones and Treloar [36] in a recent article.

There is no reason to anticipate that, in general, linear Mooney-Rivlin plots are obtained at least over a certain range of relatively small stretch ratios. Though not illustrated here, our data on the carbon-filled SBR gave the Mooney-Rivlin plots of markedly upward curvature, and again this curvature was found to be due mainly to the dependence of $\partial W/\partial I_1$ on λ_1.

Figure 28 [41] depicts the isochronal Mooney-Rivlin plots for SBR-1, where the extrapolated values of $\partial W/\partial I_1$ and $\lambda^{-1}\partial W/\partial I_2$ are represented by solid lines and the sum of them by broken lines. As above, these sums are equivalent to the Mooney-Rivlin plot of uniaxial data. We again find that the slope of the sum curves depends mainly on the λ_1 dependence of $\partial W/\partial I_1$ and therefore the slope is not equal to

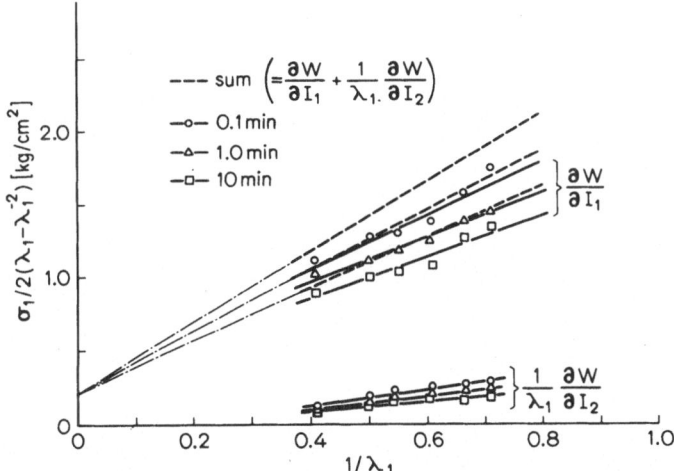

Fig. 28. Plots of $\partial W/\partial I_1$, $\lambda^{-1}\partial W/\partial I_2$, and their sum (broken lines) as function of λ^{-1} for SBR-1. These values were estimated from biaxial data in same manner as in Fig. 27. Sum curves are equivalent to Mooney-Rivlin plot, and C_1 and C_2 may be determined. Note that C_1 is apparently independent of time t, while actual values of $\partial W/\partial I_1$ are not

$\partial W/\partial I_2$. It is also observed in the figure that the C_1 value determined in the conventional manner as the intercept of the plot at $\lambda_1^{-1} = 0$ is essentially independent of the time of measurement. However, the separately evaluated $\partial W/\partial I_1$ exhibits a clear time dependence.

H. Application of the Valanis-Landel Expression for W

From a relevant biaxial extension experiment one can determine the function $w(\lambda)$ constituting the Valanis-Landel expression for W, as was done, for example, by Jones and Treloar. However, the w function so obtained is of little value unless it is tested with the stress-strain relations for other modes of deformation. This kind of test was carried out by the present authors[49] and is described below.

Figures 29 A and B show the original data from general biaxial extension measurements on the NR sample. Here the measured stresses σ_1 and σ_2 at a series of fixed λ_1 are plotted against λ_2. All these values are isochronal (10 min). The graphs have been displayed to illustrate the accuracy of our measurements. In Fig. 30, the observed σ_1 are compared with the predictions from other stress-strain relations.

First, the Mooney-Rivlin plot was prepared with the paired values of σ_1 and λ_1 taken on the uniaxial line ($\lambda_2 = (\lambda_1)^{-1/2}$) in Fig. 29 A. The values of C_1 and C_2 determined from it were taken equal to $\partial W/\partial I_1$ and $\partial W/\partial I_2$, respectively, and then

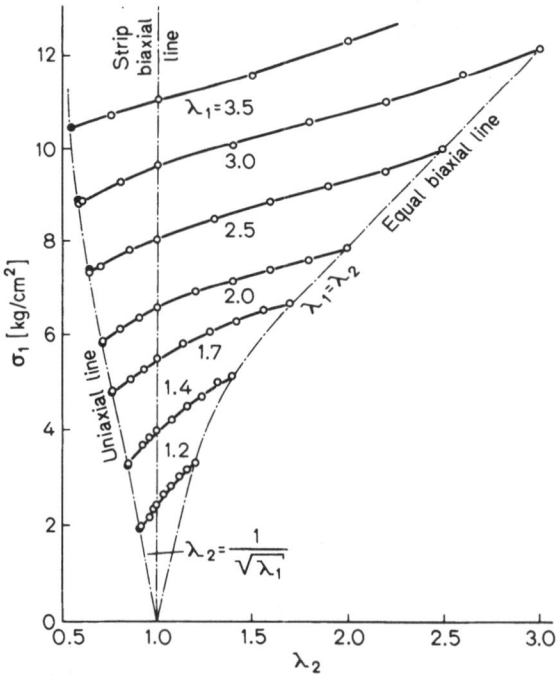

Fig. 29 A. σ_1 at fixed λ_1 as a function of λ_2 for NR at room temperature. Filled circles ($\bullet$), data obtained with ring-shaped specimen

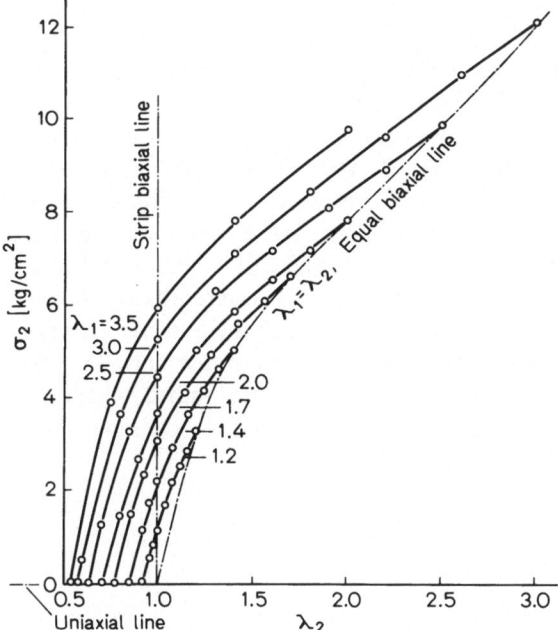

Fig. 29 B. σ_2 at fixed λ_1 as a function of λ_2 for NR at room temperature

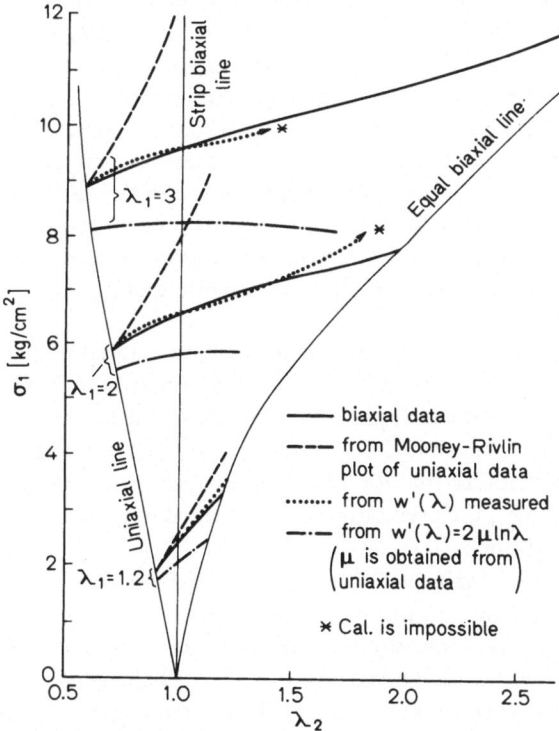

Fig. 30. Comparison of σ_1 values predicted from various methods with experimental values from biaxial extension measurements

Eq. (13) was used to compute σ_1 for given sets of λ_1 and λ_2. The broken lines in Fig. 30 show the computed values of σ_1 for $\lambda_1 = 1.2, 2$ and 3 as a function of λ_2. Their agreement with the directly observed values (solid lines) is very poor. Second, the Valanis-Landel expression for W [Eq. (8)] with $w(\lambda)$ given by Eq. (44) was substituted into Eq. (26) to calculate σ_1. The resulting σ_1 versus λ_2 curves for $\lambda_1 = 1.2$, 2, and 3 are indicated by dot-dash lines in Fig. 30. Again, they deviate markedly from the corresponding observed curves. Finally, the σ_1 versus λ_1 relation for $\lambda_2 = 1$ read from Fig. 29 A, *i.e.*, the data corresponding to strip biaxial extension measurements, was used to determine the experimental form of $w'(\lambda)$ and then σ_1 for $\lambda_1 = 1.2$, 2, and 3 were computed as a function of λ_2 by using Eq. (26). The results are shown by dotted lines in Fig. 30. They now agree fairly closely with the observed curves, except in the region approaching the equal biaxial line. We must say, however, that this similar agreement was not found with the SBR vulcanizates studied, which implies that the validity of the Valanis-Landel expression for W depends on the kind of rubber as well as on the region of deformations.

VI. Concluding Remarks

From the viewpoint of the mechanics of continua, the stress-strain relationship of a perfectly elastic material is fully described in terms of the strain energy density function W. In fact, this relationship is expressed as a linear combination of the partial derivatives of W with respect to the three invariants of deformation tensor, I_1, I_2, and I_3. It is the fundamental task for a phenomenologic study of elastic material to determine W as a function of these three independent variables either from molecular theory or by experiment. The present paper has reviewed approaches to this task from biaxial extension experiment and the related data. The results obtained so far demonstrate that the kinetic theory of polymer network does not describe actual behavior of rubber vulcanizates. In particular, contrary to the kinetic theory, the observed derivative $\partial W/\partial I_2$ does not vanish.

Kawabata [44] has pointed out on the basis of a simple network model that of the two derivatives, $\partial W/\partial I_1$ and $\partial W/\partial I_2$, the former should be related primarily to intramolecular forces such as the entropy force which plays a major role in the kinetic theory of rubber elasticity, while the latter should be a manifestation of intermolecular interactions. He predicted the possibility that $\partial W/\partial I_2$ assumes negative values in the region of small deformation. In fact, the prediction was confirmed experimentally by Becker and also by the present authors.

Finally, it is pertinent to emphasize again that either of the two derivatives mentioned above is an extremely sensitive function of its arguments I_1 and I_2 at small deformations. This fact makes precise measurement of these quantities for the arguments close to 3 a very difficult task. We may say that it is a challenging one because detailed investigations in this region are expected to give more significant information about the molecular mechanism responsible for the mechanical behavior of rubberlike materials.

Acknowledgments. The authors are indebted to Professor H. Fujita, Department of Polymer Science, Osaka University, who gave them valuable suggestions and comments for preparation of this paper. Thanks are due to their former graduate students, Dr. Y. Obata and Messrs. K. Fukuma, T. Akagi, and K. Kitai for their cooperation in experimental work. Japan Synthetic Rubber Company supplied the rubber samples used, and Mitsubishi-Monsanto Chemicals Company supported part of this research, to which their thanks are also due.

VII. References

1) Meyer, K. H., von Susich, G., Valko, E.: Kolloid-Z. *59*, 208 (1932).
2) Guth, E., Mark, H.: Mh. Chem. *65*, 93 (1934).
3) Kuhn, W.: Kolloid-Z. *68*, 2 (1934).
4) James, H. M., Guth, E.: J. Chem. Phys. *11*, 455 (1943).
5) Treloar, L. R. G.: Trans. Faraday Soc. *39*, 36 (1943).
6) Kuhn, W.: Kolloid-Z. *76*, 258 (1936).
7) Rivlin, R. S.: J. Appl. Phys. *18*, 444, 837 (1947).
8) Rivlin, R. S.: Phil. Trans. Roy. Soc. *A241*, 379 (1948).
9) Rivlin, R. S.: Phil. Trans. Roy. Soc. *A242*, 173 (1949).
10) Rivlin, R. S.: Proc. Roy. Soc. *A195*, 463 (1949).
11) Rivlin, R. S., Saunders, D. W.: Phil. Trans. Roy. Soc. *A243*, 251 (1951).
12) Rivlin, R. S., Thomas, A. G.: Phil. Trans. Roy. Soc. *A243*, 289 (1951).
13) Boussinesq, J.: Compt. Rend. Acad. Sci. Paris *71*, 400 (1870).
14) Thomson, W. (Lord Kelvin): Phil. Trans. Roy. Soc. *A153*, 586 (1894).
15) Finger, J.: Sitzber Akad. Wiss. Wien *(IIa)103*, 1073 (1894).
16) Murnaghan, F. D.: Finite deformation of an elastic solid. New York: John Wiley 1951.
17) Eringen, A. C.: Nonlinear theory of continuous media. New York: McGraw-Hill 1962.
18) Rivlin, R. S.: Finite elastic deformation. Text for the lecture given at California Institute of Technology, Pasadena, California 1953.
19) Blatz, P. J., Ko, W. L.: Trans. Rheol. Soc. *VI*, 223 (1962).
20) Valanis, K. G., Landel, R. F.: J. Appl. Phys. *38*, 2997 (1967).
21) Flory, P. J., Rehner, J.: J. Chem. Phys. *11*, 512 (1943).
22) Wall, F. T.: J. Chem. Phys. *10*, 485 (1942).
23) Kuhn, W., Kuhn, H.: Helv. Chem. Acta *29*, 1095 (1946).
24) Mooney, M.: J. Appl. Phys. *11*, 582 (1940).
25) Becker, G. W.: J. Polymer Sci. *C-16*, 2893 (1967).
26) Zapas, L.: J. Res. Nal'l Bur. Std. *8-A*, 525 (1966).
27) Sakaguchi, K., Kawabata, S., Kawai, H., Hazama, N.: J. Soc. Materials Sci. Japan *17*, 356 (1968).
28) Kawabata, S.: Proc. 1974 Symposium Mech. Behavior of Materials *2*, 299 (1974).
29) Treloar, L. R. G.: Proc. Phys. Soc. *60*, PT2, 135 (1948).
30) Thomas, A. G.: Trans. Faraday Soc. *51*, 569 (1955).
31) Gent, A. N., Thomas, A. G.: J. Polymer Sci. *28*, 625 (1958).
32) Tschoegl, N. W.: J. Polymer Sci. A-1 *9*, 1959 (1970).
33) Alexander, H.: Int. J. Engng. Sci. *6*, 549 (1968).
34) Hart-Smith, L. J.: J. Appl. Maths. Phys. *17*, 608 (1966).
35) Treloar, L. R. G.: Rep. Prog. Phys. *36*, 755 (1973).
36) Jones, D. F., Treloar, L. R. G.: J. Phys. D: Appl. Phys. *8*, 1285 (1975).
37) Treloar, L. R. G.: Physics of rubber elasticity. Oxford: Clarendon Press 1958.
38) Ogden, R. W.: Proc. Roy. Soc. *A326*, 565 (1972).
39) Blatz, P. J., Sharda, S. C., Tschoegl, N. W.: Proc. Nat. Acad. Sci. USA *70*, 3041 (1973).
40) Obata, Y., Kawabata, S., Kawai, H.: Kogyo Kagaku Zasshi *73*, 1519 (1970).
41) Yoshihara, N., Kawabata, S., Kawai, H.: J. Soc. Materials Sci. Japan *19*, 317 (1970).
42) Obata, Y., Kawabata, S., Kawai, H.: J. Soc. Materials Sci. Japan *19*, 330 (1970).
43) Obata, Y., Kawabata, S., Kawai, H.: J. Polymer Sci. A-2 *8*, 903 (1970).
44) Kawabata, S.: J. Macromol. Sci.-Phys. B-8 *(3–4)*, 605 (1973).
45) Smith, T. L., Dickie, R. A.: J. Polymer Sci. A-2 *7*, 635 (1969).
46) Kawabata, S., Akagi, T.: Proc. 16th Japan Congress on Materials Sci. *1973*, 253.
47) Kawabata, S., Kitai, K.: unpublished.
48) Kawabata, S.: Preprints, Society of Polymer Science, Japan *25*, No. 2, 416 (1976).
49) Kawabata, S.: unpublished.

Received July 7, 1976

ESCA Applied to Polymers

David T. Clark

Department of Chemistry, University of Durham, South Road, Durham City, England

Table of Contents

1. Introduction

Since solids communicate with the rest of the universe by way of their surfaces, it is a truism that structure and bonding (in the chemical sense) of the surface of solids is of fundamental importance in any detailed discussion at the molecular level of many important phenomena. Despite the obvious importance of the nature of the surface and immediate subsurface of polymeric coatings for example, there are few techniques currently available for routinely delineating the aspects of structure and bonding which are of crucial importance in determining many of the physical, chemical, mechanical and electrical properties. This is more particularly the case when the horizon is broadened to encompass not only "academic" studies under relatively idealized conditions but real situations corresponding to polymeric coatings in their working environment. For example, it may be important to study various properties as a function of ageing or weathering or surface modification in general (*e.g.* Casing prior to adhesive bonding). Techniques capable of handling such widely differing situations non-destructively (as far as the samples are concerned) have until the advent of ESCA[1] been conspicuous by their absence.

Over the past six years or so we have been applying the relatively newly developed technique of Electron Spectroscopy for Chemical Application (ESCA) to studies of structure and bonding across a broad front encompassing organic, inorganic and polymeric systems. These studies together with complementary theoretical analysis have demonstrated that ESCA is an extremely powerful tool for investigations of structure and bonding with an information content per spectrum unsurpassed by any other spectroscopic technique. This distinctive attribute confers upon ESCA wide ranging applicability and versatility in respect of studies on polymeric systems and it is the purpose of this review to enlarge upon this theme.

It will become apparent that there are areas of study in which the required information can only at present be derived from ESCA studies, whilst in others the technique nicely complements the more established spectroscopic tools. In general, however, ESCA provides data at a much coarser level than most other spectroscopic tools and information pertaining for example to conformational effects may only be inferred rather indirectly. In many areas of application ESCA does not compare favourably in terms of resolution, sensitivity etc. with more established spectroscopic tools. The fact remains, however, that this is more than compensated by the great range of information available from a single ESCA experiment such that in the future one can envisage that ESCA will be the technique of choice for any initial investigation of a polymer sample. The ability to provide information straightforwardly on uncharacterized samples is unique to ESCA and gives the technique great potential (already exploited in some areas) for tackling not only academic problems but those of an applied 'trouble-shooting' nature.

The application of ESCA to structure and bonding in polymers has largely been pioneered at Durham, however, the field is already so large that it is only possible in the space available to give a brief review of the current fields of application. It is a reflection of how rapidly the field as a whole is developing that an article is apposite at this time despite the fact that extensive reviews have been written by the present author in the past three years[2]. Since the recent developments derive almost exclusively from programmes in my own laboratory I make no apologies for presenting what might to the uninitiated appear to be biased coverage. Perusal of the literature, however, will readily confirm that this is just the way things are! The review is there-fore entirely objective in the best scientific tradition. By contrast the best literary traditions often require subjective assessments and honest authors make this implicit in their chosen titles[3].

It is obviously desirable for reviews such as this to be largely self contained such that the uninitiated can follow most of the material without resorting to extensive background reading. To facilitate this therefore a minimum of background informa-tion is provided.

2. The ESCA Experiment

ESCA involves the measurement of binding energies of electrons ejected by inter–actions of a molecule with a monoenergetic beam of soft X-rays. For a variety of reasons the most commonly employed X-ray sources are $Al_{K\alpha 1,2}$ and $Mg_{K\alpha 1,2}$ with corresponding photon energies of 1486.6 eV and 1253.7 eV respectively. In principle all electrons, from the core to the valence levels can be studied and in this respect the technique differs from UV photoelectron spectroscopy (UPS) in which only the lower energy valence levels can be studied. The basic processes involved in ESCA are shown in Fig. 1.

With the conventionally employed X-ray photon sources cross sections for core levels for most elements of the periodic table are within two orders of magnitude of that for the C_{1s} levels and the technique thus has a convenient sensitivity range for all elements. The cross sections for photoionization of core levels is generally con-siderably higher than for valence levels and this taken in conjunction with the fact that core orbitals are essentially localized on atoms, and therefore have binding en-ergies characteristic of a given element means that in ESCA the predominant empha-sis is on the study of core levels. Although core electrons do not take part in bonding they monitor closely valence electron distributions and it is this particular feature which endows the technique with such wide ranging capabilities.

The removal of a core electron (which is almost completely shielding as far as the valence electrons are concerned) is accompanied by substantial reorganization of the valence electrons in response to the effective increase in nuclear charge. This perturbation gives rise to a finite probability for photoionization to be accompanied by simultaneous excitation of a valence electron from an occupied to an unoccupied level (shake up) or ionization of a valence electron (shake off). These processes giving rise to satellites to the low kinetic energy side of the main photoionization peak, follow monopole selection rules and their measurable parameters (intensities and separation with respect to the direct photoionization peaks) enormously broadens the scope of the technique as will become apparent in the ensuing discus-sion. De-excitation of the hole state can occur via both fluorescence and Auger processes, for elements of low atomic number the latter being the more probable. The lifetimes of the core hole states are typically in the range $10^{-13}-10^{-15}$ sec.

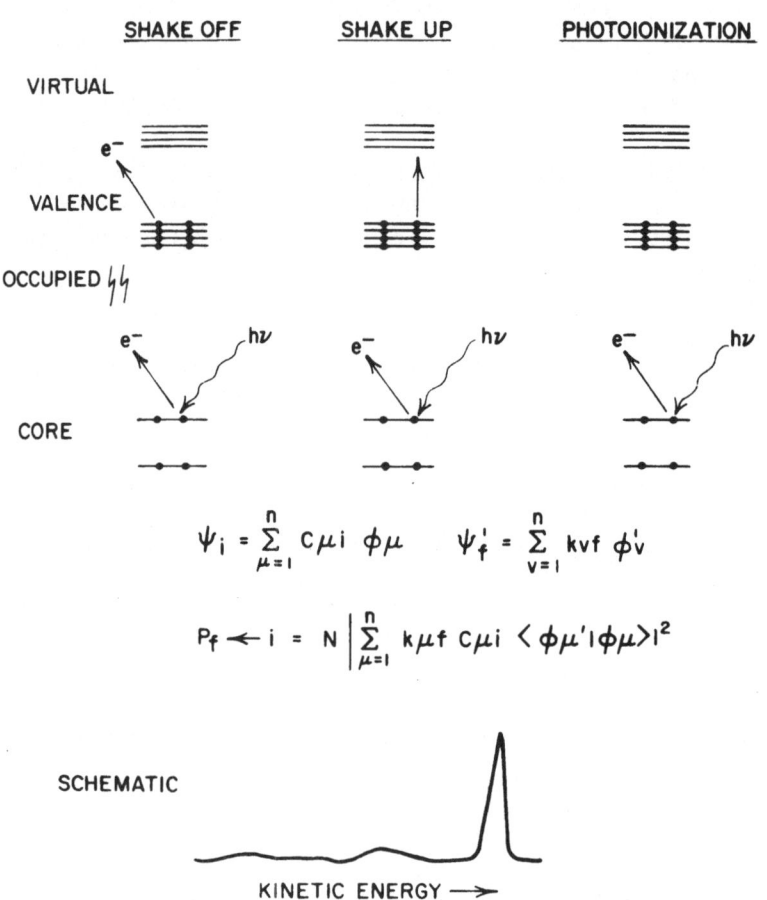

Fig. 1. Relationship between direct photoionization and shake up and shake off phenomena for core levels

emphasizing the extremely short time scales involved in ESCA compared with most other spectroscopic techniques.

The basic experimental set up for ESCA is shown in Fig. 2, and is largely self explanatory.

The most flexible of the commercially available designs employ double focussing electrostatic analyzers with retarding lens systems. This allows ready access to the sample for in situ preparation or modification and also allows angular studies to be made straightforwardly. Two alternatives to this arrangement with its relatively high constructional costs have been marketed commercially; the cylindrical mirror analyzer (Physical Electronics) and a novel design marketed by DuPont based on a low pass mirror, quadrupole lens and high pass filter. In addition to making routine angular dependent studies difficult such designs also make logical upgrading difficult. The philosophy of 'bolt on' modification and upgrading has much to commend

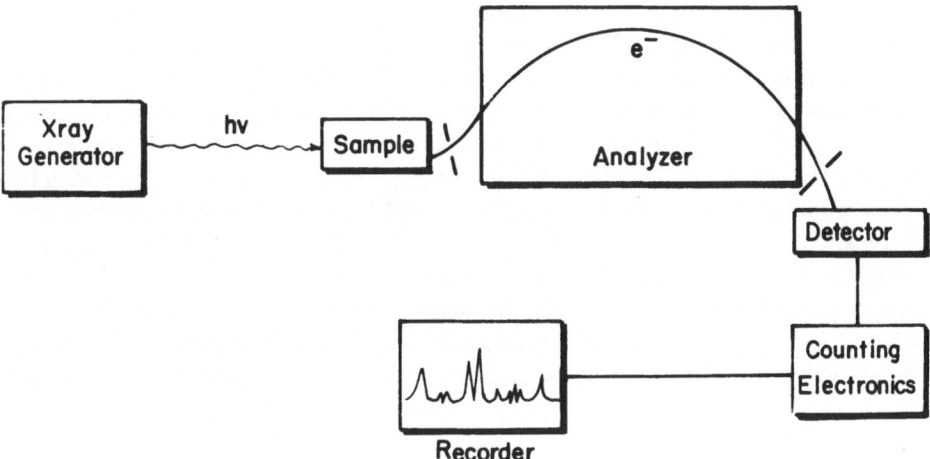

Fig. 2. Schematic of basic ESCA experiment

itself particularly in inflationary periods and there is commercial as well as scentific
logic in instrumental designs which anticipate developments over a decade or so. The
two most important instrumental developments which are already available are X-ray
monochromatization and position sensitive detectors. The former (Hewlett Packard,
Dispersion Compensation, A.E.I. Slit Filtering) based on dispersion of $Al_{K\alpha 1,2}$ radiation from spherically bent quartz discs improves signal/noise, signal/background
(by removing the bremmstrahlung) and resolution whilst the latter increases the
rate of data acquisition over that available from a single channeltron such that real
time investigations by ESCA become feasible.

To maintain the integrity of the sample during the time taken for measurement
and to obviate scattering of electrons entering the analyzer samples are maintained
in the spectrometer source at pressures of 10^{-7} torr or better. The relatively low
sticking coefficient for most small molecules comprising the extraneous atmosphere
in the spectrometer means that the vacuum requirements are somewhat less stringent
than in the study of e.g. evaporated metal films where pressures in the range 10^{-10}
torr would typically be required. Conventional cold trapped diffusion pumps backed
by two stage rotary pumps are therefore normally employed and the vacuum requirements are such that it is feasible to use pre-pumped insertion locks. Samples may
thus be introduced from atmosphere into the spectrometer and be ready for investigation at the requisite operating pressure in a matter of minutes. Samples may conveniently be studied as films or powders mounted on a sample probe which may be
taken into the spectrometer from atmosphere via insertion locks and valves. Provision
is usually made to enable samples to be heated or cooled in situ and ancillary equipment may be mounted directly onto the source of the spectrometer for in situ preparation or pre-treatment (e.g. argon ion bombardment, plasma synthesis, electron
bombardment, UV irradiation, chemical treatment etc.). Addition of a quadrupole
mass spectrometer facilitates many sample treatment studies and allows close control
to be kept of the extraneous atmosphere in the sample region.

Powders may be studied by mounting on double sided Scotch tape, films may be studied directly as such or films may be produced by solvent casting, dip coating, spin casting or by hot pressing etc. The method of presentation of samples for analysis allows considerable flexibility in terms of both size and shape. The technique therefore provides a means of direct investigation of surface coatings etc. in situ without the necessity of removal as might be required for IR investigation and definitely would be required for microanalysis or NMR investigation and this obviates many ambiguities and problems which might otherwise arise. As a simple example we might consider a block copolymer which might well have a surface domain structure different from the bulk. Taking the sample into solution and re-deposition might completely change both the bulk and surface structures. In any case for many polymeric films it will often be the case that they are insoluble (*e.g.* cross-linked system).

3. Principle Advantages and Disadvantages of ESCA as a Spectroscopic Technique

The principle advantages of ESCA as a spectroscopic tool in the study of polymeric materials are set out in Table 1.

Table 1. Principle advantage of ESCA in studying polymeric materials

(1)	Technique essentially non-destructive.
(2)	High sensitivity and modest sample requirement.
(3)	Materials may be studied in situ in their working environments with a minimum of preparation.
(4)	Large number of information levels available from a single experiment.
(5)	For solids unique capability of differentiating surface from sub-surface and bulk phenomena. Analytical depth profiling possible.
(6)	Information levels such that 'ab initio' investigations are feasible.
(7)	Data often complimentary to that obtained by other techniques. Unique capabilities central to the development of a number of important fields.
(8)	Theoretical basis well understood, results of considerable interest to theoreticians and may be quantified.

The typical X-ray fluxes employed in commercially available spectrometers are such that there are relatively few systems for which appreciable radiation damage occurs during the time taken to record a spectrum. Polythiocarbonyl fluoride depolymerizes rather rapidly whilst polyvinylidene fluoride slowly eliminates HF and cross-links. By contrast the dose rates involved in conventional Auger spectroscopy (employing an electron beam for excitation) are several orders of magnitude larger so that radiation damage poses severe problems[4]. The surface chemistry of polymers may therefore only be conveniently studied by ESCA and Auger spectroscopy is not a viable alternative. This contrasts with the situation for inorganic systems. The sample requirements are modest and the surface sensitivity of the technique is such that the technique samples ~ the outermost 100 Å or so of sample and depending on spectrometer design ~ 0.2 sq. cms. in area.

Although as will become apparent if we consider any one level of information available from ESCA and compare this with that available from the *most* competitive of the other available spectroscopic technique; for the particular case in question, it is invariably the case that ESCA compares relatively unfavourably. The most distinctive feature of ESCA as a spectroscopic tool however which sets it apart from any other is the large range of available information levels and these are shown in Table 2. For a fuller description see Ref.[2].

Table 2. Hierarchy of information levels available in ESCA

(1)	Absolute binding energies relative peak intensities, shifts in binding energies. Element mapping for solids, analytical depth profiling, identification of structural features etc. Short range effects directly longer range indirectly.
(2)	Shake up – shake off satellites. Monopole excited states; energy separation with respect to direct photoionization peaks and relative intensities of components of 'singlet and triplet' origin. Short and longer range effects directly (Analogue of UV).
(3)	Multiplet effects. For paramagnetic systems, spin state, distribution of unpaired electrons (Analogue of ESR).
(4)	Valence energy levels, longer range effects directly.
(5)	Angular dependent studies. For solids with fixed arrangement of analyzer and X-ray source, varying take off angle between sample and analyzer provides means of differentiating surface from sub-surface and bulk effects. Variable angle between analyzer and X-ray source angular dependence of cross sections, asymmetry parameter β, symmetries of levels.

It is the composite nature of these information levels which endows ESCA with such wide ranging capabilities and has seen the technique emerge as one of the most powerful shots in the chemist' s locker. The way in which these information levels may be exploited will become apparent from the discussion given below.

In studying solids the short mean free paths of electrons and their strong dependence on kinetic energy provides a means of differentiating surface from subsurface and bulk phenomena and hence analytical depth profiling by studying core levels with different escape depth dependencies. This is illustrated in Fig. 3 which shows a schematic representation of the generalized relationship between electron mean free paths and kinetic energy and hence the differing sampling depths for photoemitted electrons of differing kinetic energies. (For a detailed discussion of how this is invoked in practice readers are referred to Ref.[5].) Such studies can be considerably broadened in scope if they are coupled with angular dependent investigations.

As a preface to a discussion in relation to solids it is worthwhile briefly considering the more general possibilities. The cross section for photoionization from a given level is strongly dependent on the photon energy and exhibits angular dependence expressed in terms of the asymmetry parameter β. Angular resolved and photon energy dependent studies have been carried out on gas phase systems with particular reference to valence levels and the requisite instrumentation involves the variation in angle between photon source and electron analyzer[6]. This tends to be a specialized area and the equivalent experiment for solids has extra complications which make it non-trivial in nature[5]. For most commercially produced spectrometers the angle between photon source and analyzer is fixed so that the angular dependent studies referred to in the literature for solids are different in type for those described for gases since they do not directly involve investigations of the asymmetry parameter β. The basic philosophy behind angular dependent studies for solids is to vary the take off angle between the sample surface and the electron analyzer. Since electron mean free paths are so short it is clear that a grazing exit angle for the photoemitted electrons will enhance surface features compared with a take off angle normal to the surface and this is illustrated in Fig. 4. Angular dependent studies of a single core level can in principle therefore give rise to information concerning inhomogeneities in the surface and subsurface region of a sample. If core levels are available with different escape depth dependencies then we can combine angular dependent studies with investigations of relative peak intensities for electrons of different kinetic kinetic energies and in principle this should allow the differentiation and classification of samples into homogeneous and inhomogeneous on the ESCA depth sampling scale as illustrated in Fig. 5. If data pertaining to the bulk structure is available from microanalysis for example then we may also answer the question; is the outermost few tens of Angstroms of the sample the same in composition as the bulk?

The information levels available from ESCA studies are such that samples may be studied for which there is a minimum of prior information and in this respect ESCA is again different in character than most of spectroscopy. Having elaborated the advantages we may briefly consider the disadvantages which are surprisingly few. In terms of cost, standard ESCA instrumentation falls in the same league as continuous wave NMR although 'state of the art' instrumentation come somewhat closer to the cost of Fourier transformation NMR spectrometers. In setting up a routine ESCA facility therefore the overall costs are comparable to that for Fourier Transform IR, Laser Raman and Mass Spectrometers. The vacuum system associated with ESCA instrumentation means that *routine* sample handling requires provision of vacuum interlocks and also implies that it is not possible to switch the spectrometer on to routinely investigate a sample. In this respect the technique is comparable with mass spectrometry however it does not suffer from the same background problems. A

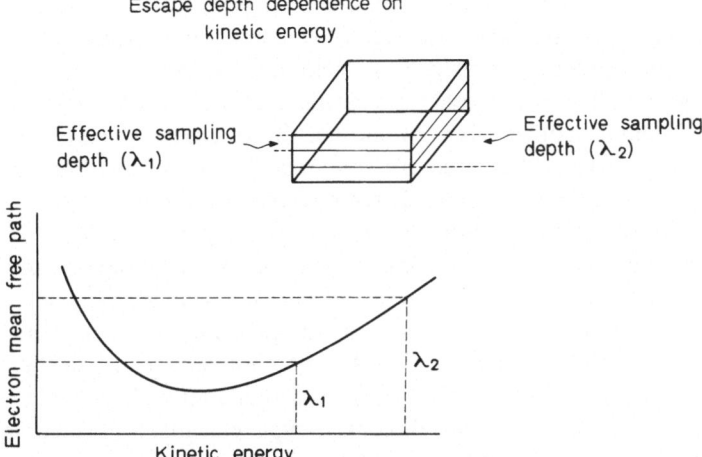

Fig. 3. Schematic of generalized form of escape depth versus kinetic energy for electrons. [Also illustrated are the different effective sampling depths for electrons corresponding to differing escape depths ($\lambda's$).]

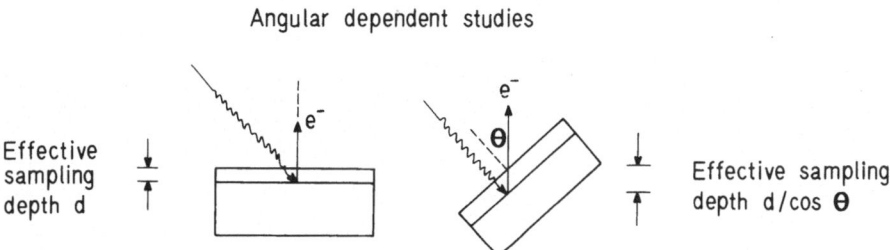

Fig. 4. Angular dependent studies for polymer samples in which spectra are studied as a function of electron take off angle θ with respect to the sample surface

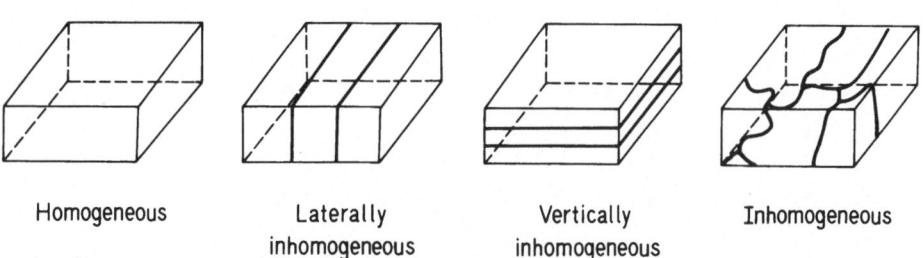

Fig. 5. Classification of samples with respect to ESCA investigation into homogeneous, laterally inhomogeneous, vertically inhomogeneous and inhomogeneous types

similar criticism may of course be levelled at NMR where the stability of the field
is such that there is usually a considerable time lag involved between switching the
instrument on and recording a spectrum. Whilst the technique has superior depth
resolution (in the range $\sim$ 100 Å) to any other, the spatial resolution is poor and
typically an area $\sim$ 0.25 cm is sampled. As a corollary of this of course we may
note that unless the outermost $\sim$ 100 Å or so of a thin organic or polymeric film of
the sample is representative of the bulk then it is not possible to say anything about
the bulk structure by means of ESCA without sectioning the sample.

    With conventional unmonochromatized X-ray sources, and slitted designs two
features are of importance in studying thick samples. The first is sample charging
arising from a distribution of positive charge over the sample surface under the conditions of X-ray bombardment. The second is that the polychromatic nature of the
X-ray source (characteristic lines superimposed on bremmstrahlung) leads to a relatively poor signal/background ratio for the technique. Sample charging for insulating
films usually amounts to no more than a few eV shift in the kinetic energy scale and
may readily be corrected for by standard techniques and we shall briefly address
ourselves to these in a subsequent section. The advent of efficient monochromatization schemes and multiple collector assemblies considerably alleviates the signal/

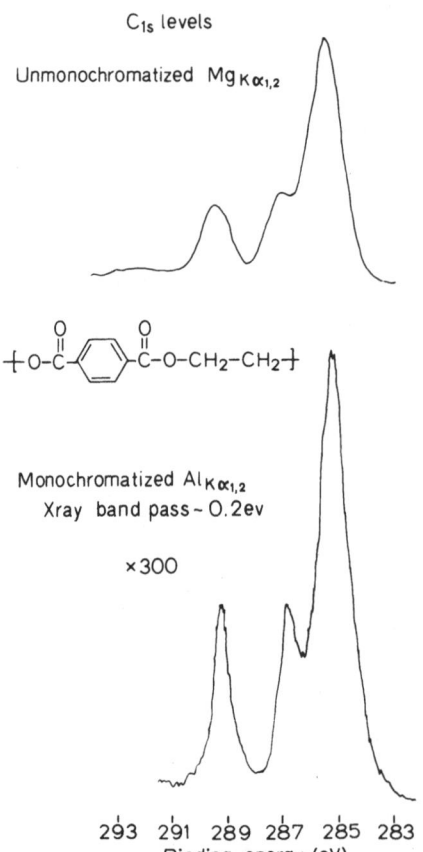

Fig. 6. C_{1s} levels for polyethylene terephthalate measured with unmonochromatized
$MgK\alpha_{1,2}$ and monochromatized $AlK\alpha_{1,2}$
X-ray sources

background and signal/noise ratios, however sample charging can be much more severe a problem for thick insulating film and needs careful consideration. Since the composite linewidths for C_{1s} levels for organic and polymeric materials studied by means of unmonochromatized X-rays are largely dominated by the inherent width of the X-ray source there is a considerable overall improvement in linewidths on going to instrumentation employing monochromatization schemes as is evident from the data in Fig. 6. It is still the case however that the overall shift to linewidth ratio for C_{1s} levels as a function of electronic environment are poor compared with say ^{13}C NMR. A particularly interesting case is that for C_{1s} levels appropriate to hydrocarbons in which carbon is formally in sp^3, sp^2 and sp hybridization whilst for the core levels the shift to line width ratio might typically be ~ 1 with an efficient monochromatization scheme, the corresponding ratio for ^{13}C is $\sim 10^3$.

Finally we may note that to take full advantage of the technique often requires a relatively high level of theoretical competence. However one of the interesting features about the technique is its capability for exploitation at many levels. Thus the technique lends itself not only to routine trouble shooting problems where often only a straightforward comparison is required without any attempt being made to understand the problem at a fundamental level and on the other hand the technique provides a powerful tool for the investigation of phenomena of considerable interest to a chemical physicist. We may note in this connection that the technique is quite competitive in terms of time taken to record a spectrum (typically ~ 10 minutes)

Table 3. ESCA applied to polymers

A. Aspects of structure and bonding (static studies)

    (i)   Gross chemical compositions
          (a) elemental compositions,
          (b) % incorporation of comonomers in copolymers,
          (c) polymeric films produced at surfaces.
   (ii)   Gross structural information
          e.g. for copolymers, block, alternating or random nature. Domain structure in block copolymers.
  (iii) Finer details of structure
          (a) structural isomerisms,
          (b) experimental charge distributions in polymers.
   (iv) Valence bands of polymers.
    (v) Identification of polymers, structural elucidation.
   (vi) Monopole excited states.

B. Aspects of structure and bonding (dynamic studies)

    (i)   Surface treatments *e.g.* Casing, plasma modification.
   (ii)   Thermal and photochemical degradation.
  (iii) Polymeric films produced at surfaces by chemical reaction *e.g.* fluorination (including the use of ESCA for depth profiling and quantitative measurement of film thickness).
   (iv) Chemical degradation of polymers, *e.g.* oxidation, nitration etc.

C. Electrical properties

    (i)   Mean free paths of electrons as a function of kinetic energy.
   (ii)   Photoconductivity of polymers.
  (iii) Statics and dynamics of sample charging.
   (iv) Triboelectric phenomena.

---

and with developments already in hand the technique should lend itself to real time applications. The areas of application of ESCA in relation to polymers which have already been delineated are shown in Table 3.

4. Energy Referencing

For samples studied as solids three situations may clearly be distinguished. In the first the sample is in electrical contact with the spectrometer. This is usually the case for films deposited in situ on a conducting substrate in the spectrometer source. Since the mean free path for the incident X-ray beam is very large it is possible, depending on the conditions, for films of the order of 1000 Å to have sufficient charge carriers to remain in electrical contact with the spectrometer. This can most readily be the case by applying a bias voltage to the sample probe. If the sample is in electrical contact the apparent shift in energy scale will exactly follow the applied bias. By shifting the position of the true zero of the kinetic energy scale it is possible to study the secondary electron distribution and this provides a direct energy reference. If the sample has been deposited on a substrate such as gold it is possible to measure the core levels of the sample whilst monitoring the $Au_{4f7/2}$ core level and this provides a very convenient means of energy referencing.

The second situation which arises for thick insulating samples. Thus it is often convenient to study samples mounted on double sided Scotch tape either as powder or as discrete films. In this circumstance there is only a fortuitous possibility that the sample will be in electrical contact with the spectrometer and in general it will be floating at some potential due to surface charging and indeed this charging process may be time dependent. If care is taken in the measurements the charge built up on a sample and its time dependence may be used to investigate electrical and chemical characteristics of samples and an example of this is given in a subsequent section. The most reliable method of energy referencing is to follow the slow build up of hydrocarbon contamination at the surface. With a base pressure of $\sim 10^{-8}$ torr the partial pressure of extraneous hydrocarbon material is such that taken in conjunction with the low sticking coefficient for most organic and polymeric systems it normally takes several hours before any signal arising from hydrocarbon (binding energy 285 eV[a] is apparent. It is of course possible to deliberately leak in straight hydrocarbon material to follow the build up at the surface. Such material almost always goes down in uniform coverage and at submonolayer coverage acquires the same surface potential as the sample. This is not necessarily the situation with regard to metals deposited on the surface since there is a marked tendency to island and as such differential charging may occur. In addition since gold is normally evaporated from a filament the possibility of surface damage, reaction or evaporation of substrate during deposition cannot be discounted. The use of the so-called 'gold decoration' technique is therefore not recommended for organic and polymeric

---

[a] This must of course be independently established for a given spectrometer. It almost certainly arises from long chain hydrocarbon material.

materials[7]. Since the factors which determine both absolute and relative binding energies of core levels may be shown to be very short range in nature it is often possible to study smaller molecules which contain the appropriate structural features as thin films in electrical contact with the spectrometer which may be straightforwardly referenced. Comparison may then be drawn between this model and the insulating sample in question and thus allow direct correction for sample charging. A further possibility which has received considerable attention of recent years is the use of electron 'flood guns'; the prime motivation being the very large sample charging for thick insulating samples in spectrometers employing monochromatic X-ray sources. The removal of bremmstrahlung as a source of secondaries can lead to shifts in the kinetic energy scale in the hundred eV range and can be compensated by flooding the sample with low energy electrons. Samples can become negatively charged however and the method needs great care to achieve an accuracy comparable with that for the other methods. An alternative source of low energy electrons is to illuminate the sample region with UV radiation from a low pressure, low power mercury lamp via a quartz viewing port in the source region of the spectrometer. Sufficient secondaries are generated from photoemission from the metal surfaces that sample charging is reduced to a low level[8].

The third situation which can arise is for thick films > 1 micron which have been built up by deposition on a conducting substrate. Such films behave as 'leaky' capacitors in that they exhibit rather striking time dependent charging and discharging characteristics and follow and applied bias potential in a particular manner. Since the dynamic equilibrium which is established under X-ray irradiation invariably produces an overall positive charge on the sample the application of a positive bias voltage causes a smaller shift in the kinetic energy scale than the applied voltage whereas a negative bias voltage produces a larger shift in the kinetic energy scale than the applied voltage. From a study of these effects and from the secondary electron distribution the energy referencing may readily be established. The investigation of such effects as a function of film thickness in the range 1–100 micron provides an interesting insight into the electrical characteristics of polymer samples[8] and the typical behaviour which is observed is shown in Fig. 7.

The energy reference in each case for the measurements   described above is the fermi level and although the exact location of this level in relation to the valence and conduction bands is generally unknown for polymers, as we have noted under the conditions of X-ray irradiation it is possible for an 'insulator' to be in electrical contact with the spectrometer i.e. their fermi levels are the same. Despite the difficulties associated with defining an analytical expression for the fermi level of an insulator, the use of the fermi level as energy reference is operationally convenient. If the work function of the insulator is known we may calculate the binding energy with respect to the vacuum level.

Although sample charging has been widely regarded as somewhat of a nuisance which must be circumvented our own view has been that the study of sample charging is of interest in its own right. Thus the study of the phenomena provides an interesting means of studying photoconductivity in polymeric films. Since the total electron flux leaving a sample surface is a function of the total cross section for photoionization we might anticipate that under a given set of instrumental conditions (e.g. sam-

Sample charging
polymer films

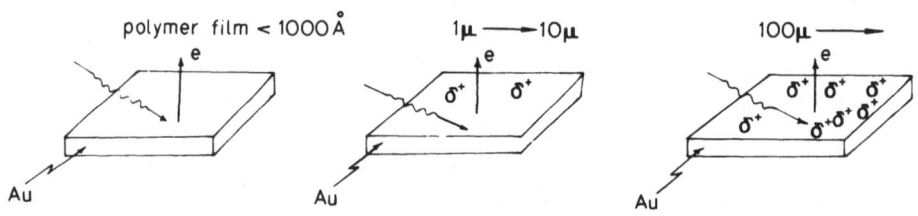

Au in electrical contact with probe.
Bias voltage characteristics
polymer core levels.

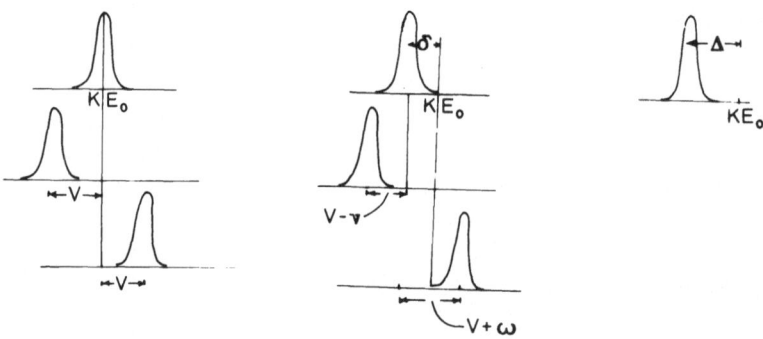

Fig. 7. Typical sample charging characteristics for polymer films

ples, either metal or polymer mounted in such a way as to be insulated from the
spectrometer), should show a relationship to sample charge measured by ESCA. Thus
if we take a typical metal such as gold and mount it on a sample probe by means of
a piece of double side electrically insulating "scotch" tape then under given set of
experimental conditions the core levels exhibit a shift to lower kinetic energy arising
from sample charging. This contrasts of course with the situation where the sample
is mounted directly in electrical contact with the spectrometer as shown in Fig. 8.
If a series of polymer samples ranging from high density polyethylene through polyvinylidene fluoride to polyhexafluoropropene are also studied under similar conditions a striking correlation is obtained between the sample charging as evidenced by
shift in kinetic energy scale and the normalized theoretically calculated total cross
section for photoionization and this is shown in Fig. 9. Sample charging is therefore
characteristic of the fine detail of the electronic structure of the sample. It is a
straightforward matter to show that sample charging essentially reflects surface structure since if we allow a hydrocarbon deposit to build up on an insulated sample of
gold there is a decrease in sample charge which exactly parallels the build up[8]. As
will become apparent in the ensuing discussion there is a considerable amount of
information encoded in sample charging which has hitherto been largely neglected.

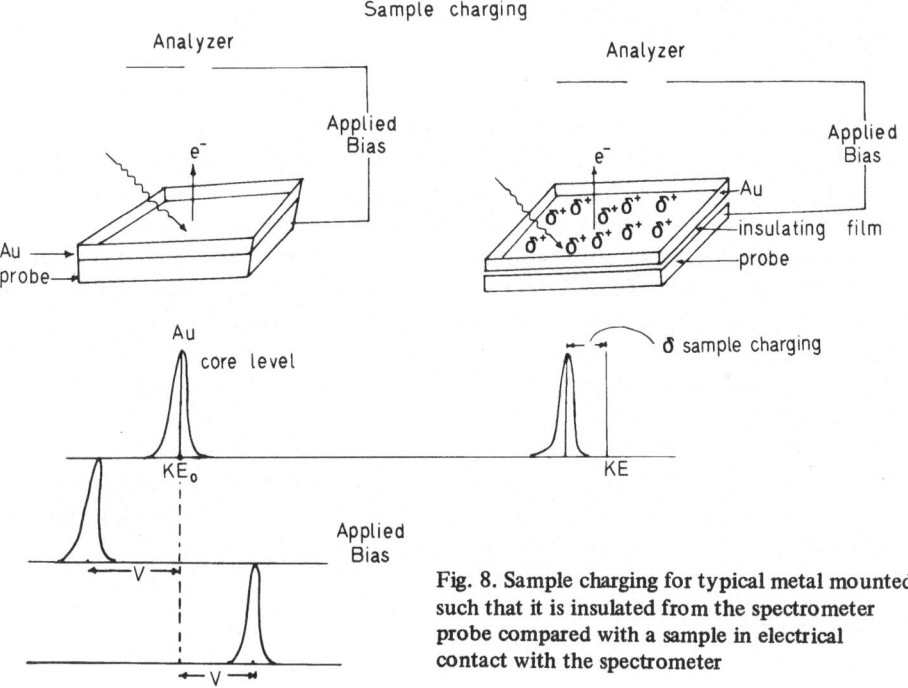

Fig. 8. Sample charging for typical metal mounted
such that it is insulated from the spectrometer
probe compared with a sample in electrical
contact with the spectrometer

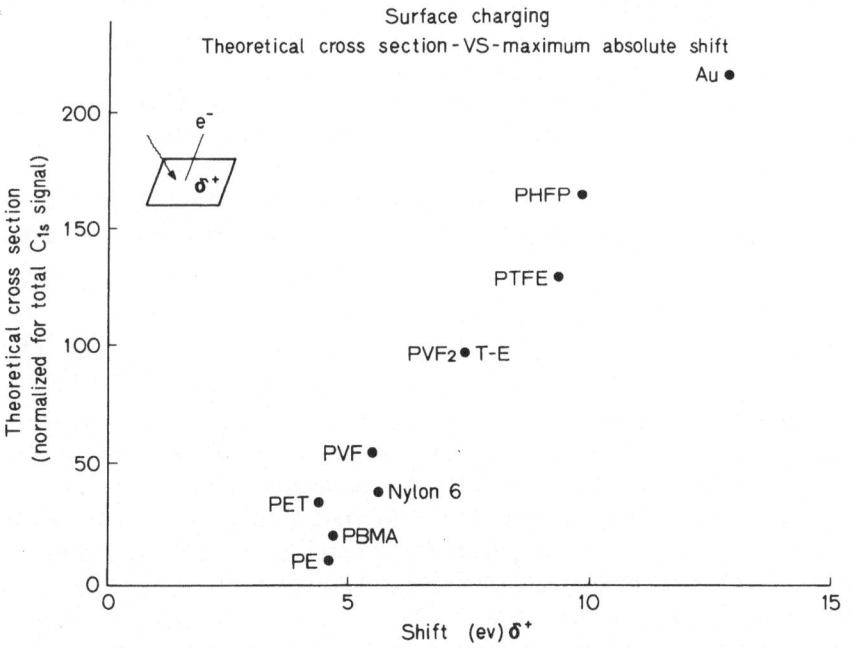

Fig. 9. Sample charging; shift in KE scale vs. theoretically calculated total cross section for
photoionization

5. Mean Free Paths of Electrons as a Function of Kinetic Energy in Polymeric Materials

Undoubtedly the most important area of application of ESCA to polymeric materials is in the study of solids where the surface sensitivity and the capability of differentiating surface from subsurface phenomena places the technique in a class of its own. Both these features are a consequence of the extremely short mean free paths (escape depths) of electrons in solids[9]. Thus in general the ESCA spectrum of a given core level consists of well resolved peaks, corresponding to electrons escaping without undergoing energy losses, superimposed on a background tailing to lower kinetic energy arising from inelastically scattered electrons. For the commonly used X-ray sources the mean free path of the photons is typically $\sim 10^5$ Å which is many orders of magnitude larger than the typical mean free paths of photoemitted electrons so that in the applications which are outlined below we may assume that the X-ray beam is essentially unattenuated over the range of surface thickness from which the photoelectrons emerge[10]. The intensity of electrons of a given energy observed in a homogeneous material may be expressed as

$$dI = F\alpha NKe^{-x/\lambda}dx$$

where F is the X-ray flux, α is the cross section for photoionization in a given shell of a given atom for a given X-ray energy, N is the number of atoms in a given volume element, K is a spectrometer factor for the fraction of electrons that will be detected and depends on geometric factors and on counting efficiency, λ is the electron mean free path and depends on the KE of the electron.

The two situations of common ocurrence are for a bulk homogeneous material for which the intensity of the elastic peak is given as in Eq. (1), and for an overlayer of thickness d for which expressions (2) and (3) are obtained.

For bulk homogeneous material A of thickness essentially infinite compared with the typical electron mean free paths the intensity of the elastic peak. I_α is given by Eq. (1).

$$I_\alpha^A = F\alpha N_A K\lambda_A \tag{1}$$

It should be noted that this expression refers essentially to a given angle (ϕ) between the X-ray source and the analyzer (this is fixed in most commercially available spectrometers), and a fixed take off angle (θ) for the electrons with respect to the sample surface. Since different core levels from different samples may well show different angular dependencies (with respect to θ and ϕ) for the absolute intensities according to (1), and since absolute intensities depend also on surface roughness and the atom density in the outermost regions of the sample as well as factors peculiar to a given spectrometer, great caution must be exercised in attempting to use ratios of absolute intensities for the same and different core levels in different samples as a means of establishing λ from Equation (1)[5].

For an overlayer (thickness d) of sample A on bulk sample B the corresponding expressions are given by Eqs. (2) and (3).

$$I^A = I_\alpha^A(1 - e^{-d/\lambda}A) \tag{2}$$

$$I^B = I_\alpha^B(e^{-d/\lambda}A) \tag{3}$$

It is clear that the difficulties associated with angular dependence, instrument factors, and atom densities of A and B involved in absolute intensity measurements can be obviated by studying ratios of intensities of the same levels as a function of angle for overlayers of different thickness d. In this way we may obtain directly escape depths for photoemitted electrons in material A and in general since the kinetic energy will be different for photoemission from the core levels of A and B this will yield two values λ_A and λ_A. If the experiment is now repeated with different core levels of sample A and with a different X-ray source we may start to build up a picture of escape depths as a function of kinetic energy.

A large body of literature has now been accumulated on escape depths in a variety of metals and semi conductors obtained predominantly from such investigations and these reveal a general dependence of escape depth on kinetic energy in which mean free paths are at a minimum of a few angstroms in the region ~ 80 eV kinetic energy and increase to higher kinetic energy roughly in a square root dependence on the kinetic energy[9]. The mean free paths also increase at a lower kinetic energy and indeed most of the experimental data falls within a relatively narrow band of the so called generalized curve of mean free paths as a function of kinetic energy[9].

Although a detailed knowledge of escape depths of electrons in organic and polymeric material is essential to the quantitative development of the technique for surface studies, until very recently there has been a divergence of opinion on the relative magnitudes with respect to typical semiconductors and metals. One of the first indications of the surface sensitivity of ESCA as a technique derives from Siegbahn's work on fatty acid double layers which indicated a sampling depth in the region of 100 Å. Escape depth studies on graphite indicated close agreement with data obtained for most other materials placing the escape depth in the range of ~ 10 Å for electrons of kinetic energy ~ 100 eV. Experiments on Langmuir-Blodgett fatty acid films on the other hand tend to suggest escape depths in the region of 50 Å for 1000 eV electron whilst recent attempts at absolute intensity measurements have suggested escape depths as high as 100 Å for typical polymers. Escape depths as long as this does not imply a very surface sensitive technique and the direct study of surface phenomena such as reactions initiated at surfaces, surface domain structure of block copolymers and comparison with techniques with surface sensitivities in the range of hundreds of angstroms indicates that these results are artefacts of the particular experiments which have been reported. The study of surface phenomena therefore suggests that mean free paths in polymeric materials are comparable with those in metals and semiconductors. The particular artefacts which undoubtedly give rise to apparently long mean free paths in attempting to study Langmuir-Blodgett films and in measuring absolute intensities are not difficult to trace. For the former, with great care, highly oriented films may be produced[11] (in most of the published work it is not clear that the criteria necessary to obtain such films has been satisfied) and if a range of fatty acid chain lengths are available then the method should be useful although it again relies heavily on absolute rather than relative intensities. With highly oriented films however it is imperative to perform a very careful and detailed angular distribution study to investigate the possibility of channeling phenomena. So far no angular resolved studies have been reported indeed it is not clear from the literature whether this is a factor which has even been considered[11]. Absolute intensity measurements[12] are fraught with difficulty as we have previously noted and again one is lead to the inescapable conclusion that this does not provide a means of obtaining escape depths in the simple minded experimentation which has been proposed.

It is clear therefore that direct measurements of escape depths in organic and polymeric materials requires the production of films of known thickness as overlayers on a given substrate and the measurement of relative intensities of a given core level as a function of angle (θ). Such studies are by no means trivial experimentally, however, we have recently completed a detailed investigation using as prototype models various polyparaxylylene polymers produced by in situ polymerization of *para*-xylylenes produced in a pyrolysis tube from paracyclophane precursors[13]. Both the parent and substituted derivatives have been studied to give a range of core

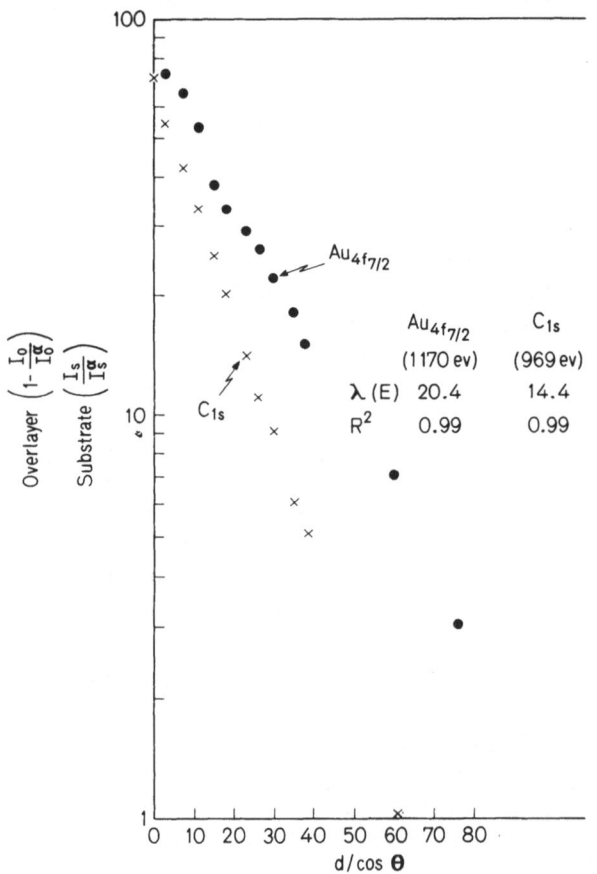

Fig. 10. ln $(1 - I_0/I_0^\alpha)$ for the overlayer and ln (I_S/I_S^α) for the substrate, versus $d/\cos\theta$, for a series of thicknesses of polyparaxylylene on a gold substrate studied at a constant θ of 10° with a Mg$K\alpha_{1,2}$ X-ray source. (Note the ordinate has been arbitrarily normalized to a ratio of 100 where $I/I_0 = 1$)

levels spanning different escape depths and allowing comparisons to be drawn for different polymer samples. In each case film thicknesses have been measured directly by means of a quartz deposition monitor and careful angular dependent studies have confirmed that the polymer films are produced uniformly and evenly in these experiments. As a typical example Fig. 10 shows data for deposited films of poly-paraxylylene on a gold substrate. By studying the C_{1s} levels of the overlayer and the $Au_{4f_{7/2}}$ levels of the substrate at a fixed take off angle of 10° with respect to the normal to the sample surface, excellent statistical correlations are obtained giving mean free paths of ~ 20 Å and 14 Å for electrons of kinetic energy 1170 eV and 969 eV respectively. If such investigations are carried out as a function of take off angle data such as that displayed in Fig. 11 are obtained. For photoemission from a

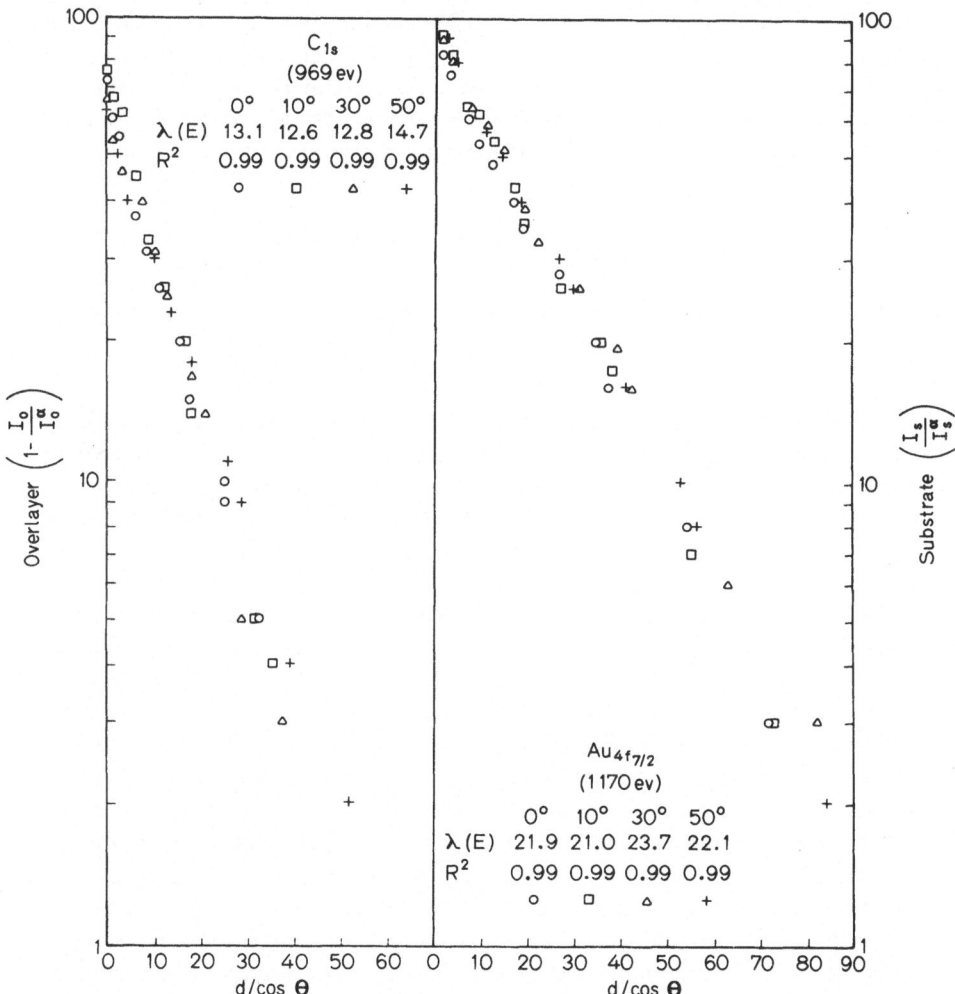

Fig. 11. ln $(1 - I_0/I_0^\infty)$ for the overlayer and ln (I_S/I_S^∞) for the substrate versus $d/\cos\theta$, for a series of thicknesses of poly *para*-xylylene on a gold substrate studied as a function of θ, 0°, 10°, 30° and 50° with a $MgK\alpha_{1,2}$ X-ray source.
(Note the ordinate has been arbitrarily normalized to a ratio of 100 where $I/I_0 = 1$)

given core level the relevant plots produce essentially parallel straight lines with a very small scatter in derived mean free path. This again provides good evidence for the uniform nature of the films produced. The data in Figs. 10 and 11 refer to a $Mg_{K\alpha_{1,2}}$ X-ray source. If similar studies are carried out using A monochromatized $Al_{K\alpha_{1,2}}$ photon source then the corresponding kinetic energies for photoemission from the C_{1s} and $Au_{4f_{7/2}}$ levels are 1202 eV and 1403 eV. The consistency of the results is shown by the close similarity of escape depths for photoemission from the $Au_{4f_{7/2}}$ substrate core levels with a $Mg_{K\alpha_{1,2}}$ photon source and the C_{1s} levels of the overlayer using an $Al_{K\alpha_{1,2}}$ source since the kinetic energies are closely similar. Studies have also been made on the monochloro and bromo substituted polypara-xylylene systems and these reveal that escape depths are within experimental error the same as for the unsubstituted system[13]. The average escape depths derived from these detailed studies are ~ 14 Å, ~ 22 Å, ~ 23 Å and ~ 29 Å for electrons of kinetic energy 969, 1170, 1202 and 1403 eV respectively. These studies therefore confirm our earlier view[2, 14] that electron mean free paths as a function of kinetic energy for polymers are closely similar to those for typical metals and semiconductors, and are only subtly dependent on the detailed electronic structure of a given class of materials[15].

6. Information Derived from Absolute and Relative Binding Energies and Relative Peak Intensities

In the previous sections an updated review has been given of the background of ESCA as a spectroscopic tool with particular reference to the study of polymers. In this and succeeding sections we review some representative examples which illustrate how the hierarchy of information levels outlined in Table 2 may be exployed to in-vestigate structure bonding and reactivity of polymeric systems along the lines set out in Table 3.

The first levels of information available derive from the measurement of abso-lute and relative binding energies and relative peak intensities. The distinctive nature of core levels means that identification of elements is straightforward as is the distinction between peaks arising from direct photoionization and Auger processes. A simple example is shown in Fig. 12 where the data refer to an ethylene-tetra-fluoroethylene copolymer. A wide survey scan such as this is an extremely useful precursor before detailed high resolution studies are attempted on individual core levels. As we have emphasized in previous reviews[2], with appropriate calibration, the relative intensities and shifts in binding energy for components of a given core level may be used to identify structural features and repeat units. Figure 13 for example shows high resolution ESCA spectra for two polymer samples and from the absolute and relative binding energies and relative peak intensities these may be identified as PVC and polyisopropyl acrylate.

Previous studies of substituent effects on core levels in simple monomeric sys-tems has shown that these are highly characteristic for a given substituent and follow simple additivity models[2]. Indeed the results may be quantified by detailed non-

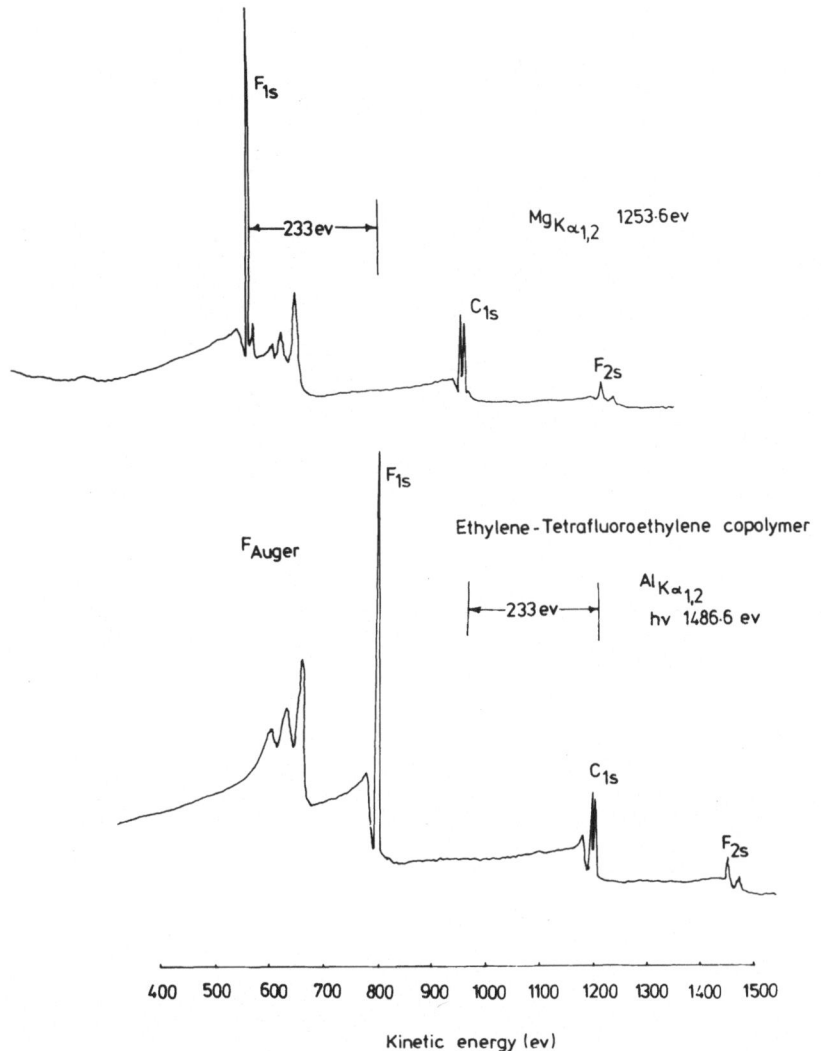

Fig. 12. Wide scan ESCA spectra for an ethylene tetrafluoroethylene copolymer obtained using $Mg_{K\alpha_{1,2}}$ and $Al_{K\alpha_{1,2}}$ photon sources

empirical calculations and this forms a sound basis for understanding the electronic factors determining both absolute and relative binding energies. This has enabled computationally inexpensive models based on an all valence electron CNDO/2 SCF MO formalism to be developed which can be extended to quantitatively describe polymers[2]. A large amount of data has previously been reviewed which relates to fluorocarbon based polymers[2]. Moreover a systematic study of a large number of homopolymers of simple vinyl monomers provides a compilation of substituent effects on $C_{1s}, N_{1s}, O_{1s}, F_{1s}, Si_{2p}, S_{2p}$ and Cl_{2p} levels. Figure 14 for example shows some of the data pertaining to substituent effects on C_{1s} levels in polymer.

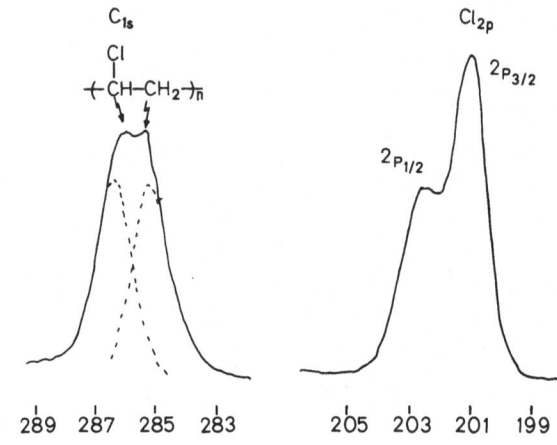

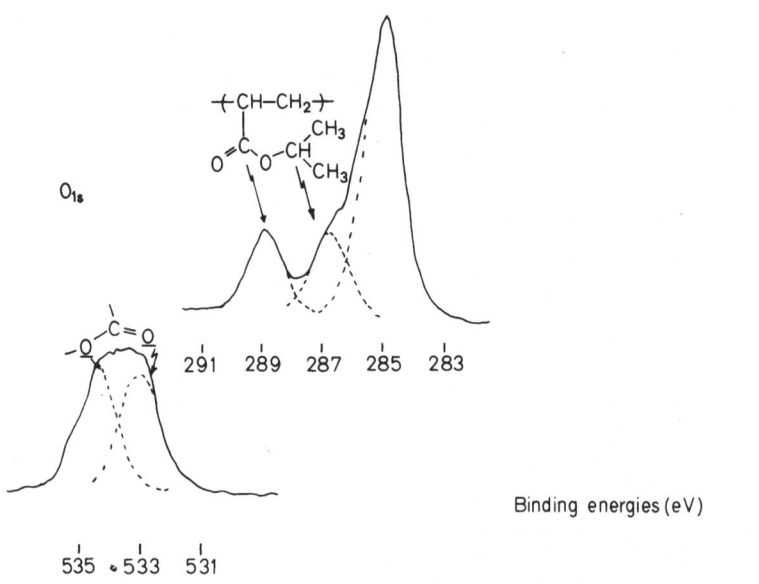

Fig. 13. Core level spectra for PVC and polyisopropyl acrylates

The characteristic nature of many substituent effects may now in fact be used as a "fingerprint" much in the same manner as one might use IR or NMR data.

Since as we have previously emphasized, the factors which determine shifts in core binding energies are short range in nature it is sometimes the case that isomeric species have core level spectra which are virtually identical. Although the spectra may therefore be used to identify the gross structures they may not allow a distinction to be drawn between isomeric species. As a simple example we might consider the isomeric polybutylacrylates[16]. The O_{1s} and C_{1s} level spectra for samples of these polymers are indistinguishable. Although not applicable in this particular case the possibility exists (and will be described in a subsequent section) of effecting a

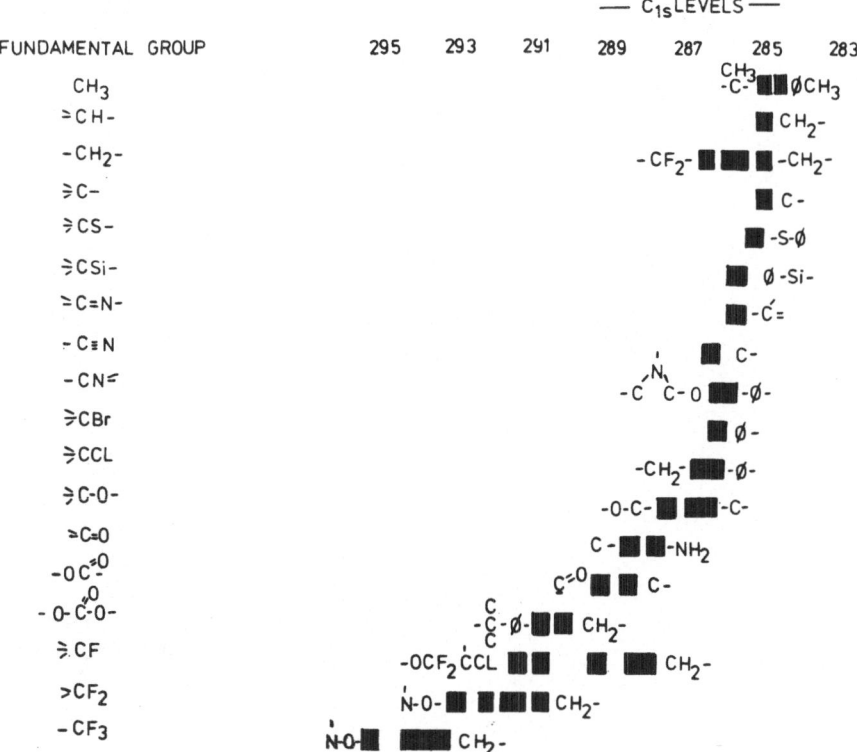

Fig. 14. Correlation diagram for C_{1s} levels in polymeric systems as a function of electronic environment. (The horizontal scale for each block is taken to indicate the typical range of binding energies found for a given structural type.)

distinction between closely related structures by investigating the low energy satellite structure arising from shake up transitions. Although the cross sections are considerably smaller for photoionization of valence as opposed to core levels we have previously outlined[2] the particular advantages of studying the valence levels of polymers by means of ESCA rather than UPS. In contrast to the core level spectra the valence levels of the isomeric polybutylacrylates are highly characteristic of the side chain structure since this forms an appreciable part of the whole[16]. This is clearly apparent from the valence levels shown in Fig. 15, and moreover comparison with appropriate model systems allows an unambiguous assignment of particular structural isomers.

The detection and semi-quantitative estimation of surface contamination of polymers is of importance in a number of fields. We have previously shown that ESCA has a unique capability for the detection of the initial stages of oxidation of polymers initiated at the surface and of adsorbed water and hydrocarbon contaminants[2]. One area where knowledge of the precise nature of the surface is of critical importance is in delineating models for the interpretation of data relating to triboelectric phenomena. Contacting polymer films from opposite ends of the so-called triboelectric series results in charge transfer such that there is a considerable build up of static charge on each component of the contacting pair. Such transfer could con-

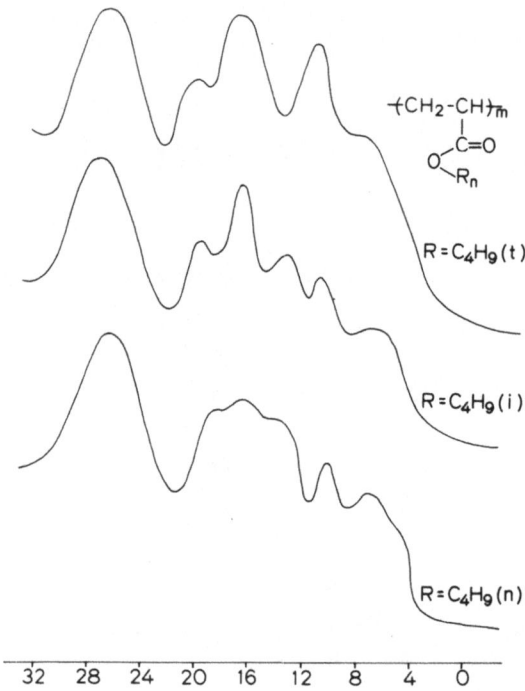

Fig. 15. Valence level spectra of isomeric polybutyl acrylates excited be $Mg_{K\alpha_{1,2}}$ radiation

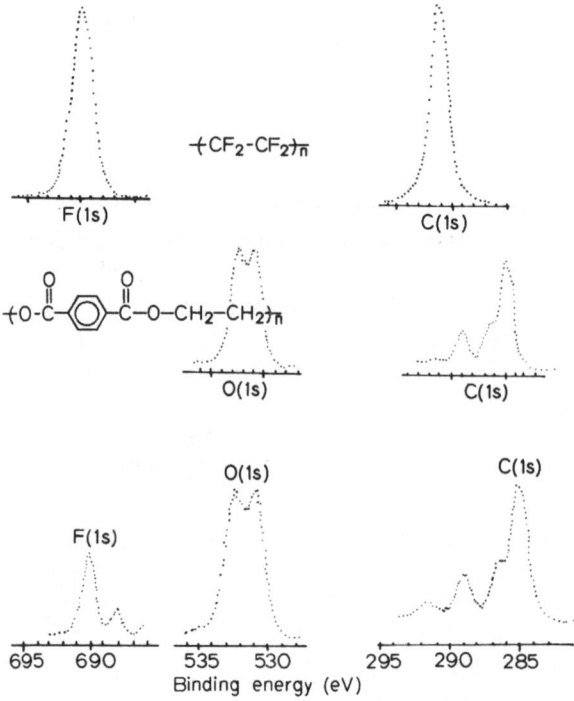

Fig. 16. Core level spectra for PTFE and PET and for the PET component after lightly contacting the two polymer films

ceivably occur via electron transfer from the material of low to that of higher work function thus equalizing the fermi levels or alternatively by mass transfer (transfer of ions) between the two components of the contacting pair. Even more likely is that both processes are of importance however the two possible mechanisms are not entirely separable since the propensity for absorption of ions at the surface of a given polymer will undoubtedly be a subtle function of its electronic structure as will the work function. It is known that triboelectro-phenomena in general are explicable in terms of a charge density of the order of 1 in 10^4 of surface sites which is probably an order of magnitude lower than can currently be detected by ESCA. Nonetheless it is of interest to see if in contacting polymer samples there is mass transfer or not between the components. If mass transfer is observed it certainly does not resolve the problem of how charging occurs but it certainly allows one to say that mass transfer cannot be ruled out as a possible mechanism.

We have therefore studied the surfaces of a variety of polymer films both before and after contacting events[17]. The great advantage of such an investigation by means of ESCA is the ability to look at *both* halves of a contacting pair. As one example Fig. 16 shows the core level spectra for PTFE, and PET films, which are highly characteristic for each polymer. In addition there is also shown the core level spectra for the polyethylene terephthalate component after lightly contacting the PTFE film. Since the polymers are from the opposite ends of the triboelectric series the films show a strong tendency to adhere to one another, even on a light contact. The observation of the high binding energy component in the C_{1s} spectrum and the observation of the F_{1s} levels shows that some PTFE transfers to the PET surface. It may be estimated that this represents fractional monolayer coverage. Of particular interest is the fact that in the F_{1s} peak in addition to the major high binding energy component associated with covalent $\underline{C}F_2$ linkages there is a lower binding energy peak attributable to fluoride ion thus providing evidence for bond cleavage accompanying the mass transfer. Examination of the other half of the component namely PTFE shows the presence of PET as evidenced by the characteristic O_{1s} and C_{1s} levels. These simple experiments illustrate the utility of ESCA in this area.

It is often the case that particular structural features may be characteristic of the end groups of a given polymer system. The direct detection of such end groups by means of their characteristic binding energies provides a convenient means of establishing DP's in relatively low molecular weight material. A particularly favourable situation arises for systems for which the terminal groups involve $\underline{C}F_3$ residues. If due care is taken to ensure that ESCA statistically samples the repeat unit (by for example considering the relative intensities of the same element with differing escape depth dependencies) then the comparison of area ratios for chemically shifted components of a given core level may be used to straightforwardly estimate DP's[18]. For example in a series of fluorocarbonate polymers of the general formulae shown in Fig. 17 it may readily be shown that the carbon 1s level appropriate to

$$\overset{\text{O}}{\overset{\|}{\text{carbonate}}}$$

carbonate $-O-\underline{C}-O$ and $\underline{C}F_2$ environments occur at approximately the same binding energy, and this can be independently corroborated by studying appropriate model systems. The C_{1s} level for the series of low molecular weight materials shown

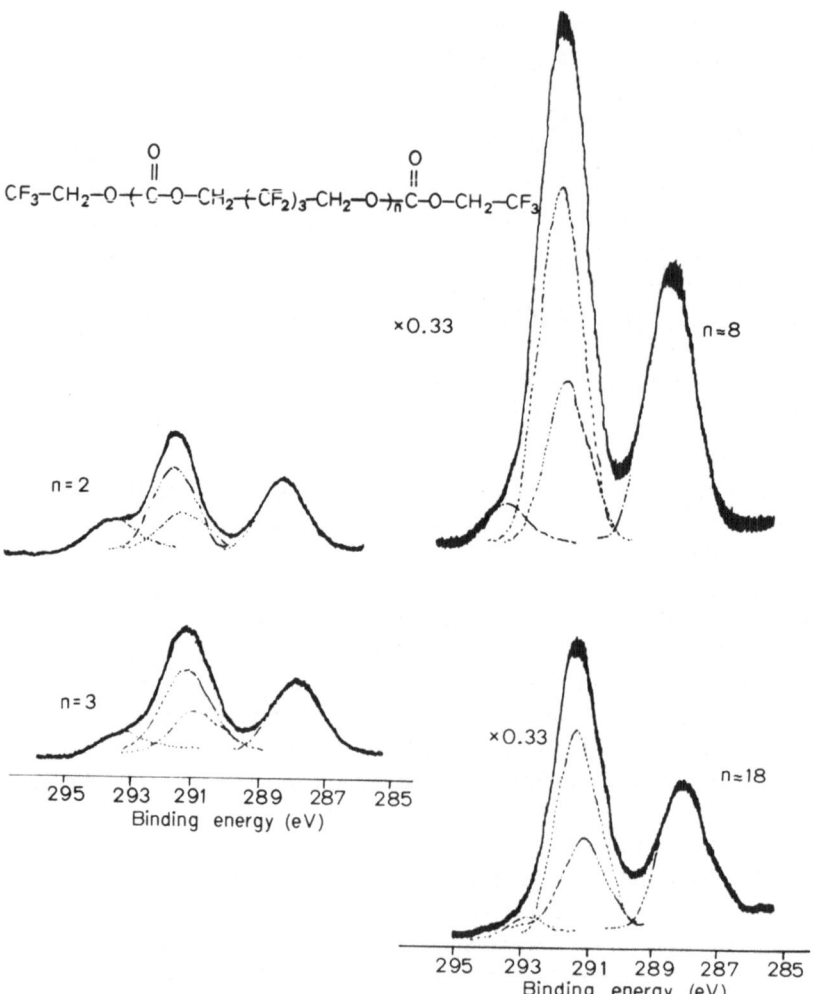

Fig. 17. C_{1s} core level spectra for a series of fluorocarbonate polymers

in Fig. 17 fall into three distinct regions and with appropriate calibration of line-widths and lineshapes for individual components from the study of model compounds the lineshape and analysis produces the component analysis indicated by the dotted curves. From the relative areas of the CF_3 carbons to the $\underline{CF_2}$ and carbonate carbon peaks DP's may be elaborated as in Fig. 18.

The two methods of elaborating DP's give slightly different results which may indicate specific orientation effects, however the two are within ~ 10% and show an excellent correlation with DP's determined by vapour pressure osmometry and this is shown in Fig. 19.

By contrast DP's determined by ^{19}F NMR do not agree with those determined by these two techniques although the reason for this discrepancy is not clear.

Although there are well developed techniques for studying chemical compositions and features of structure and bonding pertaining to the bulk of polymer

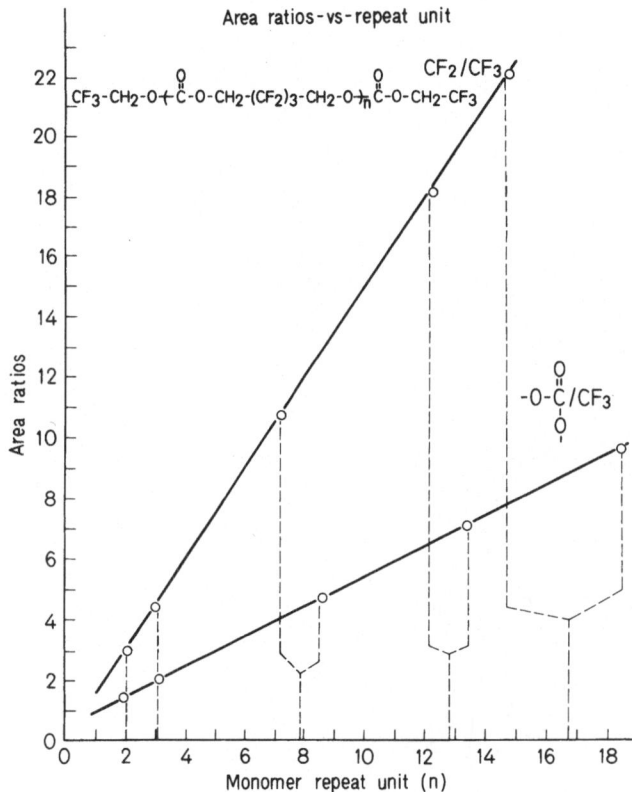

Fig. 18. DP's for fluoro-carbonate polymers obtained from CF_3/CF_2 and $CF_3/-O-\overset{\overset{O}{\|}}{C}-O$ core level intensities

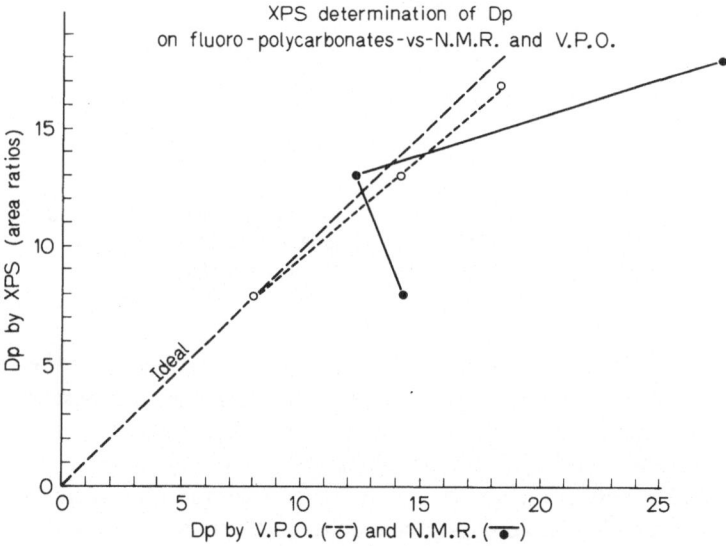

Fig. 19. Correlation between DP's for fluorocarbonate polymers determined by ESCA, NMR and VPO

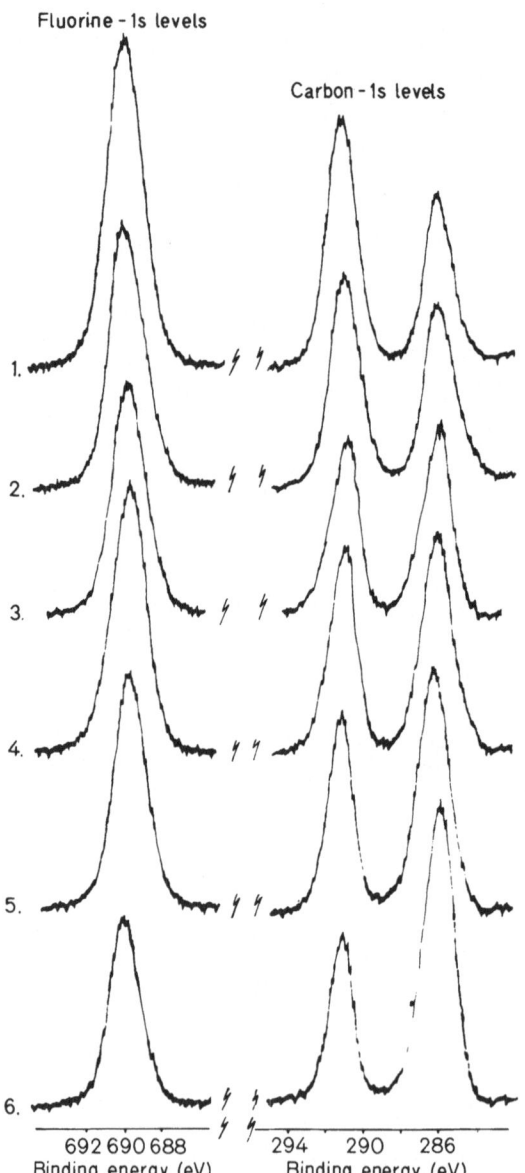

Fig. 20. Core level spectra for a series of ethylene tetrafluoroethylene copolymers

samples, until the advent of ESCA, information with regard to surface compositions could only be inferred rather indirectly by for example surface free energy measurements. Since any solid communicates with the rest of the world primarily by ways of its surface such information is important in many areas. ESCA may therefore be conveniently employed to answer the question 'is the surface composition typical of the bulk'. As a simple example Fig. 20 shows the core level spectra for a series of ethylene-tetrafluoroethylene copolymers, which we have previously discussed in some detail[19]. This particular system serves as a classic example of how to establish compositions on the basis of relative peak intensities and hence serves as an introduction to the study of polymeric systems in general.

The higher binding energy component of the C_{1s} levels arise from $\underline{C}F_2$ groups whilst the lower binding energy component is appropriate to CH_2 components. The shifts in binding energy between the two component peaks also establishes that the copolymers are largely alternating in character. Much of our early work in ESCA applied to polymers involved the study of homopolymers of simple fluorocarbon monomers and this has allowed the accumulation of an extensive background of information on absolute and relative binding energies as a function of structural feature and on the relative intensities of core levels[2]. By comparison with these simple homopolymers the compositions of the copolymers may readily be established by two independent means.

Firstly from a comparison of the integrated area ratios for the F_{1s} and C_{1s} levels and secondly from the individual components of the C_{1s} levels. This readily establishes that the materials are copolymers of ethylene and tetrafluoroethylene which are largely alternating in character and that the outermost surface sampled by ESCA is identical in composition to the bulk. This is shown in Table 4 where a comparison is drawn with compositions determined by standard microanalysis (carbon by combustion, fluorine by potassium fusion). ESCA is highly competitive as a routine means of establishing compositions for fluoropolymers in particular, in terms of accuracy, nondestructive nature and speed.

We may also use the data in Fig. 20 to illustrate how theoretical models have reached a level of sophistication where absolute binding energies for core levels may be accurately calculated and also to show how we may take the measured absolute binding energies to compute directly charge distributions in polymers[20]. Figure 21 shows a comparison of experimental and theoretically calculated binding energies for the roughly 50:50 ethylene-tetrafluoroethylene copolymer system. Also shown are the charge distributions obtained from the experimental data by inversion of the charge potential model and by direct computation.

As an example of more complicated systems studied at a somewhat lower level mention might be made of the application of ESCA to the routine evaluation of protein quality and quantity in legume and cereals[21].

Table 4. Analysis of ethylene/tetrafluoroethylene comonomer incorporations

Sample	Composn. monomer mixture mol% C_2F_4	Copolymer composition (mol % C_2F_4)				
		Predicted from monomer reactivity ratios	Calc. from C analysis	Calc. from F analysis	Calc. from area ratio C_{1s} peak: F_{1s} peak	Calc. from C_{1s} (CH$_2$ peak): C_{1s} (CF$_2$ peak)
1	94	63	61	61	63	62
2	80	53	52	54	52	52
3	65.5	50	49	48	47	46
4	64	50	47	45	44	45
5	35	45	41	40	42	40
6	15	36	–	–	32	31

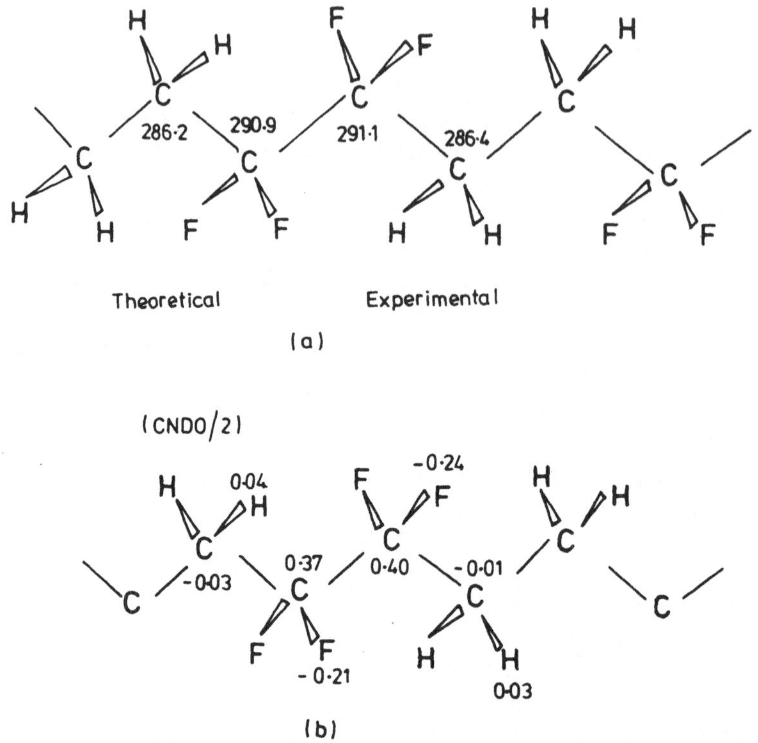

Fig. 21. Theoretically calculated and experimentally determined C_{1s} core binding energies for 50/50 alternating copolymer of ethylene and tetrafluoroethylene. Also shown are charge distributions obtained by direct SCF computation and from the experimental binding energies by inversion of the charge potential model

Since the protein content of legume is substantially larger than for cereal there is a considerable research effort amongst various organizations in the developed countries to select the best strains (defined in terms of protein quantity and quality) for introduction in the under-developed nations in order to improve nutritional diet. This essentially comes down to a routine screening operation in which the total nitrogen, ratio of side chain to backbone nitrogen and sulphur containing amino content need to be estimated with the minimum of sample preparation. The particular advantages of employing ESCA for such an operation compared with routine wet analyses are that the technique is non-destructive, requires small quantities of material, is fast and readily distinguishes between different oxidation states of a given element, important in the particular case of sulphur. By recording the core level spectra of ball milled samples of various varieties of legume, and cereal both nitrogen and sulphur content may routinely be established. For example Fig. 22 shows the correlation between the relative area ratios of N_{1s} to C_{1s} core levels with total nitrogen determined by Kjeldahl analysis. The high nitrogen containing legumes are clearly distinguished for the barley and wheat of lower nitrogen content. It should be noted that the ball milled material is a mixture of protein and carbohydrate (cell wall ma-

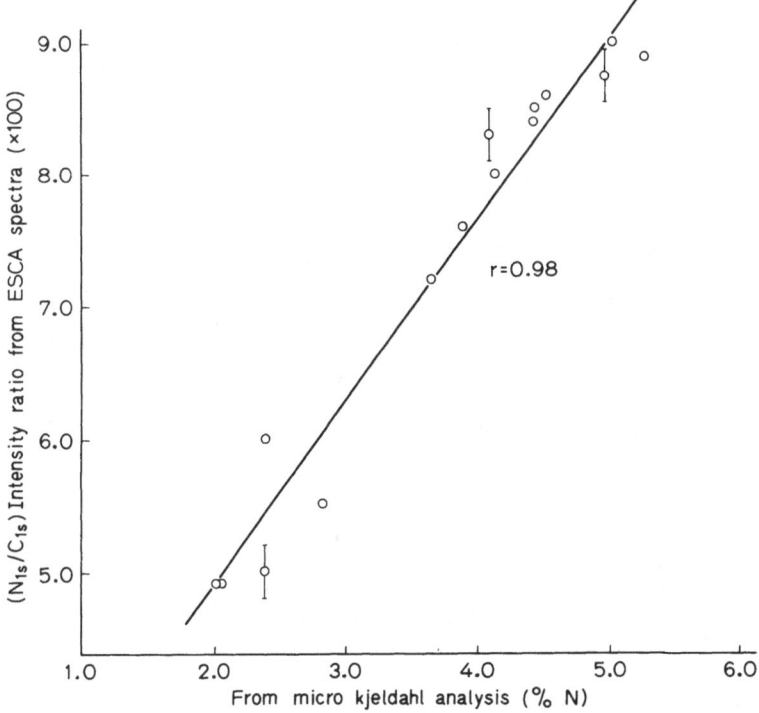

Fig. 22. Correlation of core level intensity ratios with percentage nitrogen determined by micro-keldahl for a series of ball milled, cereals and legumes

terial etc.) and that the ESCA analysis may therefore be directly related to nutritional value. In studying a series of barleys of varying lysine content the N_{1s} core levels show a higher binding energy component as shown in Fig. 23. Comparison with model systems shows that this arises from protonated side chain nitrogen and the relative intensities of the two components may be related to the amount of protonated side chain nitrogen. The sulphur $2p$ core levels inevitably show a doublet structure with peaks centred at $\sim$ 164.0 eV and 168.7 eV the latter being characteristic of inorganic sulphate which is inevitably present. The lower binding energy component arising from the sulphur containing amino acids may be used to quantify the amount present and for different strains of legume this can vary over a significant range. This is important since a high sulphur containing amino acid content is essential to a high quality diet. Nitrogen to sulphur peak area ratios from core level spectra are thus of considerable utility in routinely establishing the main features of the protein quantity in biologically important systems[21, 22].

The application of ESCA to the elaboration of chemical composition is well established in the case of fluorocarbon based systems for which the span in shift range for the C_{1s} level is particularly favourable consequent upon the large electronic effect of replacing hydrogen by fluorine. In many cases comparable information may be derived from the more familiar spectroscopic techniques such as IR and NMR; one area in which ESCA comes into its own, however, is in the analysis of polymeric

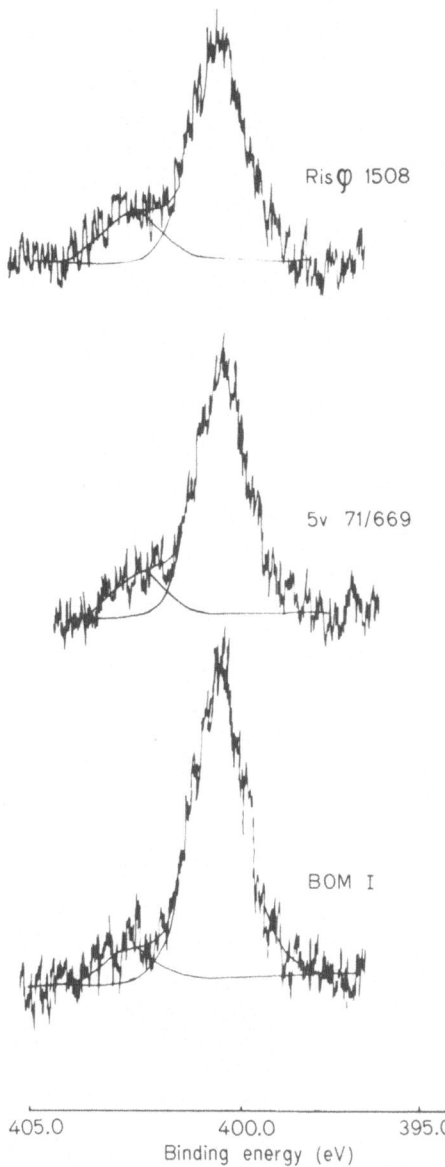

Fig. 23. N_{1s} levels for barleys of varying lysine content

films produced by glow discharge polymerization techniques. The desirable properties of such films are already apparent in many applications particularly in the coatings and electronics industry, however, the nature of the processes involved means that such films are highly insoluble and therefore inconvenient if not impossible to study directly by either high resolution NMR or transmission IR. There has therefore been considerable interest over the past year or so in the application of ESCA to studies of structure and bonding of glow discharge synthesized films with the predominant emphasis being on those based on fluorocarbon percursors[23–25].

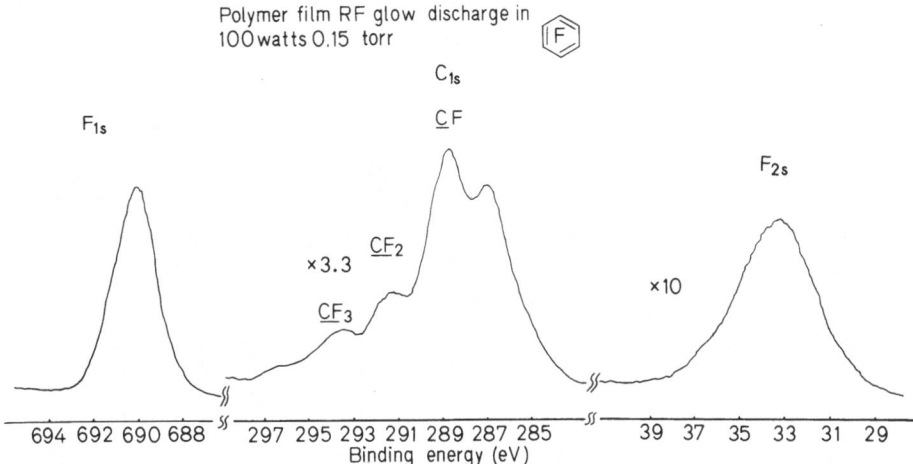

Fig. 24. ESCA spectra of polymer film produced in an inductively coupled RF plasma excited in perfluoro benzene

Figure 24 for example shows the core level spectra for polymer films produced by RF glow discharge polymerization of perfluorobenzene[25]. The complex nature of the C_{1s} spectrum with components of binding energy appropriate to $\underline{CF}_3$, $\underline{CF}_2$ and $\underline{C}F$ sites as well as carbons not directly attached to fluorine is evidence of the rearrangements occurring in the monomeric species prior to polymerization. A wealth of detail may be extracted from such data. Angular dependent studies and the investigation of appropriate area ratios readily establish the homogeneity of such polymer films on the ESCA depth profiling scale and indeed the synthesis of such films of known thickness provides a further means of studying escape depths as a function of kinetic energy over a considerably larger range than those described previously for poly *para*-xylylene films[25]. It can be shown that the stoichiometry of such films determined by ESCA is identical to that obtained by microanalysis of the bulk and the fine structure of the C_{1s} levels reveals considerable detail of the structure of such films. It is interesting to note in this connection that a shoulder in the C_{1s} levels centred ~ 296.5 eV is too high in binding energy to be associated with a C_{1s} level shifted by attached electronegative substituents (cf. Fig. 14). It arises in fact from $\pi \to \pi^*$ shake up transitions accompanying photoionization of C_{1s} levels associated with $\underline{C}F$ groups thus showing that the polymer structure encompasses unsaturated components involving vinylic type $\underline{C}F$ groups. Clearly ESCA has an important role to play in the elaboration of the structure of such polymeric films and detailed studies have been made as a function of partial pressure, power, residence time etc. of the structure of such films produced by inductively coupled RF plasmas for a variety of monomer precursors[25].

If we are merely interested in overall stoichiometries and comparison with microanalytical data for the bulk then as we have noted in the case of biological polymers all that is required is the measurement of area ratios for the various core levels, (once appropriate reference standards have been established). If however, we wish to obtain information on structure and bonding then the first line of informa-

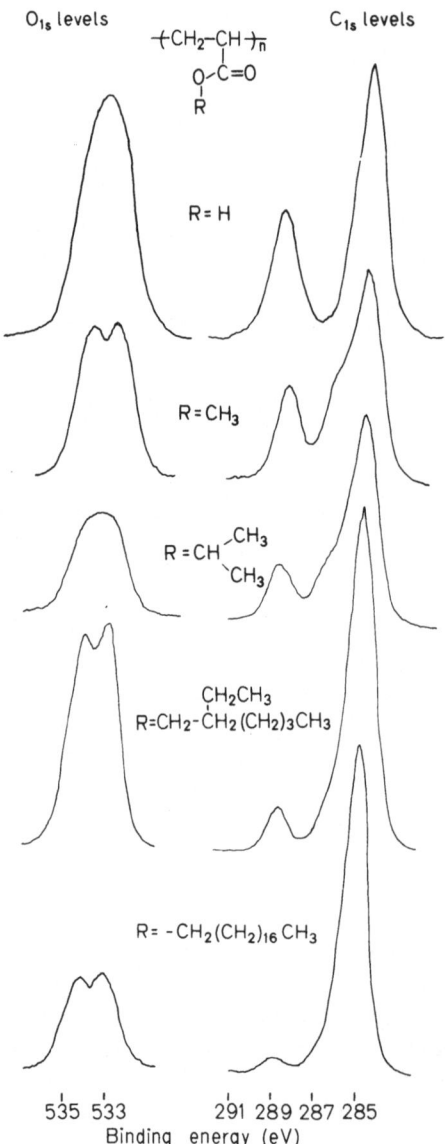

O_{1s} levels C_{1s} levels

R = H

R = CH_3

R = CH $\stackrel{CH_3}{\diagdown}$ CH_3

CH_2CH_3
R=CH_2-$\stackrel{|}{CH_2}$ $(CH_2)_3CH_3$

R= -$CH_2(CH_2)_{16}CH_3$

535 533   291 289 287 285
Binding energy (eV)
O_{1s} levels C_{1s} levels

Fig. 25. Core level spectra for a series of polyalkyl acrylates

tion is from the investigation of absolute and relative binding energies for the components of the individual core levels if these may be resolved. Fluorinated systems
as we have noted are particularly useful in this respect since the shift range for C_{1s}
levels is rather large (cf. Fig. 14). With appropriate calibration, however, the type
of analysis illustrated for the ethylene-tetrafluoroethylene polymer system may be
extended to systems for which the shift range is somewhat less favourable. As examples we may consider recent work on a series of polyalkyl acrylates and methacrylates[16].

Typical core level spectra for a series of polyalkylacrylates are shown in Fig. 25.

The O_{1s} levels show a doublet structure (somewhat obscured in the case of poly-acrylic acid because of hydrogen bonding effects); the binding energies for the two components being characteristic for an ester group (cf. polyethylene terephthalate). The C_{1s} levels in each case show a high binding energy component attributable to

$-C\overset{\displaystyle\diagup O}{\diagdown_O}$ type environments shifted ~ 4 eV from the main peak which in each case

arises from CH_3 $\underline{CH_2}$ and $\underline{CH}$ type environments from the backbone carbons and carbons of the alkyl group not attached to oxygen. In going from polyacrylic acid to the polymethylacrylate a shoulder to the high binding energy side of the low binding energy component develops, shiftet by ~ 1.6 eV and attributable to the carbon attached to the ester oxygen. This shoulder gradually decreases in relative intensity with respect to the main peak as the chain lengths of the alkyl group increases. The assignment of core levels is readily confirmed by reference to simple model compounds for which detailed theoretical analyses have been performed[16].

In previous studies on polymeric systems we have shown that the substantial differences in escape depth dependence for deep lying valence levels which are core like in character in character (viz. F_{2s}) with respect to tightly bound core levels (e.g. F_{1s}) may usefully be employed for analytical depth profiling[14]. A study of the valence levels for simple model compounds reveals that in esters the O_{2s} levels are well separated from the remainder of the valence band and are essentially core like in nature. The approximate kinetic energies pertaining to photoemitted electrons from O_{1s} and O_{2s} levels using a $Mg_{K\alpha 1,2}$ photon source are ~ 720 eV and 1227 eV respectively and from a consideration of the generalized curve of escape depth versus kinetic energy these should correspond to significant differences in electron mean free

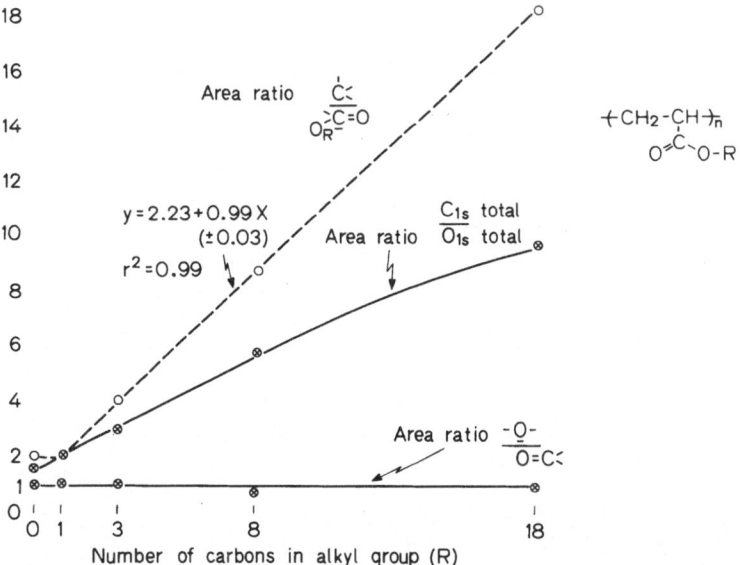

Fig. 26. Plot of area ratios for the core levels versus number of carbons on the alkyl group of a series of polyalkyl acrylates. (These have been corrected for differences in cross section and instrumental sensitivity.)

paths. By studying a series of simple oxygen containing organic molecules as condensed films such that ESCA statistically samples the molecules it can be shown that with the particular experimental arrangement peculiar to our spectrometer the apparent O_{1s} to O_{2s} area ratio is 11 ± 1. By measuring the area ratios of the O_{1s} to C_{1s} peaks versus the stoichiometric ratios for these molecules it is also possible to derive the instrumentally dependent apparent sensitivity ratios for oxygen with respect to carbon for unit stoichiometry. The instrumental dependent factors may then be employed to interrogate the data derived from Fig. 25 to answer the question: On the ESCA depth profiling scale do we statistically sample the repeat unit of the polymer? Figure 26 shows an analysis of the area ratios for the core and O_{2s} levels which are also essentially core like (cf. Fig. 15). For polyacrylic acid the measured areas for the various structural features for the O_{1s} levels and C_{1s} levels and the overall ratios for the C_{1s} to O_{1s} levels (corrected for differing sensitivity factors) are 1.0, 2.0 and 1.6 respectively, in excellent agreement with the theoretical values of 1.0, 2.0 and 1.5 based on a statistical sampling of the polymer repeat unit. The area ratios for the individual components for the C_{1s} levels show an excellent correlation with the number of carbon atoms in the alkyl groups[b]. (Slope; Experimental 0.99, Theoreti-

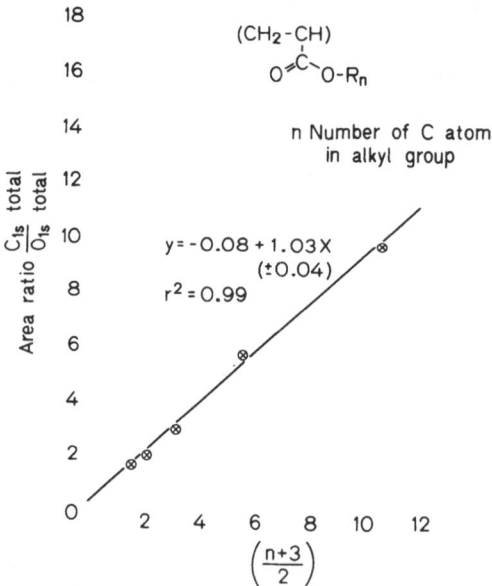

Fig. 27. Plot of area ratios for the C_{1s} and O_{1s} levels of a series of polyalkyl acrylates as a function of chain length of the alkylgroup

---

[b] It is convenient in this correlation to plot the area ratios for the two best resolved peaks, namely the carbonyl carbon which although of low relative intensity for the longer chain systems nonetheless is well removed from the main component arising from the backbone and carbons not directly attached to oxygen. This obviates any error due to deconvoluting the signal arising from the other carbons directly attached to oxygen (viz. of the ester group) since for the long chain systems it is obviously preferable to have a small error in a large than a small quantity. Since the area ratio does not therefore include the carbons of the ester group which are directly bonded to oxygen this leads to an obvious break in the curve for polyacrylic acid.

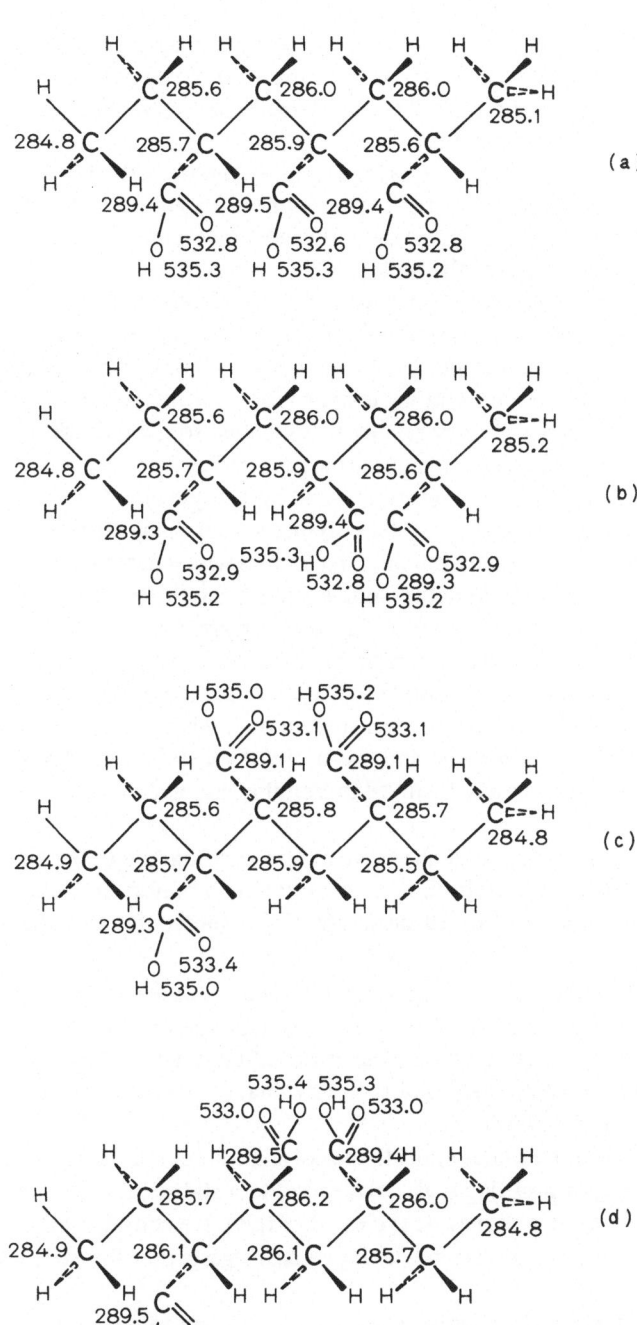

Fig. 28. Conformational models of polyacrylic acid with calculated binding energies from the charge potential model shown for (a) HT-HT 'isotactic' model, (b) HT-HT 'syndiotactic' model, (c) HH-HT 'isotactic' model and (d) HH-HT 'syndiotactic' model. (HH head to head, HT head to tail)

cal 1.0). By contrast the plot of Total C_{1s}/O_{1s} area ratios (corrected) against the chain length for the alkyl group fall on a smooth curve, however replotting the data in a different form, as shown in Fig. 27 reveals the underlying linear correlation with the appropriately derived theoretical parameter.

Within experimental limits the slope is unity as required by theory if the ESCA experiment statistically samples the repeat units of the polymers. In sum total therefore the ESCA data shows that for these systems the outermost few tens of Angstroms of the samples are representative of the bulk and that compositions, integrity of the immediate surface, and homogeneties may routinely be established. The analysis also strongly suggests that there are no specific orientation effects of side chain alkyl groups at the surface. The theoretical models previously developed to quantitatively describe absolute and relative core binding energies for fluoropolymers based on the charge potential model[2, 20] may readily be extended to the polyalkyl acrylates once appropriate values for the charge potential parameters k and $E^°$ are established for each core level. This is readily accomplished by studying model compounds and absolute binding energies for model polymer systems may then be directly calculated from appropriate model systems. The factors which determine absolute binding energies are short range in nature and as such it should be clear from the data in Fig. 28 that the calculations are all in excellent agreement with the experimental data irrespective of structural isomerism or tacticity. It is therefore the case that only in special circumstances is it possible to obtain information on structural isomerism[26] and tacticities from the core level spectra themselves. It will become clear however, that shake up phenomena and the study of valence energy levels since they both involve electrons intimately involved in bonding are much more sensitive probes in this respect.

Whilst for the polymer samples discussed above, structure and bonding in the outermost few tens of Angstroms sampled by ESCA corresponds to that in the bulk this is not always the case. For example Fig. 29 shows the ESCA spectra for a further series of polyalkyl acrylates.

A distinctive feature clearly evident in all of the spectra is the obvious inequality in intensity of the two component peaks of the O_{1s} levels. A similar analysis to that presented in a previous section provides the following information. Figure 30 for example shows a plot of the ratio of intensities of the individual components of the O_{1s} levels and also the total O_{1s}/O_{2s} ratios.

For comparison purposes the dotted lines indicate the correlations expected for samples which on the ESCA depth profiling scale correspond to a statistical sampling of the appropriate repeat unit in the polymer. It is clear that there are considerable deviations from such correlations in a direction which overall suggests that the samples are oxidized.

If we consider the polydecyl acrylate for example, the O_{1s}/O_{2s} ratio is significantly higher than for the reference compounds suggesting that since the mean free path for the O_{1s} levels is considerably shorter than for the O_{2s} level that the oxidation is largely confined to the surface. The absolute binding energies in each case for the O_{1s} component levels which have apparently increased in intensity corresponds to $\underline{C} = 0$ structural features, as is apparent from a comparison with data for the model systems. It is interesting to note that high resolution infrared studies

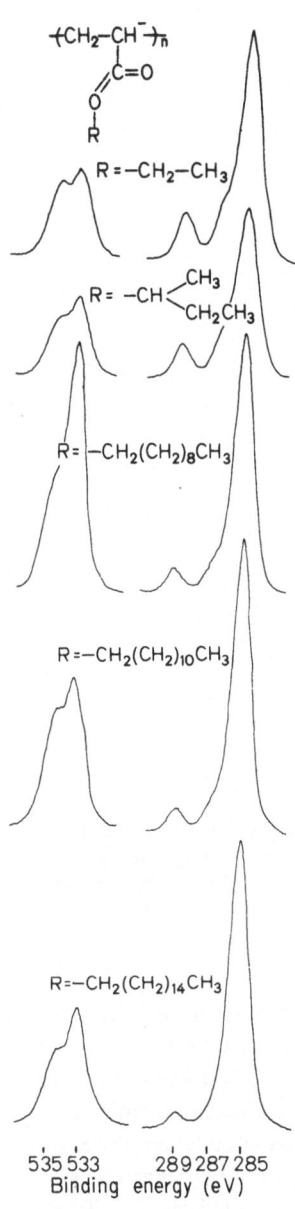

R = −CH₂−CH₃

R = −CH with CH₃ and CH₂CH₃

R = −CH₂(CH₂)₈CH₃

R = −CH₂(CH₂)₁₀CH₃

R = −CH₂(CH₂)₁₄CH₃

535 533   289 287 285
Binding energy (eV)

O₁ₛ levels   C₁ₛ levels

Fig. 29. Spectra for O_{1s} and C_{1s} core levels for a series of surface oxidized polyalkyl acrylates

revealed no major distinction of the type clearly evident from the ESCA spectra and the carbonyl region for all of the samples showed only a single peak in the range

1734 ± 6 cm^{-1} consistent with $-C\underset{\diagdown O-R}{\overset{\diagup O}{}}$ structural features. This is readily under-

standable since the infrared data pertains essentially to the bulk. Further evidence for the oxidized nature of the poly-n-decyl acrylate surface is provided by the

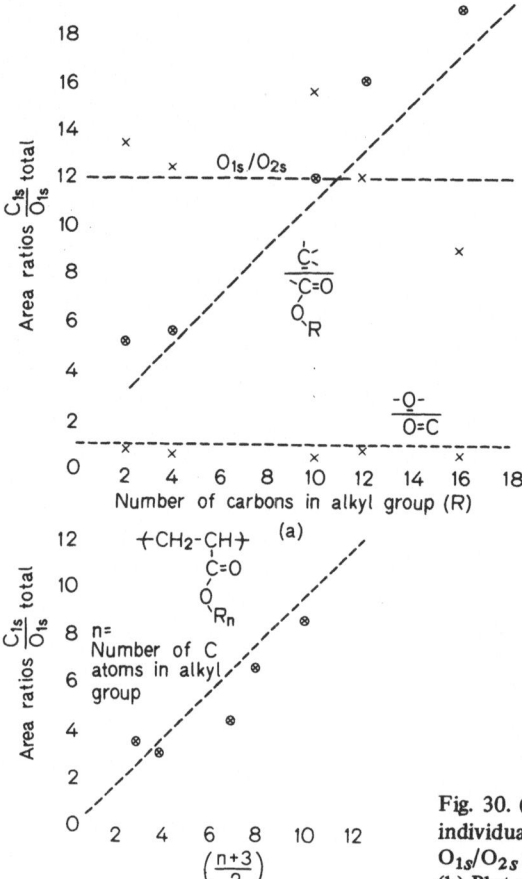

Fig. 30. (a) Plot of the intensity ratios for the individual components of the O_{1s} levels and also the O_{1s}/O_{2s} ratios for a series of polyalkyl acrylates. (b) Plot of the C_{1s} and O_{1s} area ratios versus the number of carbons in the alkyl groups

greatly increased wettability with respect to water compared with polyisopropyl acrylate as a representative example of the unoxidized samples. A comparison with poly-2-ethylhexyl and polyoctadecyl acrylates shows that the latter has a contact angle (with water) closely similar to polyisopropyl acrylate whilst the former is intermediate between that of polyisopropyl acrylate and poly-n-decyl acrylate. It is interesting to note that although the data for the poly-2-ethylhexyl acrylate generally fits well into the overall analysis previously outlined as is evident from Figs. 25–27 a close inspection of the relative intensities of the component peaks of the O_{1s} levels reveals some evidence for a small extent of oxidation (cf. Fig. 26).

If the surface oxidation inferred from the inequality of the component peaks of the O_{1s} levels is attributable to surface carbonyl features then this should also be manifest in the carbon 1s levels. It should, however, be emphasized that since the escape depth depth dependence for photoemitted electrons in the energy range considered is such that the mean free path increases with increasing kinetic energy then any surface feature will be relatively more prominent for the more tightly bound O_{1s} levels than for the C_{1s} levels. A detailed examination of the C_{1s} spectra

for the series of surface oxidized samples (Fig. 29) shows that the overall line profiles can only be quantitatively fitted with the addition of a small peak in the C_{1s}
spectrum appropriate to isolated carbonyl features as might arise from oxidation.

The predominant emphasis in this section thusfar has been the elaboration of
composition and structure for polymer systems and comparison with data pertaining to the bulk. The technique comes into its own however in the investigation of
inhomogeneous samples. Thus the capability of elaborating features of structure
and bonding nondestructively in inhomogeneous samples on the tens of Angstroms
scale is unique to ESCA. We have previously noted that the great surface sensitivity
of ESCA in the study of solids is associated with the extremely short mean free path
for electrons and the strong dependence on kinetic energy. There are thus two broad
categories of experiments which may be carried out, although inevitably in any defi-
nitive study the two are inextricably linked. The first category of experiment is to
sample levels of the same or different elements having different escape depth dependencies[14, 16]. If we consider a fluorocarbon based material for example, photoemission from the F_{1s}, C_{1s} and F_{2s} levels with $Mg_{K\alpha 1,2}$ radiation corresponds to electrons with kinetic energies of ~ 560 eV, 960 eV and 1220 eV respectively. This wide
span in kinetic energies is reflected in the substantial differences in escape depth and
hence sampling depth. Inhomogeneities in the surface and sub-surface compositions
of a sample will therefore be reflected in differing intensity ratios for these levels
compared with those for a homogeneous material. The first application of this approach was in the study of the initial stages of the surface fluorination of polyethylene[14]. By studying simple homopolymers such as polyvinylfluoride, polyvi-
nylene fluoride, polyvinylidene fluoride, polytrifluoroethylene and PTEE in which
ESCA statistically samples the repeat unit, intensity ratios for the F_{1s}/F_{2s}, F_{1s}/C_{1s}
and F_{2s}/C_{1s} levels may be established. Since the fluorination reaction is initiated at
the surface and is diffusion controlled we may anticipate that the fluorine content
will be higher at the surface. Since a greater proportion of the overall signal intensity
for the elastic peaks derives from the outermost surface the shorter the mean
free path, a heterogeneous sample with progressively lower fluorine content
into the bulk will have larger F_{1s}/F_{2s} and F_{1s}/C_{1s} intensity ratios than for a homogeneous sample whilst the F_{2s}/C_{1s} ratio will be smaller. This is in fact the case and a
careful analysis of the data pertaining to the surface fluorination of polyethylene
allows one to follow the changes in composition of the first monolayer as a function
of time[14].

The second category of experiments is to investigate relative peak intensities as
a function of electron take off angle. We can illustrate how a convolution of these
two types of experiment can provide a rather complete picture of inhomogeneties
arising in an initially homogeneous material by surface modification.

The surface modification of polymers for improvement of adhesive bonding,
and altering surface properties in general without concomitant modification of bulk
properties is an active area of research in both industrial and academic laboratories
and has been accomplished by a variety of means ranging from Corona discharge
treatment, direct chemical modification and by interaction with plasmas excited in
inert gases either capacitively or inductively[27].

The possibility of specifically improving properties associated with the outermost few tens of Angstroms of polymer surfaces under conditions in which the bulk

is left intact is particularly attractive and has led to a good deal of research activity particularly in the application of RF inductively coupled plasmas. The requisite instrumentation provides considerable flexibility in the design of apparatus and allows a close investigation of the parameters which determine the extent and rate of reaction. For plasmas excited in *e.g.* Agron the reactive species which might be expected to be involved in surface modification are: Argon ions, and metastables and electrons. In addition since plasmas are copious sources of electromagnetic radiation particularly in the vacuum ultraviolet it is clear that modification could also arise from absorption of short wavelength radiation. Indeed since the techniques which have previously been brought to bear in this area have not been noted for their surface sensitivity it is not surprising that the available evidence which essentially involves modification extending well into the bulk has suggested that the modification is dominated by radiative energy transfer processes[27]. With a surface sensitive technique such as ESCA it is possible to follow the initial stages of reaction and to differentiate the surface from subsurface and bulk reactions.

The primary objective of our investigations in this area has therefore been to investigate the relative importance of direct and radiative energy transfer processes in the interaction of various polymers with plasmas excited in inert gases[28].

The basic ESCA instrumentation allows considerable flexibility in the design of equipment for ancillary experiments and it is possible to construct a reaction chamber such that samples may be directly passed into the source region of the spectrometer subsequent to plasma treatment and a typical design is shown in Fig. 31. Typical of the results of such an investigation are those pertaining to a largely alternating copolymer of ethylene and tetrafluoroethylene (48%:52%) treated in a

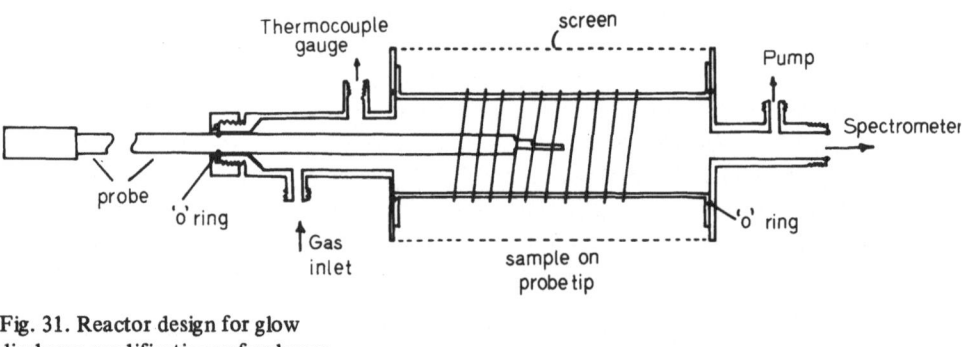

Fig. 31. Reactor design for glow discharge modifications of polymer surfaces monitored by ESCA

pulsed RF discharge at an average power loading of 0.1 watt. The core level spectra as a function of reaction time are shown in Fig. 32. The salient features are (i) a decrease in intensity of the F_{1s} levels and $\underline{CF_2}$ component of the C_{1s} levels; (ii) the appearance of structure at intermediate binding energy for the C_{1s} levels appropriate to the production of $\underline{CF}$ structural features; (iii) an overall increase in the total integrated intensity of the C_{1s} levels as a function of reaction time. Before briefly interpreting these results we may note that it is a relatively trivial matter to show that

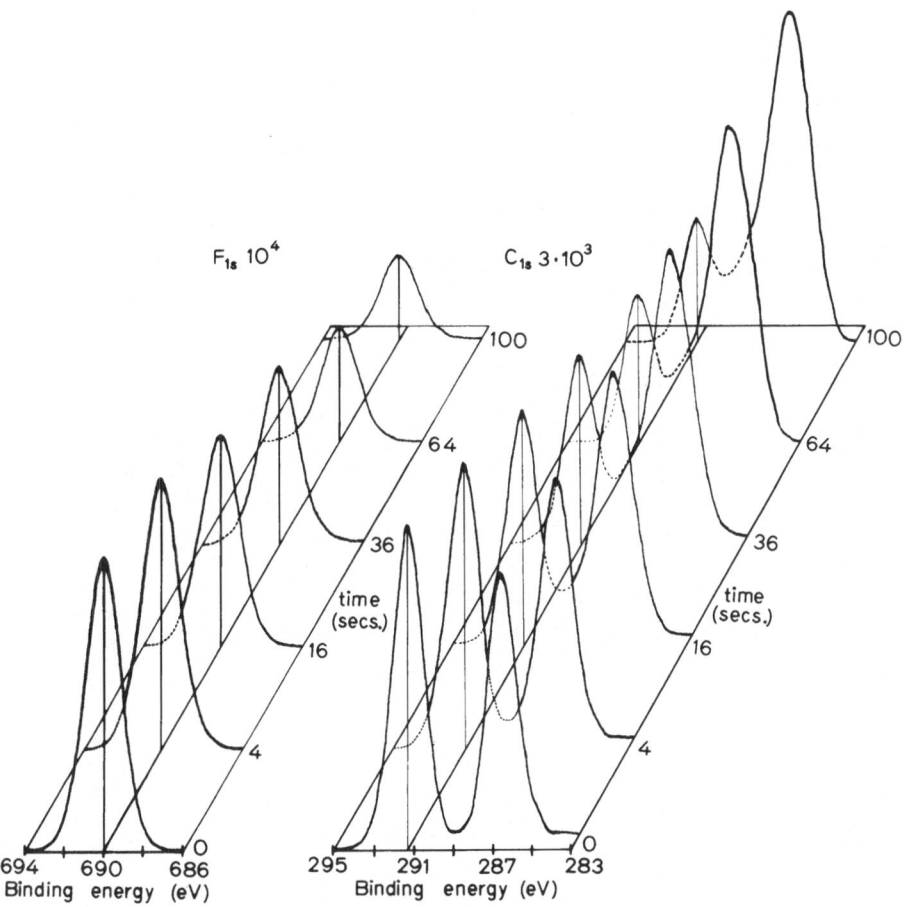

Fig. 32. Core level spectra as a function of treatment time for an ethylene tetrafluoroethylene copolymer treated in an argon plasma

the reaction is confined to the surface and subsurface of the samples and which are therefore inhomogeneous on the ESCA depth profiling scale. Figure 33 for example shows the typical form of the integrated relative intensity ratios of the F_{1s} and F_{2s} levels as a function of reaction time for treated samples. Since the kinetic energy of electrons photoemitted from the latter level is considerably larger than for the former the sampling depth is also larger and the decrease in intensity ratio compared with the homogeneous structure of the starting material is a manifestation of the inhomogeneity of the treated samples. A further indication of the substantial changes in structure and bonding in the surface regions is provided by measurements of the shift in kinetic energy scale arising from surface charging phenomena under the conditions of the ESCA experiments. As we have previously noted (Fig. 9) the equilibrium charge which a sample acquires under a given set of instrumental conditions is determined by the surface composition. Figure 34 shows the charging characteristics for typical discharge treated samples. The decrease in overall charging effect may be

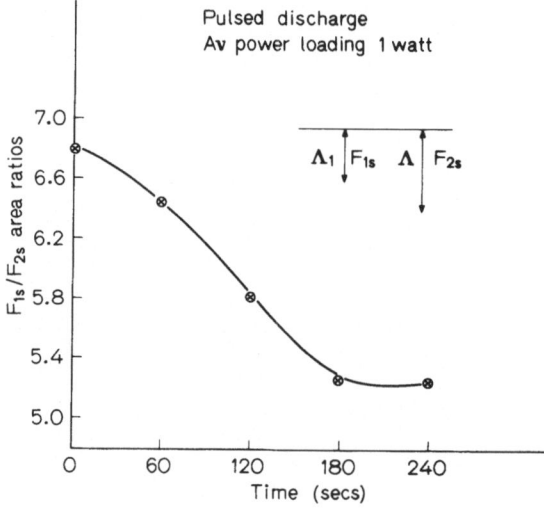

Fig. 33. F_{1s}/F_{2s} area ratios for plasma treated samples

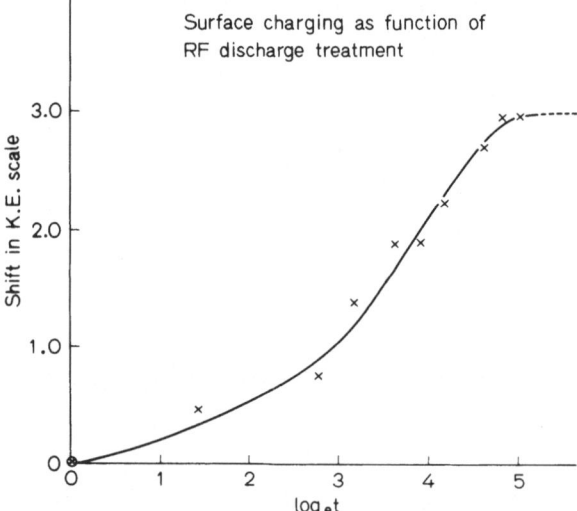

Fig. 34. Sample charging as a function of discharge treatment

directly related to the fluorine content of the surface regions and it is interesting to note that whilst the initial charging approximates to that of polyvinylidene fluoride the composition of which is quite close to that of the copolymer, the plateau value for the charging effect for the discharge modified samples is somewhat similar to polyethylene. The study of sample charging characteristics thus adds an extra information level which is complimentary to that obtained from relative peak intensities and absolute and relative binding energies.

A further dimension may be added by carrying out angular dependent studies. For a homogeneous material, although as we have previously noted the absolute intensity of a given core level will depend on take off angle because of instrumentally dependent factors, if we look at ratios of intensities for core levels of the same sym-

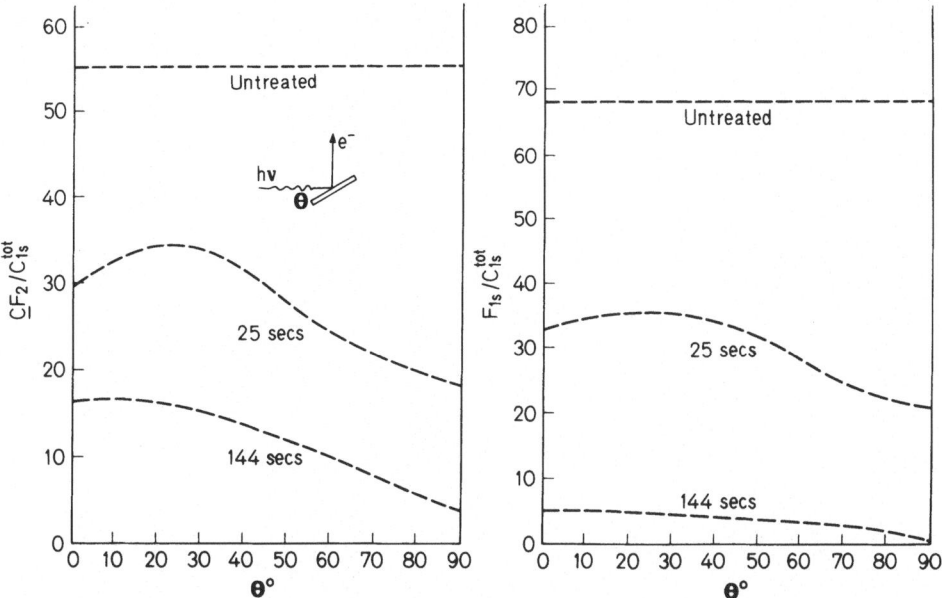

Fig. 35. Plot of intensity ratios of components of C_{1s} levels and of the total F_{1s} to C_{1s} levels as a function of take off angle for untreated and Argon glow discharge treated samples of an ethylene-tetrafluoroethylene copolymer

metry then these should be independent of take off angle. That this is the case is shown in Fig. 35. Thus the ratios of total integrated intensity ratios for the F_{1s} and C_{1s} levels and the components of the latter are independent of angle for the initial ethylene-tetrafluoroethylene copolymer.

For samples which have been treated at low power (0.2 watt.) for 25 secs and 144 secs the corresponding ratios are strongly dependent on take off angle in a sense which indicates that the surface is deficient in fluorine compared with the bulk. This can more readily be appreciated by considering a specific example as shown in Fig. 36.

The spectra correspond to a treated sample studied at two different take off angles θ of 18° and 80°. The differences are quite striking. The decrease in fluorine content in the surface regions is revealed by the large relative decrease in intensity of the component arising from $\underline{CF}_2$ structural features and concomitant increase in the lower binding energy components associated with $\underline{CF}$ and C-$\underline{C}$-C. The calculated relative intensities for the components of the C_{1s} levels are:

$(CF_2 : \underline{CF} : \underline{C}$ 1 : 0.23 : 1.86) and

$(CF_2 : \underline{CF} : \underline{C}$ 1 : 0.88 : 3.1)

for take off angles of 18° and 80° respectively. Also shown in Fig. 36 are the F_{1s} levels for the discharge treated sample and since the mean free path of the photo-emitted electrons is now considerably less than for the C_{1s} levels, the lowered fluorine content of the surface region is manifest in a marked decrease in the relative in-

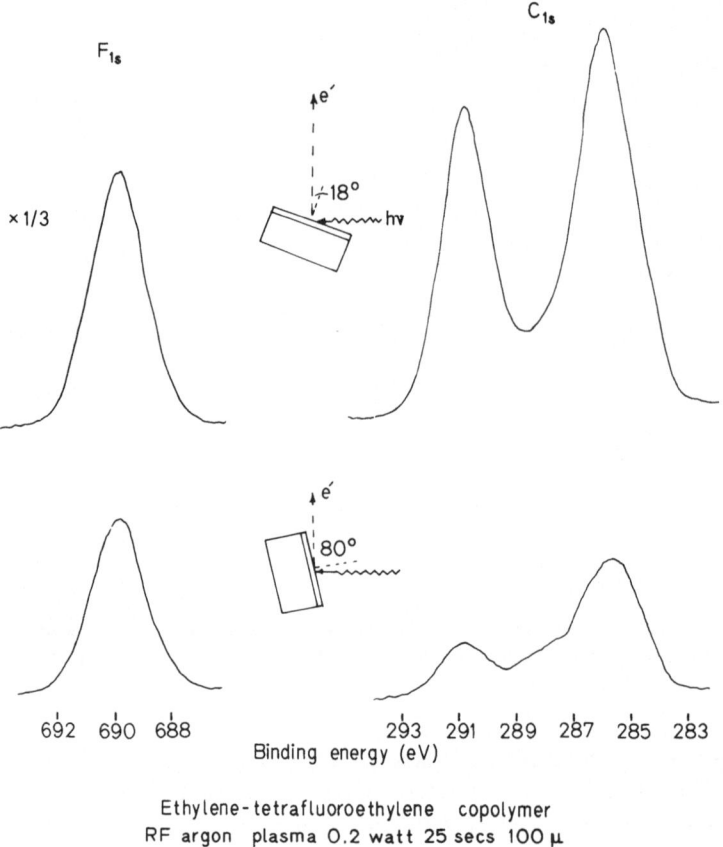

Ethylene-tetrafluoroethylene copolymer
RF argon plasma 0.2 watt 25 secs 100 μ

Fig. 36. Angular dependence of core level spectra for glow discharge treated sample of ethylene-tetrafluoroethylene copolymer

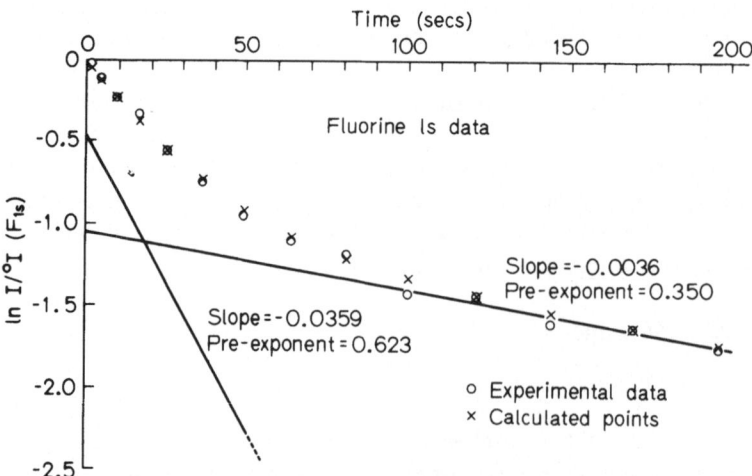

Fig. 37. Composite rate constants for interaction of Argon plasmas with ethylene tetrafluoro-ethylene copolymers obtained from analysis of the F_{1s} levels

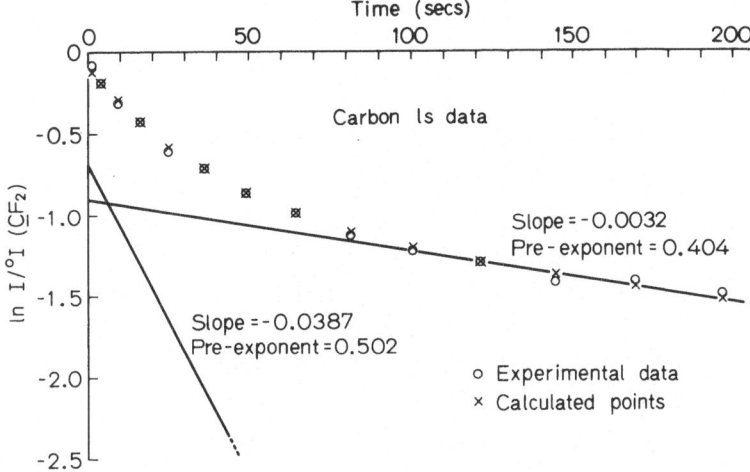

Fig. 38. Composite rate constants for interaction of Argon plasmas with ethylene tetrafluoro-ethylene copolymers obtained from analysis of the $\underline{CF}_2$ component of the C_{1s} levels

tensities of the F_{1s} levels with respect to that from the C_{1s} levels appropriate to carbons having no fluorines directly attached from 1.94 at 18° to 1.03 at 80°. The angular dependence of core level spectra can thus add considerably to our overall picture.

The intensities variations as a function of time of the F_{1s} levels and CF_2 structural features at a given take off angle forms the basis for monitoring reactions occurring at the surface and immediate subsurface and a kinetic scheme involving a separation into two components may be elaborated. An analysis in terms of such a scheme is shown in Figs. 37 and 38. Standard techniques may be used to evaluate psuedo first order rate constants for the surface and bulk reactions with the intercepts providing the pre-exponential factors which relate directly to the depth d to which the surface reaction is confined. The surface reaction is considerably faster than that for the subsurface and bulk and the most ready interpretation is in terms of a surface reaction dominated by direct energy transfer and a subsurface reaction almost exclusively arising from radiative energy transfer. The pre-exponential factors taken together with escape depths for photoemitted electrons of a given kinetic energy suggest a value for d of ~ 5 Å consistent with the direct energy transfer mechanism dominating for the first monolayer. This is most reasonable in terms of the short mean free paths for Argon ions and metastables produced in the plasma. The fact that the exponents for the faster surface reaction obtained from Figs. 37 and 38 are virtually the same reveals that the modification corresponds to conversion of CF_2 sites to $\underline{CF}$ sites with the latter undergoing little further modification. Detailed studies have been made of the effect of varying both the power and pressure and typical results are shown in Fig. 39. It is clear from this that the rate increases with both increasing power and decreasing pressure. Indeed over a considerable range of both power and pressure the rate is constant for a fixed ratio of power/pressure. This is entirely consistent with a reaction scheme in which reaction at the very sur-

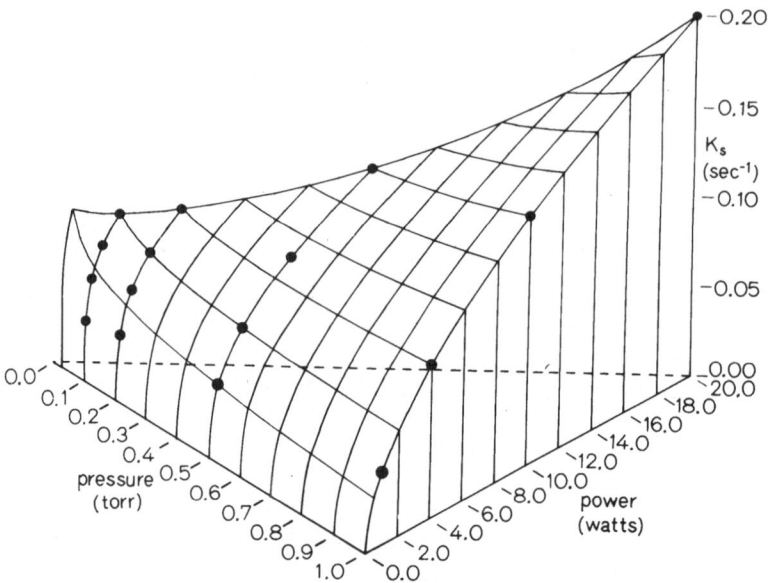

Fig. 39. Three dimensional plot of composite rate constants for surface reaction as a function of power and pressure in an industsively coupled RF plasma

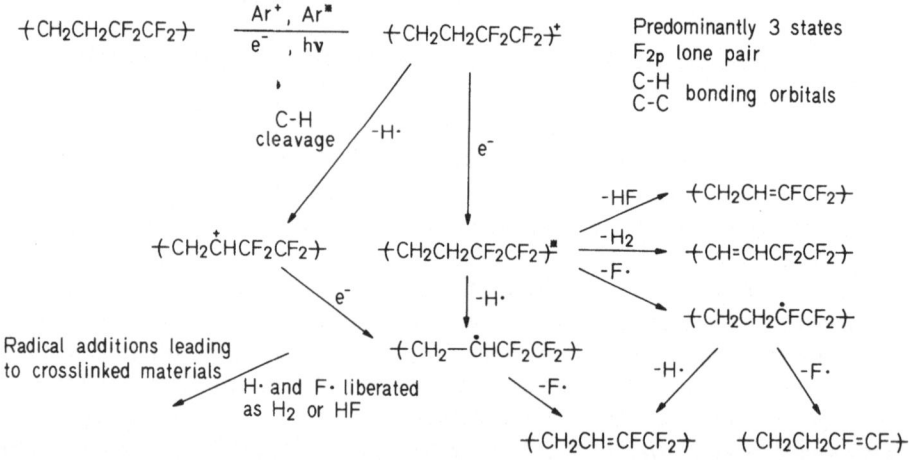

Fig. 40. Possible reaction sequence for initial stages of glow discharge modification of polymer films of an ethylene-tetrafluoroethylene copolymer

face is dominated by direct energy transfer from ions and metastables in the plasma[29].

The overall increase in the total integrated C_{1s} level intensity as a function of reaction time is readily interpreted in terms of a cross linking mechanism since the number of carbon atoms per unit area increases in the surface regions consequent upon the decrease in interchain distance on crosslinking. Indeed the kinetic scheme may be interpreted in terms of a mechanistic scheme involving energy transfer involving Argon ions and metastables and electrons as indicated in Fig. 40. Detailed studies using other inert gas plasmas (He, Ne, Kr) are entirely consistent with this overall picture and it seems a general feature that reaction in the first mono-layer or so of a given polymer is dominated by direct energy transfer processes whilst for the subsurface and bulk the dominant mechanism is by radiative energy transfer predominantly from the vacuum ultra violet output of the plasma[30].

7. Shake Up Phenomena in Polymers; the Study of the Excited States of Core Hole States

If the hierarchy of information levels available in ESCA were limited to those discussed in detail in the preceeding sections the technique would clearly be extremely useful and versatile in many applications. It is evident, however, that for systems in which only a single core level is available for study and for which no chemical shifts are apparent that the range of applications of the the technique to such systems would be limited. Fortunately ESCA is a much more interesting and subtle technique and such systems are encompassed when we consider the information available from the direct study of low energy shake up satellites.

The removal of a tightly bound core electron which is almost completely screening as far as the valence electrons are concerned leads to substantial electronic reorganization accompanying core ionization and this is sufficient perturbation such that direct photoionization leading to the ground state of a core ionized species is accompanied by simultaneous excitation of an electron from a higher occupied to lower unoccupied valence level (shake up) and in the limit leads to a doubly ionized state (shake off). Shake up and shake off processes therefore give rise to satellite peaks to the low kinetic energy side of the direct photoionization peak and may be thought of as excited states of the core hole state species as indicated in Fig. 1. It may be shown that excitation to these states follow monopole selection rules as indicated in Fig. 1 and in the sudden approximation, transition intensities are directly related to the sums of one centre overlap terms involving the occupied orbitals of the initial system and virtual orbitals of the hole state species. These monopole excited states are analogues of the more familiar dipole allowed excited states of the neutral molecule studied in conventional electronic spectroscopy. There are subtle differences however and this can be readily appreciated by the schematic in Fig. 41.

If we consider electronic transitions for a closed shell system as depicted on the left hand side of Fig. 41 in a simple orbital model we may generate a singlet and triplet state from the same excitation configuration, the latter being lower in energy than the former, the energy gap being given by twice the exchange integral. Except under special conditions the only transitions observed with substantial intensities are the dipole allowed singlet to singlet transitions. Consider now excitation involving a core hole state in the doublet manifold as depicted on the R.H.S. of Fig. 41. We may consider two possibilities for a given excitation configuration. Firstly if we

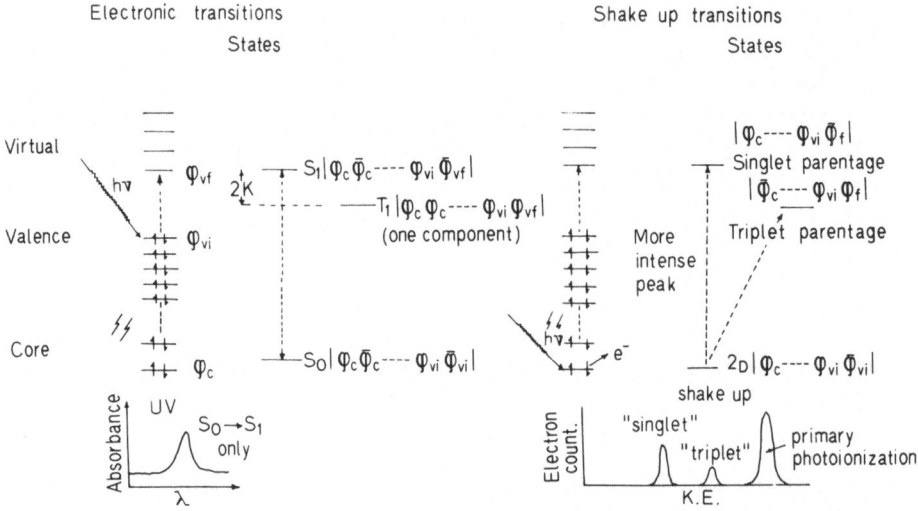

Fig. 41. Schematic illustrating the relationship between the dipole excited states of a neutral system with singlet and triplet states formally of the same excitation configuration and the monopole shake up states for a core hole state

consider the shake up transition to be such that the unpaired electron in the valence level and that excited to the virtual orbital have opposite spins we may consider this doublet state to be of 'singlet origin' by analogy with the excited singlet state of a given excitation configuration, previously discussed. However we may also generate a doublet state by having the electrons in the valence level and that promoted to the virtual orbital with the same spin whilst the remaining core electron has opposite spin. This by analogy corresponds to the shake up state of triplet origin. Again the 'triplet' state is lower in energy than that of singlet origin, however, since both re-present doublet states, transitions from the ground state of the core hole state may be viewed as both being allowed. In principle therefore we should have as experimental observables the energy separations and intensities for the components of the shake up states of given excitation configuration. We might anticipate naively that the shake up state of singlet origin would be the more intense and in the following discussion this will largely be implicit. The detailed theoretical treatment of shake up states in general is by no means as simple as we have portrayed, however, the simplistic model presented here is conceptually very useful and forms a good starting point for more sophisticated treatments[31]. Fortunately as will become apparent if we restrict ourselves to the interpretation of trends and differences particularly of shake up intensities a rather good description may be obtained at relatively modest computational expense[31, 32].

Before considering the experimental aspects of shake up phenomena in polymers it is worthwhile briefly considering some of the theoretical background since this provides considerable insight into the processes involved.

We have previously emphasized that the binding energy is characteristic of a given core level and varies within narrow limits (chemical shifts).

Relaxation energies (associated with contraction of the valence electron cloud consequent upon core ionization) are also characteristic of a given core level and also vary within narrow limits as a function of the bonding environment of the atom on which the core level is located[33]. For C_{1s} levels for neutral systems for example, binding energies measured with respect to the fermi level as energy reference, fall in the range 285–295 eV whilst relaxation energies might typically fall in the range 12 ± 2 eV[3, 30]. The direct relationship between shake up and shake off processes and relaxation energies may be readily understood from theoretical relationships first established by Manne and Aberg[34]. They showed that the weighted average over the direct photoionization and shake up and shake off peaks corresponds to the binding energy appropriate to the unrelaxed systems and this is shown schematically in Fig. 42. Since relaxation energies fall within such a narrow range for a given core

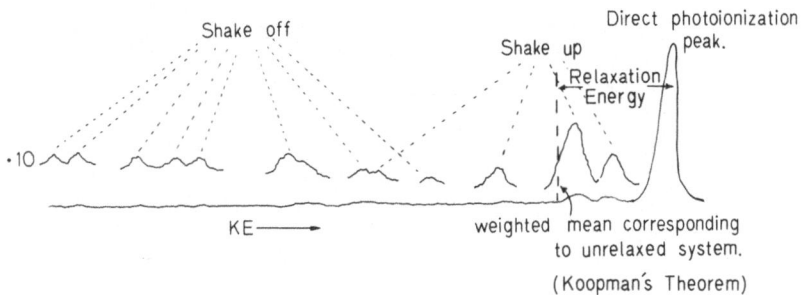

Fig. 42. Relationship between relaxation energies, Koopmans' Theorem and the relative intensities of direct photoionization and shake up and shake off transitions

level it is clear that shake up and shake off are perfectly general phenomena which are present in every system, the feature which changes from one system to another being the weighting coefficients (probabilities) for each transition. It is clear that transition probabilities for high energy shake off processes should be relatively small and that transitions of highest probability should fall reasonably close to the centroid. In principle relaxation energies should be available from experiment provided all of the relevant shake up and shake off processes can be estimated in terms of energies and intensities. In practise this is not a feasible proposition particularly for solids since the overall situation is considerably complicated by the presence of the general inelastic tail (arising from photoemission from a given core level followed by energy loss by a variety of scattering processes) which provides a broad energy distribution usually peaking for organic systems $\sim$ 20 eV below the direct photoionization peaks. This generally obscures any underlying high energy shake off processes such that it is only for systems exhibiting relatively high intensity low energy shake up peaks that information derived from this source can conveniently be exploited. Fortunately such a situation generally obtains for polymer systems which contain either unsaturated backbones[35] or pendant groups[32c, 32e] since low energy $\pi \to \pi^*$ shake up transitions are available.

As a typical example Fig. 43 shows the C_{1s} spectra for typical saturated polymers polyethylene (high density) and polydimethylsiloxane which represent prototype systems with saturated backbones and unsaturated pendant groups[32c, 32e].

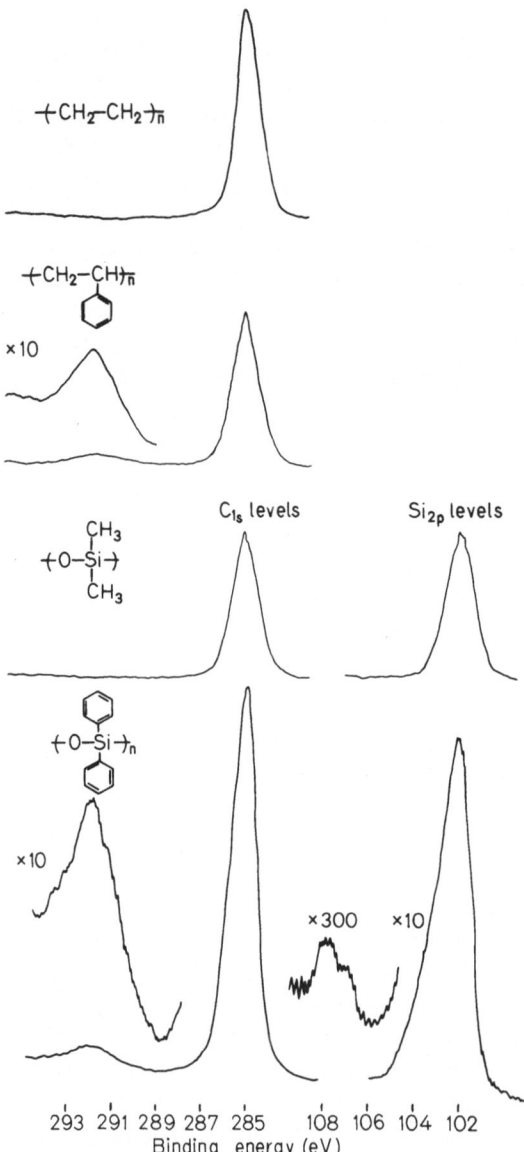

Fig. 43. Core level spectra of polyethylene, polystyrene, poly-dimethylsiloxane and pholydiphe-nylsiloxane showing shake up structure

For the latter, well developed shake up structures are apparent which clearly distinguishes them from the saturated systems although the differences in lineshape and linewidths for the main photoionization peaks are closely similar. (For the siloxanes of course, the relative intensities of the C_{1s} with respect to the O_{1s} and Si_{2p} levels may be used to effect a ready distinction between the two siloxanes). It will become apparent that the transitions giving rise to the satellite structures in PS and PDPS are due to $\pi \rightarrow \pi^*$ transitions as might indeed be inferred from the much smaller shake up peak associated with the Si_{2p} levels and the lack of any low energy shake up structure accompanying the O_{1s} levels for PDPS. To elucidate the nature of these shake up transitions as a preliminary to utilizing such data for structural studies, we have made a systematic study of para substituted polystyrenes the objective being to study the transition energies and peak intensities as a function of the electronic demand of the substituents in the classic mould of physical organic chemistry[32c, 36]. To complement the experimental studies, theoretical computations of shake up probabilities have been made within the sudden approximation employing the equivalent cores concept and a semi empirical all valence electron SCF MO formalism. Within this framework the calculation of shake up probabilities involves summation over weighted overlap terms involving the occupied and virtual orbitals involved in the transitions.

For the parent polystyrene the marked asymmetry of the satellite structure to the C_{1s} levels strongly suggests that at least two transitions are involved. The substituents investigated included ring nitrogen (poly-4-vinyl pyridine), *para* CH_3, $(CH_3)_3C$, Cl, Br, NH_2 and OMe and thus ranged from an extremely powerful π electron acceptor to strong π electron donors. The effect on the sense of asymmetry of the satellite peaks for the C_{1s} levels in this series depends markedly on the electronic demands of the substituent being in an opposite sense for the two extremes of donor and acceptor properties. The centroids for the satellite structures increase in energy separation with respect to the direct photoionization peaks as the *para* substituent changes from being an overall pi electron donor to a pi electron acceptor. The most striking feature however is the strong dependence (on the substituent) of the overall intensity of the shake up satellites, with respect to the direct photoionization peak. Table 5 shows the relevant data. By contrast the satellite structure accompanying

Table 5. Core level binding energies, and transition energies and intensities for low energy shake up structures in poly *para*-substituted styrenes

Substituent X	Core level	Binding Energies[a] (eV)		Mean shake up[b] energie (eV)	Total shake up[c] intensity (%)
(N)	C_{1s}	285.0	285.9	7.1	5.8
H	C_{1s}	285.0		6.6	8.1
tBu	C_{1s}	285.0		6.5	7.5
Me	C_{1s}	285.0		6.5	7.0
Cl	C_{1s}	285.0	286.3	6.7	6.3
Br	C_{1s}	285.0	286.2	6.6	5.9
OCH_3	C_{1s}	285.0	286.7	6.7	3.7
NH_2	C_{1s}	285.0	286.3	6.4	3.3
(N)	N_{1s}	399.5		6.5	7.1 (7.3)
Cl	Cl_{2p}	200.5	201.9	6.5	2.7 (3.0)
Br	Br_{3d}	69.8	70.9	6.8	1.9 (2.3)
OCH_3	O_{1s}	534.2		6.6	2.0 (3.8)
NH_2	N_{1s}	400.4		6.4	5.8 (12.3)

[a] Binding energies relative to hydrocarbon at 285.0 eV.

[b] Measured to centroids of asymmetric satellite peaks.

[c] Intensities expressed as a percentage of the total intensity due to the core level from atoms in the ring and directly attached to the ring. Figures in brackets refer to the relative substituent/carbon shake up intensity for equal numbers of atoms.

core ionization from the *para* substituent is symmetrical in nature and the centroid corresponds to that for the lower energy component of the C_{1s} satellite structure if this is deconvoluted into two components.

The trend in intensities for the substituent shake up satellites is in the opposite sense to that for the C_{1s} levels and this becomes more evident on consideration of the ratio corrected for equal numbers of atoms. (The intensity ratio for the methoxy derivative is almost certainly a lower limit since there is some evidence from the O_{1s} spectrum that there is a small amount of water and/or oxidation at the surface.)

Comparison of the UV spectrum of polystyrene in the 2600 Å region with that of toluene shows a close relationship in terms of both extinction coefficients and vibronic fine structure. The effect of *para* substituents is most conveniently characterized by the shift in the band corresponding to the ν_{0-0} transition. The comparison of substituent effects on the electronic excited states of the *para* substituted polystyrenes parallels those for the corresponding *para* substituted toluenes. Such a correlation would only be expected if the $\pi \rightarrow \pi^*$ transitions were effectively localized within a given pendant group of the polymer system. This conclusion is reinforced by the observation that polystyrene and toluene show similar shake up structure in their ESCA C_{1s} spectra with respect to both band profiles and intensities (when due allowance has been made for the differing number of carbon atoms in the repeat unit). It is evident that *para* substituted toluenes are good model system for the dipole excited states for the poly *para* substituted styrenes and it will become apparent that this carries over to the monopole excited (shake up) states. Two parameters relating to substituent effects on the low lying $\pi \rightarrow \pi^*$ excited states of substituted benzenes. Platt's spectroscopic moment[37] relates to the intensity changes

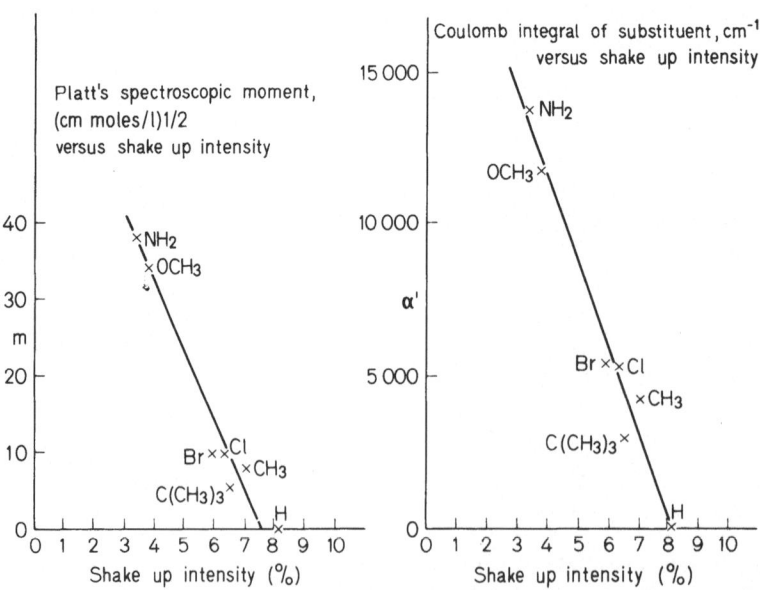

Fig. 44. Correlation of low energy shake up intensity accompanying C_{1s} photoionization (for a given parasubstituted polystyrene) with Platts' spectroscopic moment and the coulomb intergral of the substituent

consequent upon introduction of a given substituent. Whilst the coulomb integral of a given substituent derives from a localized orbital description of the dipole excited states of substituted alternants and non alternants and is a measure of the change in potential of an electron in a $2p_z$ orbital on an adjacent carbon atom[38]. This may be derived from experimental data by analyzing the first order inductive shift in non alternants and second order shift in alternants. Both substituent constants are therefore intimately related to substituent effects on the $\pi \to \pi^*$ dipole excited states. Figure 44 shows the correlation with the shake up intensities. The trends displayed are quite striking and leave little doubt that the satellites arise from $\pi \to \pi^*$ excitations.

In a review such as this it is inappropriate to present a detailed theoretical interpretation of the results since this is available elsewhere[36]. Suffice it to say that it may readily be shown that the $\pi \to \pi^*$ shake up transitions involve the highest occupied and lowest unoccupied virtual orbitals of the pendant group.

It is a comparatively straightforward matter to establish that on the ESCA depth scale the surface morphology of the polymers studied are such that the repeat units are statistically sampled and it is therefore appropriate to use the corresponding *para* substituted toluenes as model systems. Figure 45 shows the four one electron transition involving the two highest occupied and lowest unoccupied orbitals (symmetries designated with respect to approximate local C_{2v} symmetry of the π electron system). Of the four transitions those arising from the $a_{2\pi} \to b_{1\pi}$ and $b_{1\pi} \to a_{2\pi}$ excitations are formally monopole forbidden. However, the strong perturbation consequent upon removal of a core electron effectively removes the symmetry restriction except for hole states corresponding to photoionization from core levels associated with atoms located on C_2 axes (*e.g.* Cl, C2, C5 and X). The calculated shake up

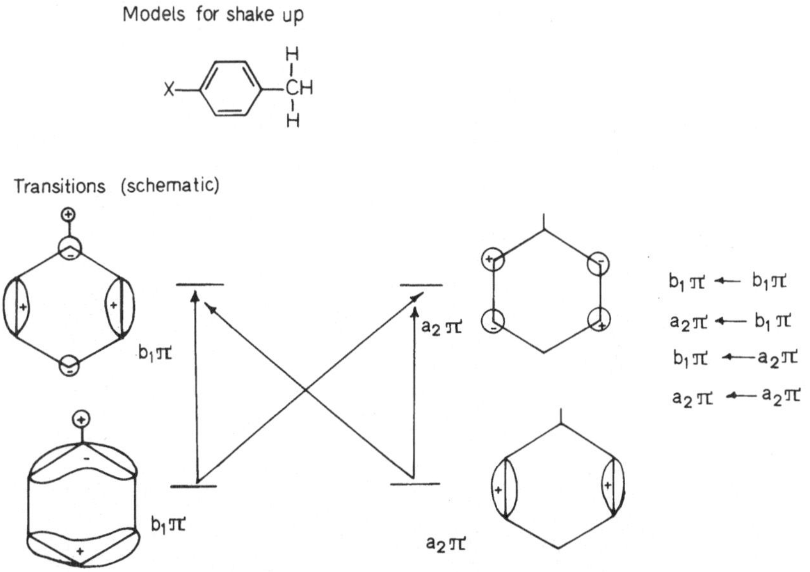

Fig. 45. The four one electron transitions involving the highest occupied and lowest unoccupied orbitals of substituted benzenes (in local C_{2v} symmetry)

Table 6. Calculated shake up probabilities for low energy $\pi \to \pi^*$ transitions in model compounds[26]

% Shake Up

Transition	Atom 1	2	3	4	5	6	7	8	9	Total C_{1s}	O_{1s}
$b^*_{1\pi} \leftarrow b_{1\pi}$	0.90	9.13	3.40	2.23	9.61	2.23	3.40	—	—	4.41	—
$a^*_{2\pi} \leftarrow b_{1\pi}$	0	0	0.06	0	0	0	0.06	—	—	0.02	—
$b^*_{1\pi} \leftarrow a_{2\pi}$	0	0	5.81	6.80	0	6.80	5.81	—	—	3.60	—
$a^*_{2\pi} \leftarrow d_{2\pi}$	0.11	0.01	0	0	0.02	0	0	—	—	0.02	—

Transition	Atom 1	2	3	4	5	6	7	8	9	Total C_{1s}	O_{1s}
$b^*_{1\pi} \leftarrow b_{1\pi}$	1.15	10.20	3.02	4.07	10.24	3.59	4.08	1.39	0.45	4.60	1.39
$a^*_{2\pi} \leftarrow b_{1\pi}$	0	0	0.06	0.10	0	0.10	0.06	0.48	0.71	0.13	0.48
$b^*_{1\pi} \leftarrow a_{2\pi}$	0	0	6.06	4.84	0	5.36	4.72	0.05	0	2.62	0.05
$a^*_{2\pi} \leftarrow a_{2\pi}$	0.10	0.04	0	0	0.07	0	0	0.03	0.10	0.04	0.03

probabilities for the parent system and for a prototype system with a good pi electron donor are shown in Table 6. Two features are evident. Firstly, the intensity for
the low energy shake up satellites to the C_{1s} spectra largely derives from two transi-
tions ($b_{1\pi} \rightarrow b_{1\pi}$ and $a_{2\pi} \rightarrow b_{1\pi}$) and indeed the distinct asymmetry of the satellites
would tend to confirm this as we have already pointed out that more than one tran-
sition is involved. Secondly when hydrogen is replaced by a π electron donating
substituent, whilst the intensity of the $b_{1\pi} \rightarrow b_{1\pi}$ transition is predicted to remain
essentially the same, that for the $a_{2\pi} \rightarrow b_{1\pi}$ transition is calculated to decrease. The
computed net overall decrease in shake up intensity predicted by the model is therefore qualitatively in agreement with experiment. As an interesting side light it might
be mentioned that computations on the orthogonal conformation of the methoxy
derivative ($2p_z$ lone pair on oxygen perpendicular to the ring) shows a substantial
difference in shake up probabilities[36]. This suggests that shake up structure should
be sensitive to conformational preference which contrasts markedly with the behaviour of the direct photoionization peaks.

The analysis of the shake up satellites in terms of a two component structure
leads to the correlation shown in Fig. 46 where for convenience the data has been
analyzed in terms of the coulomb integrals of the substituents.

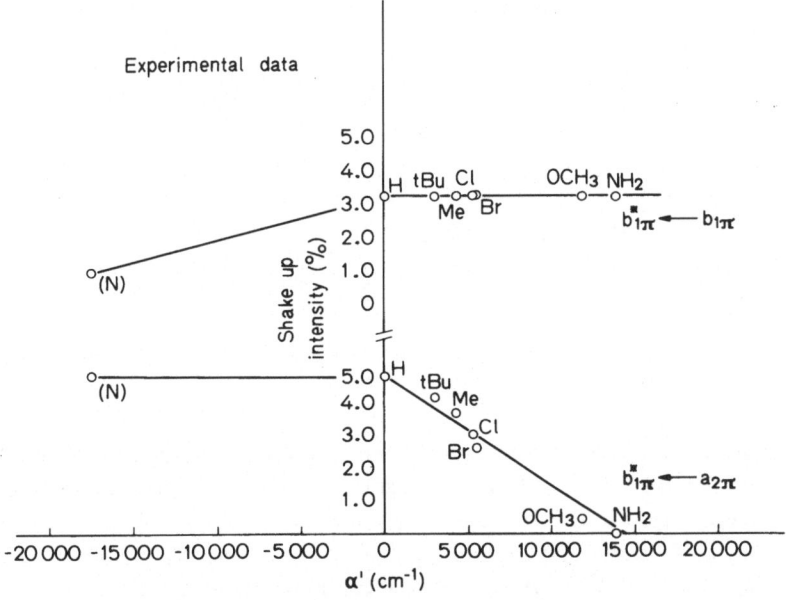

Fig. 46. Experimentally derived shake up probabilities for components contributing to the low
energy shake up satellites in para substituted polystyrenes

It is gratifying to note that the theoretical calculations on model systems repro-
duce the trends shown in Fig. 46 providing strong confirmation for the overall validity of the interpretations[32e, 36]. Low energy shake up satellite structures are often
highly characteristic of the π electronic structure of the pendant group as is clear
from a comparison of Fig. 43 and Fig. 47. In each case theoretical analysis indicates

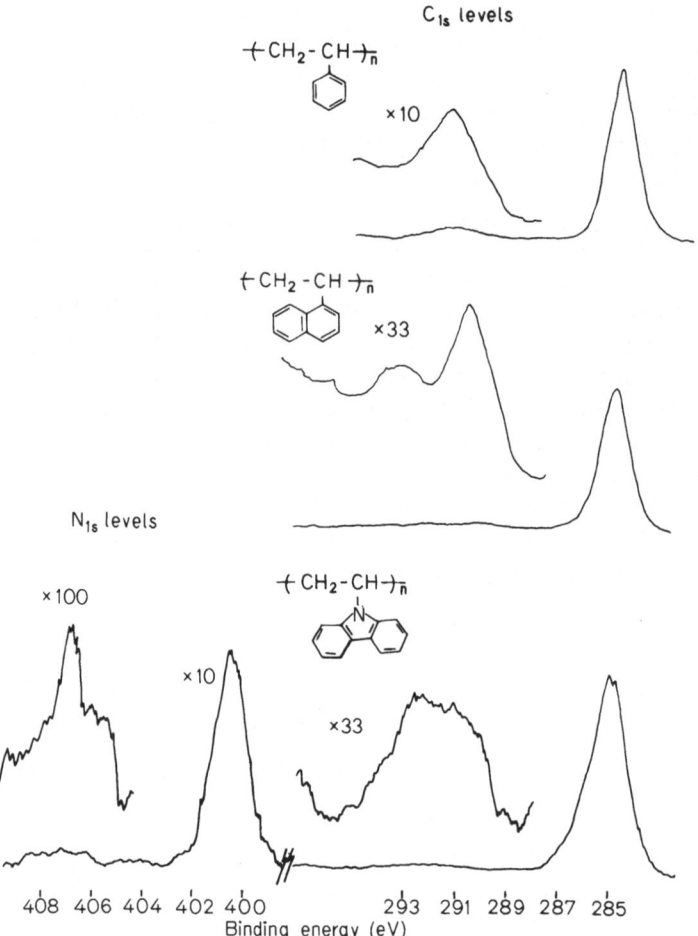

Fig. 47. Core level spectra for polystyrene, polyvinylnaphthalene and polyvinylcarbazole showing low energy shake up structures

that the transitions involve the highest occupied and lowest unoccupied orbitals and this is indicated in Fig. 48 [32e, 32a]. Clearly the distinctive nature of shake up satellites can add an extra dimension to the utility of ESCA as a structural tool. As one straight-forward example Fig. 49 shows the C_{1s} and O_{1s} levels for poly-n-hexylmeth acrylates and polyphenylmeth acrylates [16]. The core level spectra are essentially the same, however, the distinctive nature of the shake up structure for the unsaturated pendant group allows a ready distinction to be made.

Having identified and qualitatively understood the main features of the low energy satellite structures arising predominantly from shake up processes[c], we may

---

[c] Discrete energy loss peaks will also contribute to the overall structure. Comparison of model systems studied in the gas and condensed phases shows however, that the low energy structures in the systems in question are dominated by shake up processes.

Shake up transitions (schematic)

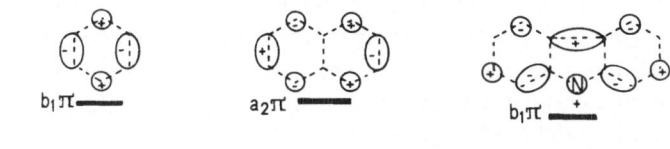

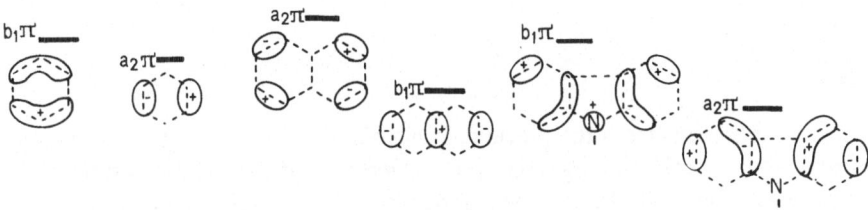

Fig. 48. Orbitals involved in the low energy shake up satellites accompanying core ionization in polystyrene, poly-1-vinylnaphthalene and polyvinylcarbazole

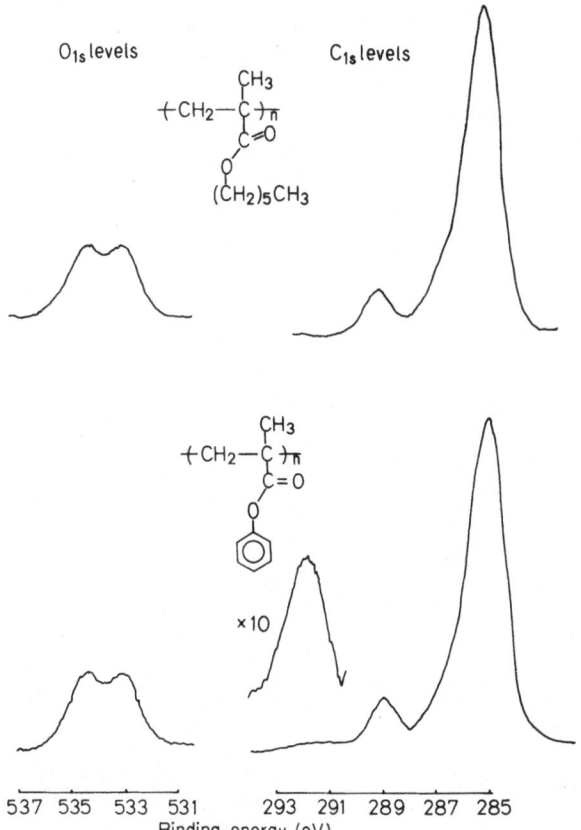

Fig. 49. Core level spectra for poly-n-hexylmethacrylate and polyphenylmethacrylate showing the shake up structure for the latter

now consider the implications with regard to enlarging the scope of the technique. As a particular example in this section we outline an investigation of some alkane-styrene copolymers of general formula which illustrate the utility of ESCA for studying copolymer composition in systems for which the primary sources of information are of themselves insufficient and for which the extra dimension is provided by observation of shake up satellites[39].

$$[CH-CH_2CH_2-CH-(CH_2)_n]_m$$

$$n=0,1,3,5,6,10$$

It should be evident from the previous discussion that characteristic low energy shake up structure accompanying direct photoionization of the C_{1s} levels should be specifically associated with the styrene component. Figure 50 shows the measured C_{1s} levels and shake up satellites for the series of alkane-styrene copolymers and it is evident by visual inspection that the relative intensities of the shake up satellites with respect to the main photoionization peaks decrease with increasing chain length of the alkane component. The measured intensities and energy separations are given in Table 7. A clear trend exists between shake up intensities and the chain length of the alkane component, and the structure of the shake up satellites and the energy

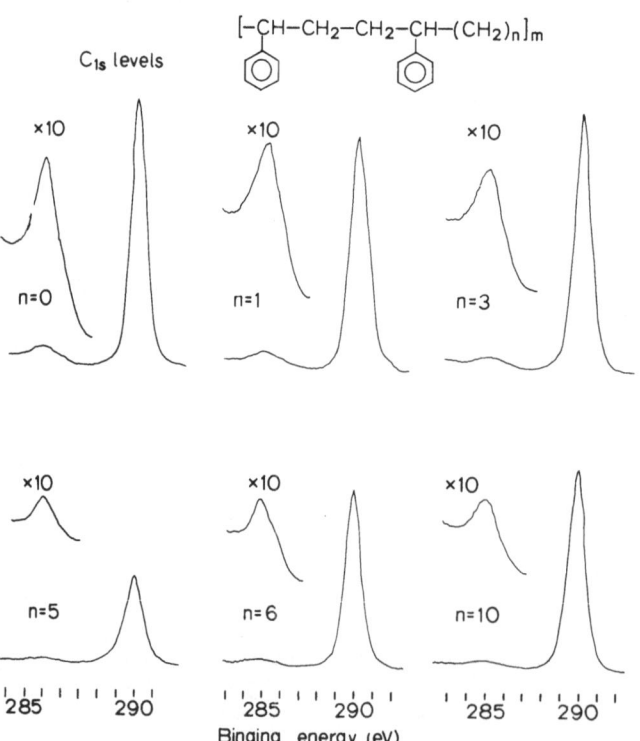

Fig. 50. C_{1s} levels for a series of alkane-styrene copolymers showing shake up structure

Table 7. ESCA data for the alkane-styrene copolymers binding energies/eV[a]

n	C_{1s}	C_{1s}^s	Δ	Area ratios $(C_{1s}/C_{1s}^s)^b$
0	285.0	291.6	6.6	13.4
1	285.0	291.6	6.6	14.5
3	285.0	291.6	6.6	18.5
5	285.0	291.6	6.6	22.7
6	285.0	291.6	6.6	23.8
10	285.0	291.6	6.6	32.2

[a] Relative to C_{1s} at 285.0 eV, Δ given with respect to centroid of asymmetric shake-up peak.
[b] Ratio of main photo-ionization peak to shake-up peak.

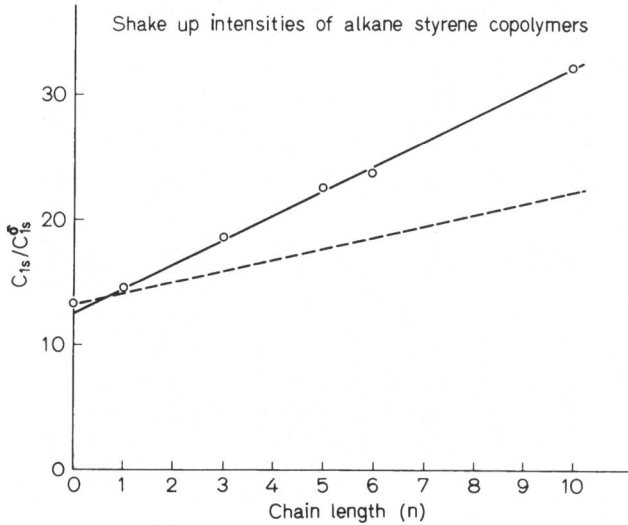

Fig. 51. Plot of ratio of area ratios for direct photoioniza- tion peak and low energy shake up satellite for a series of alkane styrene copolymers as a function of chain length n of the alkane component

separations remain essentially constant. This becomes clearer from a graphical representation of the data as shown in Fig. 51. Also shown is the correlation ex- pected on the basis that the repeat units of the polymers are statistically sampled. It is clear that copolymer compositions may be established from the measurement of shake up intensities and the least squares plot of the intensity ratio of the direct photoionization peak to shake up satellites versus n (the chain length of the alkane component) gives a correlation coefficient of 0.997 the slope being 1.91 and inter- cept 12.91. The latter may be compared with the measured value of 13.4 for the parent system (*i.e.* polystyrene). The calculated slope assuming statistical sampling of the repeat unit is 0.90. The fact that an additive model applies to the experimental data but with a much larger dependence of intensity on n than predicted theoretically would strongly suggest that there are specific orientation effects of the polymer chains in the surface regions sampled by ESCA. An alternative possibility is that the samples are contaminated with hydrocarbon. Such contamination would contribute to the C_{1s} peak at 285 eV but not to the low energy satellite structure. Even if the extent of contamination were such as to produce the linear correlation found experimentally (which is most unlikely on the basis of the method of preparation) on the basis of the likely escape depth dependence on kinetic energy for C_{1s} levels employing $MgK\alpha_{1,2}$ radiation such an explanation is untenable. What is clearly required is a model in which the repeat unit is not statistically sampled in such a sense that the phenyl groups are discriminated against. As n becomes large we might reasonable expect a structure based on the folded chain structure of polyethylene[40]. With this in mind and with the aid of models we have considered possible structures which would lead to the results illustrated in Fig. 51. Two such models which exhibit an increased gradient with respect to that expected if the data correspond to statistically sampling the repeat unit are shown in Fig. 52 for the particular case of $n = 3$. The model with a phenyl group specifically oriented at the surface comes remarkably close to the experimental data and it may fairly be claimed that for this system in addition to providing information on composition; shake up satellites also

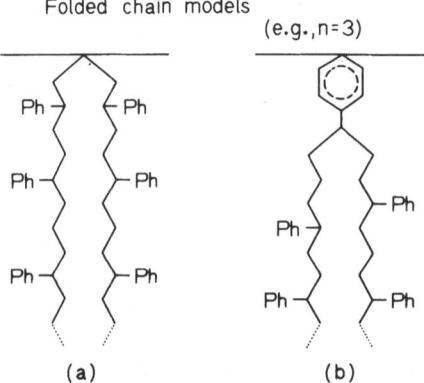

Folded chain models (e.g., n=3)

(a)   (b)

Fig. 52. Possible structures for alkane styrene copolymers which would lead to nonstatistical sampling of the repeat unit (for the particular case of $n = 3$)

provide an interesting insight into the possible surface morphology.

Other examples of the exploitation of shake up phenomena includes for example the investigation of the relationship of surface to bulk domain structure in AB block copolymers[41]. These studies considerably extend the scope of ESCA as one of the most important shots in the polymer chemist and physicists' locker for studying aspects of structure, bonding and reactivity relating to the surface regions of polymers.

Acknowledgement. I would like to acknowledge the contributions made to the development of the ESCA programme in Durham by a succession of willing and able research workers. In the most recent installment of the ESCA applied to polymers programme, particular mention should be made of Jim Peeling, Ron Thomas, Alan Dilks and Derek Shuttleworth, I would also like to thank the Science Research Council and Institute of Petroleum for providing financial support for these research programmes.

8. References

[1] a) Siegbahn, K., *et al.*: Nova Acta R. Soc. Sci., Uppsala Ser. IV, 20 (1967).
   b) Siegbahn, K., *et al.*: ESCA Applied to Free Molecules. Amsterdam: North Holland
   Publishing Co., 1969.
[2] a) Clark, D. T.: Chemical applications of ESCA in electron spectroscopy. Dekeyser, W.
   (ed.) Dordrecht, Holland: D. Reidel Publishing Co. 1973 (NATO Summer School
   Lectures, Ghent, September 1972).
   b) Clark, D. T.: plenary lectures ACS Symposium, Advances in Polymer Friction and Wear,
   Los Angeles, March 1974, Lee, L. H. (ed.) 5 A, New York: Plenum Press 1975.
   c) Clark, D. T., Feast, W. J.: J. Macromol. Sci. Reviews in Macromol. Chem. *C12*, 191 (1975).
   d) Clark, D. T.: Structure and bonding in polymers as revealed by ESCA in electronic
   structure of polymers and molecular crystals. (ed) Ladik J. Andre, J. M. New York:
   Plenum Press, 1975. (NATO Summer School Lectures, Namur, September 1974).
   e) Clark, D. T., in structural studies of macromolecules by spectroscopic methods, Chapter 9.
   Ivin, K. (Ed.). London: J. Wiley and Sons 1976.
[3] cf. Tedder, A. W.: With prejudice, Cassel, London 1966.
[4] cf. Wagner, C. D.: Disc Faraday Soc. *60* (1975).
[5] cf. Fadley, C. S., Baird, R. J., Siekhaus, W., Novakov, T., Beigstran, S. A. L.: J. Electron
   Spectroscopy *4*, 93 (1974).
[6] cf. Carlson, T.: Photoelectron and auger spectroscopy. New York: Plenum Press 1975.
[7] cf. Swingle II, R. S., Riggs, W. M.: CRC Crit. Rev. Anal. Chem. *5*, 267 (1975).
[8] Clark, D. T., Dilks, A., Thomas, H. R., Shuttleworth, D.: to be submitted.
[9] a) cf. Powell, C. J.: Surface science *44*, 29 (1974).
   b) Lindau, I., Spicer, W. E.: J. Electron Spectroscopy *3*, 409 (1974).
[10] Handbook of Spectroscopy, Robinson, J. W. (Ed.). Vol. 1, Cleveland: CRC Press 1974.
[11] a) Henke, B. L.: Adv. X-ray Analysis *13*, 1 (1969).
   b) Henke, B. L.: J. Phys. (Paris) *C4*, 115 (1971).
[12] Cadman, P., Evans, S., Scott, J. D., Thomas, J. M.: J. Chem. Soc. Faraday II *71*, 1777 (1975).
[13] Clark, D. T., Thomas, H. R.: J. Polymer Sci. Chem. Ed., in press (1977).
[14] Clark, D. T., Feast, W. J., Musgrave, W. K. R., Ritchie, I.: J. Polymer Science, Chem. Ed. *13*,
   857 (1975).
[15] cf. Penn, D. R.: J. Electron Spectroscopy *9*, 29 (1976).
[16] Clark, D. T., Thomas, H. R.: J. Polymer Science, Polymer Chem. Ed. *14*, 1671 (1976), *14*,
   1701 (1976).
[17] Clark, D. T., Salaneck, W., Paton, A.: J. App. Phys. *47*, 144 (1976).
[18] Clark, D. T., Thomas, H. R., Feast, W. J., Tweedale, P.: to be submitted.
[19] a) Clark, D. T., Feast, W. J., Ritchie, I., Musgrave, W. K. R., Modena, M., Ragazzini, M.:
   J. Polymer Science, Polymer Chem. Ed. *12*, 1049 (1974).
   b) cf. Chujo, R., Maeda, K., Okuda, K., Murauama, N., Hoshino, K.: Makromol. Chem.
   176, 213 (1975).
   c) Clark, D. T., Adams, D. B., Dilks, A., Peeling, J., Thomas, H. R.: Makromol. Chem. *177*,
   2139 (1976).
[20] cf. Clark, D. T.: Chemical aspects of ESCA. Electron emission spectroscopy. Dekeyser, W.
   (ed.). Dordrecht, Holland: D. Reidel Publishing Co. 1973, pp. 373–507.
[21] Peeling, J., Clark, D. T., Evans, I. M., Boulter, D.: J. Sci. Fd Agric. *27*, 331–340 (1976).
[22] Clark, D. T., Peeling, J., Boulter, D., Evans, I.M.: J. Amer. Chem. Soc. in press.
[23] Yasuda, H.: J. Macromol. Sci. Chem. *A10*, 383 (1976).
[24] Okane, D. F., Rice, D. W.: J. Macromol. Sci. Chem. *A10*, 567 (1976).
[25] Clark, D. T., Shuttleworth, D.: J. Polymer Sci. Chem. Ed. (1977).
[26] Clark, D. T., Kilcast, D., Feast, W. J., Musgrave, W. K. R.: J. Pol. Sci. Al *10*, 1637 (1972).
[27] Hudis, M.: Techniques and applications of plasma chemistry. Hollahan, J. R., Bell, A. T.
   (ed.). New York: J. Wiley and Sons 1976.

28) a) Clark, D. T., Dilks, A.: ACS Centennial Meetings New York, April 1976 International
      Symposium on Advances in the characterization of polymer and metal surfaces, Lee, L.H.,
      (ed.). Academic Press, 1976, Vol. 2, p. 101.
      b) Clark, D. T., Dilks, A.: J. Polymer Science, Polymer Chem. Ed. (1976) in press.
29) Clark, D. T., Dilks, A.: J. Polymer Sci. Chem. Ed., in press (1977).
30) cf. Sampson, J. A. R.: Techniques of vacuum ultraviolet spectroscopy. New York: J. Wiley
      and Sons 1967.
31) Clark, D. T., in: Progress in theoretical organic chemistry, Vol. 2, Csizmadia, I. G., (ed.).
      Elsevier 1976 (in press).
32) a) cf. Clark, D. T., Adams, D. B., Scanlan, I. W., Woolsey, I. S.: Chem. Phys. Letters 25,
      263 (1974).
      b) Clark, D. T., Adams, D. B.: J. Electron Spectry 7, 401 (1975).
      c) Clark, D. T., Dilks, A., Peeling, J., Thomas, H. R.: Faraday Soc. Disc. 60, 183 (1975).
      d) Clark, D. T., Adams, D. B.: Theoretica chim. Acta 39, 321 (1975).
      e) Clark, D. T., Adams, D. B., Dilks, A., Peeling, J., Thomas, H. R.: J. Electron Spectroscopy
      8, 51 (1976).
33) a) Clark, D. T., Scanlan, I. W.: J. Chem. Soc. Faraday Trans. II 7, 1222 (1974).
      b) Clark, D. T., Müller, J., Scanlan, I. W.: Theoretica chim. Acta 35, 341 (1974).
      c) Clark, D. T., Müller, J.: Theoretica chim. Acta 41, 193 (1976).
34) Manne, R., Aberg, T.: Chem. Phys. Letters 7, 282 (1970).
35) Chambers, R. D., Clark, D. T., Kilcast, D., Partington, S.: J. Polymer Sci., (Polymer Chem.
      Edn.) 12, 1647 (1974).
36) Clark, D. T., Dilks, A.: J. Pol. Sci. Polymer Chem. Ed., 15, 15 (1977).
37) a) Platt, J. R.: J. Chem. Phys. 19, 263, 1148 (1951).
      b) Jaffe, H. H., Orchin, M.: Theory and applications of ultraviolet spectroscopy. New York:
      John Wiley and Sons Inc. 1962.
38) Murrell, J. N.: The theory of the electronic spectra of organic molecules. London: Methuen
      and Co. Ltd. 1963.
39) Clark, D. T., Dilks, A.: J. Polymer Science Polymer Chem. Ed. 14, 533 (1976).
40) Wunderlich, B.: Macromolecular Physics, Vol. 1. New York: Academic Press 1973.
41) Clark, D. T., Peeling, J., O'Malley, J. M.: J. Polymer Science Polymer Chem. Ed. 14, 543
      (1976).

Received October 11, 1976

Polymer Separation and Characterization by Thin-Layer Chromatography

Hiroshi Inagaki

Institute for Chemical Research, Kyoto University, Uji, Kyoto 611, Japan

Table of Contents

I. Introduction

Separation and purification are the most fundamental procedures in chemistry.
Chemistry of low-molecular-weight substances has been recognized as exact natural
science largely by the reason that a given substance could be made free from impurities to a great extent by separation and purification procedures, as exquisitely
stated by L. Pauling as follows[1]:

> The concept "pure substances" is of course, an idealization; all actual substances
> are more or less impure. It is a useful concept, however, because we have learned
> through experiment that the properties of various specimens of an impure sub-
> stance with different impurities are nearly the same if the impurities are present
> in only small amounts. These properties are accepted as the properties of the
> ideal substance − − −

The concept "pure substances" in synthetic polymer chemistry is not immediately clear. One can purify a given polymeric product to remove impurities or
contaminants, such as unreacted monomers, oligomers, initiator fragments, and
solvent, etc. This procedure merely means the removal of non-polymeric substances
from the product. Every object in this field has a nature which has been expressed
as "polydispersity" or "polymolecularity". It follows that the molecular parameters
to be assigned to a given polymeric substance can only be given in terms of quantities
averaged over all the constituent polymer chains. There was a period during which
polymer chemists were satisfied with characterizing polymeric substances in terms
of the average quantities. At present we are, however, compelled to recognize the
fact that even polyethylene resins, which consist of the simplest monomer units,
cannot be characterized uniquely by molecular parameters, such as the average
molecular weight, the molecular weight distribution, the average degree of branching,
and so forth. One of characteristics of present polymer industry is the growing interest in "tailor-made" polymers as new polymeric materials. The properties of such
materials can hardly be interpreted by the average quantities. This is the case with
even simple statistical copolymers, as experimentally demonstrated, *e.g.,* by Kollinsky
and Markert a decade back[2].

Characterization of copolymers encounters more difficult complexities than
that of homopolymers. The heterogeneity in chemical composition and monomer
arrangement along the polymer chain should be taken into account in addition to
that in molecular weight. Cross fractionation thus becomes indispensable for the
characterization. In other words, separations according to molecular weight and
composition should be done previous to characterization. Though cross-fractionation technique has recently been reexamined extensively by Teramachi *et al.*[3], its
time-consuming nature is unavoidable and, in addition, this technique appears not
sensitive enough to the monomeric arrangement in copolymer chains. Another
separation possibility for the same purpose consists of a counter-current distribution method, which has been worked out by v. Tavel and co-workers[4]. However,
this method requires not only tedious operations but also the determination of an
appropriate set of solvents which are immiscible with each other to form a multi-
phase system. For the latter reason, this method might not be widely applicable.

A method similar to the above was proposed recently by Kuhn, which uses a pair of solvents immiscible at a lower temperature but becoming miscible with elevating temperature[5]. To the present authors this method seems more easily applicable and perhaps more promising than the counter-current distribution, since no special instrumentation is necessary for the former. At any rate, it is obvious that these three methods do not allow fractionation of copolymers only by the composition without interference of the molecular weight, because they all are based on the solubility difference among constituent species.

As discussed above, the most important requisite for fractionating a given copolymer product by the difference of either composition or monomeric arrangement is to minimize the interference of molecular weight. There are some papers which suggest that adsorption of polymer chains at liquid-solid interface is almost insensitive to the chain length[6], except for the case of oligomers. This is one of the reasons why we tried to utilize thin-layer chromatography (TLC), whose separation mechanism is largely due to adsorption-desorption processes. Application of TLC to polymer chemistry had previously been restricted only to fractionation of oligomers[7] and to analysis of additives used in polymer industry, such as plasticisers[8] and stabilizers[9]. The pioneers who first introduced the same idea as that above may be Longford and Vaughan[10], though they employed not TLC but paper chromatography to separate high polymers of differing chemical constitution. In this connection it should be mentioned that Belenkii and Gankina independently reported the feasibility of polymer separation by TLC[11] already in 1968.

This article presents recent advances in application of TLC to polymer separation and characterization. Although most polymer scientists may still be unfamiliar with TLC, this article does not contain its history and experimental procedure in detail. For the details of TLC itself refer to monographs published by E. Stahl[12] and others[13, 14]. A review on experimental results and problems of TLC specific to polymers, which have been reported during the years up to 1973, will also be seen in a monograph edited by H. L. Tung[15]. Some experimental problems experienced recently by our research group will be described and discussed in the last part of this article (Chapter VI).

II. Theoretical Background

II.1. Principle of TLC Separation

A TLC system comprises three elements. As its name implies, one of the elements is thin layer coated mostly on a glass or plastic plate, which acts as a stationary phase. Such a plate is termed "chromatoplate". The most popular materials (adsorbent) used to form the thin layer are silica gel and alumina powder. The other elements are a sample, and a solvent or solvent mixture (developer) which acts as a mobile phase to develop (elute) the sample. A stock solution of sample is prepared in advance of the TLC experiment and applied to the thin layer to form a spot. After removal of solvent contained in the spot, the sample is eluted with an appropriate

developer to separate it into final spots which will appear corresponding to component species involved in the sample. This procedure is called "development", which
can usually be completed within a time shorter than one hour. This rapidness in the
measurement and also the simplicity in the instrumentation are the most favorable
features of TLC. The development is stopped when the developer moves up to a
certain distance, L_0 from the starting point at which the sample was applied. Then
the final spots are visualized mostly by staining to determine the distance by which
each final spot migrated. Thus it becomes possible to characterize each final spot by
a measure termed "rate of flow" (R_f), which is defined as:

$$R_f = (\text{migration distance of sample})/L_0 \tag{1}$$

As implied by the above description of TLC experimental procedures, the separation is effected because the migration rate of each component species relative to
that of the developer is retarded preferentially by the stationary phase during the
development procedure. Therefore, the principle of TLC separation does not differ
from that of other chromatographic separations. The retarding action of the stationary phase may be classified into different mechanisms on the basis of molecular
interactions. "Adsorption" should firstly be mentioned, which concerns attractions
between the sample molecule and the active site of stationary phase. The next two
are "partition" and "molecular sieving". The former concerns the solubility of
sample in a liquid retained by the stationary phase; while the latter the preferential
exclusion of sample by porous materials used as the stationary phase, depending on
molecular size of sample. The last one is "ion-exchange", which concerns the ability
of sample to exchange with inorganic ions attached to the stationary phase and will
therefore have no close relation to TLC separation of neutral polymers.

As will be shown later, the former three mechanisms mentioned above are applicable to TLC separation of polymers. From the standpoint of TLC applied to polymer separation, the partition mechanism may be better expressed by "phase-separation" or "precipitation" mechanism, as will be explained in Section II.3. It should
be noted that all these mechanisms are generally present during a chromatographic
separation. Therefore, one mechanism should be made to be predominant for a
given separation aim. This can, in principle, be done by properly selecting the
developer and adsorbent. However, such a selection is the major problem in application of TLC, especially, to polymer separation, and the following three sections
will be devoted to describing the rules that have been established to solve this problem.

II. 2. Polarity-Controlled Adsorption Mechanism

It is now well known that this mechanism leads to separation largely according to
the chemical composition of sample polymers practically without interference of
the molecular weight so that the compositional distribution of copolymers can be
determined[11, 16]. This means, in turn, that the R_f value is almost independent of
the sample molecular weight, and its validity condition is that the sample molecular weight exceeds, e.g., 5×10^4.

The adsorption mechanism may be argued in terms of three types of binary interactions acting among the adsorbent (A), the developer (D), and the sample polymer (P), which are abbreviated as the interaction [A–D], [A–P], and [D–P]. The retarding action of the stationary phase is dependent on both [A–D] and [A–P], and, hence, the migration rate of the sample may be determined by competitive balance of these two kinds of interactions. One can change the R_f, value of a given sample, because [A–D] and [A–P] depend on the polarity of the developer used and on the activity of the adsorbent, which is adjustable, *e.g.*, by heat treatment[12]. This is the reason why a given copolymer with a compositional heterogeneity was separated by adsorption TLC into component species having different composition[16].

In connection with the foregoing statement, the important role of [D–P] should be mentioned. The interaction [D–P] concerns the sample solubility in a given developer. When a developer used behaves as a poor or bad solvent toward the sample polymer, no migration occurs due to precipitation. In other words, the dissolving power of the developer should be so adjusted that no precipitation of sample polymer takes place during the development. Thus the primary requisite of a developer to be used for adsorption TLC is that the developer should at least be a moderately good solvent for sample polymer[15]. This requirement narrows the variety of solvents to be selected as developer for adsorption TLC. However, it is now known that some homopolymers can be fractionated by molecular weight if a poor solvent for the sample polymer is used as developer[17–19], as will be stated in the next section.

In summary: Provided the sample polymer is soluble in the developer used, a qualitative rule established to design a TLC system appropriate for a given low-molecular-weight compound[12] is applicable in the case of polymers as well. The rule may be described as follows: If a given sample polymer has a low polarity, the polarity of solvent and the activity of adsorbent should be low and high, respectively, and vice versa.

A quantitative rule for the same purpose as above was deduced on the thermodynamic basis of adsorption equilibria by Snyder[20]. The essential part of this rule may be represented by a following equation:

$$\log K^\circ = \log V_a + \alpha(S^\circ - A_s\epsilon^\circ), \tag{2}$$

where K° is a distribution coefficient defined as a ratio of the concentration of adsorbed sample molecules to that of unadsorbed; V_a is a constant specific to a used pair of developer and adsorbent; α is a parameter indicating the activity of adsorbent surface (relative degree of the adsorbent activity); S° and $A_s\epsilon^\circ$ are dimensionless parameters introduced to describe the adsorbed phase energy related to the sample and developer molecules, respectively, and A_s is a parameter describing the area which is required by an adsorbed sample molecule on the adsorbent surface.

Equation (2) predicts that the concentration of adsorbed sample molecules is decided by the difference between adsorptive affinities of the sample and developer molecules onto the adsorbent surface; and this is just the same as the aforementioned statement that the rate of sample migration is decided by a competitive balance of the interactions [A–D] and [A–P]. On the other hand, the distribution coefficient K° is related to the R_f value by

$$R_f = [1 + (W/V^\circ)K^\circ]^{-1} \tag{3}$$

W and V° being the total weight of stationary phase and the volume of mobile phase, respectively. From the above two equations it is easily deduced that the R_f value increases from zero to unity with decrease in the K° value, viz., with decrease in the concentration of adsorbed sample molecules. Further it is obvious that Eqs. (2) and (3) confirm the aforementioned qualitative rule for designing a TLC system appropriate for a given sample, *i.e.*, selection of the developer and adjustment of the adsorbent activity.

Snyder elucidated S°, A_s and ϵ° empirically in terms of the molecular structures of the samples and solvents, and introduced a nomenclator, especially to ϵ°, which is "solvent-strength" parameter. It should be noted that an arrangement of developers according to their increasing "solvent-strengths" describes the eluotropic series, with much fewer exceptions than that according to their dielectric constants. Thus, he proved that Eq. (2) is valid for a variety of chromatographic systems involving low-molecular-weight compounds as the sample[20].

Kamiyama and Inagaki tested the validity of Eq. (2) for a number of vinyl polymers[15, 21]. Polymers tested were poly(butene-l), poly(α-methyl styrene), poly(p-chloro-styrene), polystyrene, polyvinyl chloride, poly(butyl methacrylate), poly-(ethyl methacrylate), poly(methyl methacrylate), poly(methyl acrylate), polyvinyl acetate and polyacrylonitrile. In this study it was assumed that the adsorption behavior for a given polymer be represented by that for its constituent monomeric unit. Under this assumption they calculated S° and A_s using an additivity rule proposed by Snyder[20]. It was found that a plot between S°/A_s and ϵ° for the sample polymers, except for the crystalline polymers, *i.e.*, polyvinyl chloride and poly-acrylonitrile, was approximated well by a straight line under such a condition that non-zero R_f values, *e.g.*, $0 < R_f < 0.2$, were just observed for each polymer. This implies that Eq. (2) holds largely for polymers. The different chromatographic behavior exhibited by the crystalline polymers may be attributed to a fact that it takes always fairly a long time to dissolve these polymers (cf. Section VI.1.). However, Eq. (2) will be valid for copolymers containing these monomers as comonomer.

II.3. Solubility-Controlled Phase-Separation Mechanism

As stated briefly in the foregoing section, TLC can be utilized to fractionate homopolymers according to the molecular weight. Properties required of developers used for this purpose are: The developer must behave as a poor solvent toward the sample polymer, and have a higher polarity than that of the sample polymer so that no preferential adsorption of the sample onto the stationary phase occurs. Exactly speaking, the first requirement can be fulfilled only if the developer is a θ solvent or solvent-mixture[22] for the sample, whose critical immiscible temperature is not greatly different from room temperature, *e.g.*, in a range from 15 to 80 °C. This is only the experimental requirement, however.

The above requirement implies that polymer fractionation by TLC is caused not by simple polymer precipitation but by thermodynamic phase-separation of sample polymer, which would take place during the developing process. The situation of such phase-separation will be argued below. For the sake of simplicity, it is assumed that a θ solvent for the sample polymer is used singly as developer, and TLC operation is made at the θ temperature for the sample. Polymer molecules involved in an applied spot will first face to penetration of the θ solvent when the development is started, and a phase equilibrium will be attained by which the polymer molecules are distributed in two phases, a sol- and a gel-phase, according to their molecular weights. Then, polymer molecules in the sol phase will be carried past by the developer, while those in the gel phase will be strongly retarded to migrate. In other words: The phase-separation retards sample migration at a different rate which is higher for a component with a greater molecular weight, thus giving a smaller R_f value for a higher-molecular-weight component. This is the simplest interpretation of the reason why TLC allows polymer fractionation by molecular weight.

The interpretation given above is, however, limited to that for an initial stage at which fractionation process begins, since TLC experiments are usually operated at a constant temperature. In the absence of, *e.g.*, a temperature gradient, no successive fractionation will further be achieved. Inagaki *et al.*[15, 23, 24] made a series of experiments to solve this problem and arrived at a conclusion that the phase-separation in TLC is related mainly to a change in the polymer concentration. On the basis of a binodial (cloud point) curve for a system of a sample polymer and a θ solvent used as developer, it is obvious that the phase separation is realized even at a constant temperature if the polymer concentration becomes higher by some reason. The above authors attributed the cause of this concentration effect largely to an experimental fact that the phase ratio decreases steeply with increase in the distance between the solvent front and starting point, so far as the ascending development technique[12] is employed. Here the phase ratio is defined as the weight of developer retained by unit weight of the stationary phase. Otocka and his colleagues[25] supported qualitatively the above conclusion but pointed out that an actual concentration of solution composed of the applied sample and the developer might not be high enough for the phase-separation.

II.4. Molecular Sieving Mechanism

Polymer fractionation due to this mechanism by TLC has a fairly long history. The purpose of applying this method has, however, been confined mainly to the relative determination of molecular weight of proteins, for which hydrophilic gels, such as Sephadex, are used as macroporous adsorbents[26, 27]. Halpaap and Klatyk are the pioneers who studied the fractionation possibility of hydrophobic polymers by TLC with macroporous silica gel[28]. Later, Belenkii and Gankina reported the appearance of molecular sieving effects in conventional TLC with macroporous silica gel[11]. More recently, Otocka and his colleagues[25] investigated this separation method and proposed a new name, thin-layer gel permeation chromatography (TLGPC). A similar work was also presented by Donkai and Inagaki[29]. On the other hand, a model calculation has been made by White and Kingry for chromatographic systems, in which superimposed effects of molecular sieving and adsorption are operative[30].

The principle of TLC separations by this mechanism does not differ from that of gel permeation chromatography (GPC). From the standpoint of instrumentation, however, TLC belongs to chromatography using open gel bed systems; while GPC to that using column systems. This difference specifies the operation condition under which the molecular sieving mechanism is effected in TLC, as will be mentioned below. Fractionation by GPC can be attained by eluting the sample with a solvent which can completely suppress the preferential adsorption of the sample onto the stationary phase. This requisite for the eluting solvent is the same for TLC[28, 29] but the stationary phase (thin layer) must be pre-eluted with an appropriate developer before the sample development, in order to obtain appreciable separations by the molecular sieving mechanism in TLC[25, 28, 29].

The necessity of pre-elution for TLC may suggest that the phase ratio (see Section II.3.) plays a decisive role in this type of separation; under a lower phase ratio, which is attainable by the usual ascending development, the polymer molecules in question might not be distributed completely into the inside and outside of macropores. In accordance with this consideration, Belenkii and Gankia pointed out that molecular sieving effects could be observed without the pre-elution if a horizontal development technique[12], which will make the phase ratio higher than an ascending one, was applied properly[11]. This observation has also been recognized by Otocka et al.[25].

From the experimental viewpoint it may be said that the resolution in the separation with respect to molecular weight, which is attainable by the molecular sieving mechanism in TLC, is not quite as high in comparison with that by the phase-separation mechanism. An attempt has been made to improve the resolution by increasing the thickness of thin layer from 0.25 mm to 1.0 mm[29]. An improvement of the resolution was really observed with increase in the thickness. Because of the technical difficulty, however, the thickness of thin layer was limited up to some 1.0 mm so that the best resolution was lower than that observed by GPC which was performed under analogous conditions to the TLC experiment.

An interesting version for improving the resolution in TLC was found by Belenkii and his colleagues[31]. They employed a macroporous silica gel as the stationary phase, and a binary solvent as the developer, to which a small amount of a third highly polar solvent was admixed. An appreciably large molecular-weight dependence of R_f was observed by conducting the development at a critical concentration of the third solvent, at which the molecular sieving effect just ceases to act while the adsorption effect appears incipiently. In this connection our two different observations should be pointed out. As already mentioned in Section II.2., the one was that the molecular-weight dependence of R_f in adsorption TLC was usually so small as to be neglected[16, 23]. In contrast to this observation, it was found that the R_f value depended distinctly on the molecular weight, even in adsorption TLC if a binary developer was used and the dielectric constants of its constituent solvents differed from one another by quite a small magnitude (cf. Fig. 14 in Section V.3.). The result of Belenkii et al.[31] might be related more closely to the latter observation rather than to some combined effects of molecular sieving and adsorption process. At any rate, these observations cannot be interpreted in terms of the Snyder theory i.e., Eq. (2).

III. Determination of Compositional Heterogeneities

III.1. General Experimental Directions

A most important and promising use of TLC in polymer chemistry is undoubtedly the determination of compositional distribution curves for statistical copolymers. The advantage of TLC in this respect rests in the fact that this method does not largely concern the polymer solubility which depends on the sample molecular weight. So far as constituent monomers of a given copolymer sample have an appreciably large difference in their polarities, TLC separation of the sample according to chemical composition is achievable by the adsorption mechanism without great experimental difficulties. This was true even for an azeotropic copolymer of styrene and methyl methacrylate, which has only a small heterogeneity in chemical composition[11, 33].

Developers to be employed for this purpose may be selected generally in accordance with the Snyder theory, *i.e.*, Eq. (2) given in Section II.2. However, the following precaution should be remembered on selecting the developer. A single solvent can be used effectively as developer for TLC separation of copolymer samples having greatly different composition, for it is a rule that the highest resolution with respect to chemical composition is attainable when the solvent is used singly as the developer (cf. Section VI.2.). An experimental result[16] is shown to illustrate the rule. Two statistical copolymer samples of styrene and methyl acrylate, Sm 76-1 and Sm 25, which contain 77.6 and 25.8 mol % of styrene, respectively, and have small compositional heterogeneities, were chromatographed using single solvents on activated silica gel thin-layer together with the parent homopolymers, PST and PMA. The result is summarized in Table 1. When chloroform was employed singly as developer, the final spots of PST and Sm 76-1 were in the proximity of solvent front and starting point, respectively. This means that complete separation is attainable for a fairly small composition difference such as ca. 20 mol % styrene.

The rule mentioned above is related to the so-called demixing effect[20, 24], which is explained as follows: When the developer is composed of, *e.g.*, two solvents with different polarity, the composition of the developer, hence, the solvent-strength, varies with proceeding of the development as a result of preferential adsorption of the more polar solvent component onto the adsorbent. Consequently, a concentra-

Table 1. TLC developments of styrene-methyl acrylate copolymers with single solvents

Developer	Dielec. const.	R_f			
		PST	Sm 76-1 $(77.6)^a$	Sm 25 $(25.8)^a$	PMA
CCl$_4$	2.22	0	0	0	0
Chloroform	4.62	>0.90	0	0	0
Ethyl acetate	6.02	>0.90	>0.90	>0.90	0

^{a} Styrene mole %.

tion gradient of developer is spontaneously formed during the development (cf. concentration-gradient development). In contrast to the situation with binary or multi-component developers, a single solvent keeps its polarity constant throughout the development process, and this may concern the development characteristics of single solvents.

When a given sample has a large compositional heterogeneity, a single solvent can no longer be used as developer for the heterogeneity determination. Thus, solvent mixtures are generally used. If an adopted solvent mixture is too sensitive to a small difference in chemical composition, a concentration-gradient development technique[12, 15, 16, 34] will often be useful, which continuously changes the composition of developer by adding another solvent during the course of the developing process. The purpose of using this technique is thus to depress resolutions that are too high with respect to composition and also to minimize unfavorable tailing-phenomena in chromatograms. This problem will be considered again in Section VI.2. The developer selection for nonpolar copolymers will be discussed later.

III.2. Polar Copolymers

This section is devoted to presenting a variety of TLC results on the compositional heterogeneity of "polar" copolymers which have been reported to date. For the sake of simplicity, sample copolymers to be discussed were classified into two categories, which are designated "polar" and "nonpolar" copolymers. The former refers to copolymers for which the polarities of comonomers are either relatively high or appreciably different from one another, while the latter means those composed of comonomers with lower polarities.

a) **Styrene-acrylate Copolymers**. An example for the category "polar copolymer" are statistical copolymers of styrene (ST) and methyl acrylate (MA), for which the compositional distribution was first determined with adsorption TLC by Inagaki et al.[16]. The study was begun with preparation of copolymers samples with different compositions at low conversions of monomer to polymer ($<$ ca. 10 wt.%). The copolymers having sufficiently small compositional heterogeneities were used as reference samples to decide the development condition appropriate for copolymer fractionation by chemical composition. After some preliminary tests, it was found that a concentration-gradient development on activated silica gel thin-layer of 0.25 mm thickness allowed fractionation according to the difference in chemical composition. First, chloroform and ethyl acetate were used as the initial and second solvents, respectively, in order to produce a concentration gradient during the development. A chromatogram obtained under the above condition showed that with increasing concentration of ethyl acetate, the composition of migrating components becomes increasingly rich in MA content. Quantitatively speaking, the reference samples with ST mole % higher than 50 to 100 (PST) were characterized by different R_f values from zero to unity. Under this development condition it was also affirmed that the R_f values were almost independent of molecular weight.

On the basis of the above result, the compositional distribution of a sample, SM 35H (34.2 ST mole %), which was prepared by polymerization up to a conver-

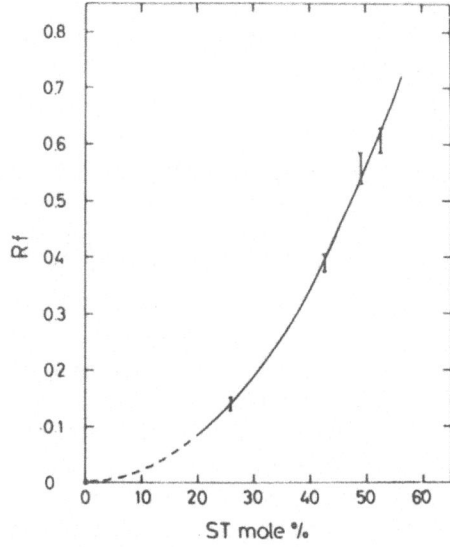

Fig. 1. Styrene-content dependence of R_f observed for statistical copolymers of styrene and methyl acrylate (reproduced from Ref.[16] by permission of the American Chemical Society)

sion of 97.5%, was determined under another condition of gradient development. To this end, a binary methyl acetate + carbon tetrachloride (1 : 5 by vol.), and methyl acetate were employed as the initial and second solvents, respectively, and the development was conducted in such a way that the rate of second solvent addition was increased with time. The sample SM 35H was chromatographed together with PMA and four reference samples whose composition ranged from 25.8 to 52.7 ST mole%. Figure 1 shows a relation between the ST-content and R_f established by the above-mentioned experiment, which served subsequently as a calibration curve in the compositional distribution determination. The chromatogram obtained for SM 35H indicated that this sample was developed not as the usual spot but as a band extending from the starting point, as shown in Figure 2. This smear (band) was analyzed to

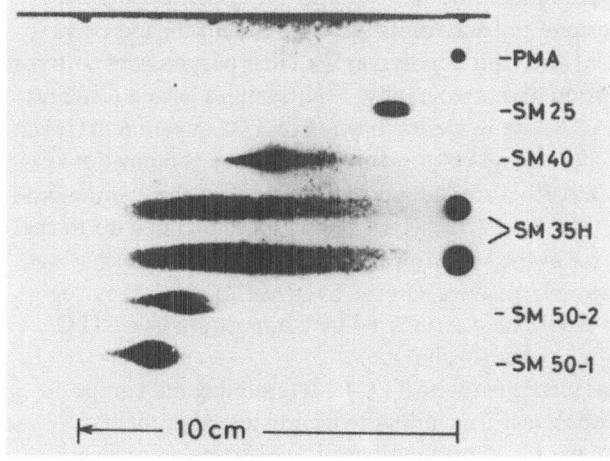

Fig. 2. TLC-smears obtained for a statistical ST-MA copolymer, SM 35H, which was prepared at a high conversion (97.5 wt. %). The upper and lower smear correspond to the sample size of 50 and 100 µg, respectively; each immobile spot for SM 35H appears because this sample contained homo-PMA (reproduced from Ref. [16] by permission of the American Chemical Society)

describe the compositional distribution by introducing some auxiliary relations deduced from other experiments. Thus it was finally shown that the compositional distribution curve deduced from TLC was in good agreement with that calculated on the basis of classical copolymerization kinetics.

b) **Styrene-methacrylate Copolymers.** Statistical copolymers of styrene and methyl methacrylate (MMA) belong to the category "polar copolymers". The development condition and chromatographic behavior of this copolymer in adsorption TLC were investigated extensively by Belenkii and Gankina[11]. They classified various solvents, applicable to the separation purpose, into two groups, "displacer" and "true solvent". The former involves solvents, such as diethyl ether, acetone, 2-butanone, and dioxane, etc., which contain oxygen capable of hydrogen-bond formation with silanol hydroxyl groups of silica gel, while the latter involves some chlorinated hydrocarbons, such as dichlorobenzene, dichloroethane and chloroform etc. It was shown that the separation by composition could be achieved using binaries, prepared by admixing a small volume of "displacer" to a large volume of "true solvent", without applying any concentration-gradient technique. The typical examples were binaries chloroform + acetone (12:2.4 by vol.) and chloroform + 2-butanone (12:0.6), which exhibited a high and low resolution with respect to chemical composition, respectively.

One of the other findings reported by Belenkii and Grankina was that the molecular-weight dependence of R_f appeared only at molecular weights below 1×10^5, so far as the adsorption mechanism in TLC was operative. This is in agreement with the finding obtained for ST-MA copolymers[16], and also with the data for ST-MMA azeotropic copolymers, which were observed using a binary chloroform + ethyl acetate as developer[15, 33]. They investigated the influence of silica-gel porosity upon the molecular-weight dependence of R_f. It was thus suggested that the use of microporous silica gels was suitable for TLC of copolymers having molecular weights lower than 10^5; while the use of silica gel with large pores for those having the higher molecular weights.

c) **Styrene-acrylonitrile Copolymers.** Another example for "polar copolymer" are statistical copolymers of styrene and acrylonitrile (AN). Teramachi and Esaki recently made a TLC experiment with this copolymer[36]. Their purpose was to transfer TLC data to column adsorption chromatography. Four samples having different AN-content ranging from 15 to 31.2 wt.% were subjected to TLC experiment. It was found that binaries of methyl acetate and chloroform with various compositions gave an R_f different from the samples with different AN-contents: when the volume ratio of the above solvents was 1:20 (MeOAc:CHCl$_3$), all the samples migrated up to the solvent front, but with a decrease in the MeOAc concentration, the samples having higher AN-contents indicated lower R_f values. On the basis of this result they suggested that ST-AN copolymers would be fractionated by composition using TLC without the aid of concentration-gradient technique.

More recently Wälchli et al. succeeded with TLC in determining the compositional distribution of ST-AN copolymers, including two commercial products[37]. The separation by composition was based on the adsorption-desorption process because of the great polarity difference in these comonomers. Concentration-gradient development was carried out, for which tetrachloroethane and ethyl acetate were employed as the initial and second solvents, respectively. However, it was noticed that this binary was applicable only to the copolymers with AN-contents below 60 mole% because of the insoluble nature of this copolymer in the binary. The compositional distribution for a sample (AS70-H) having a 28 mole% AN-content, which was prepared by radical polymerization in bulk at a high conversion of ca. 70 wt.%, was determined in the same manner as in a previous work[16], and compared with that calculated theoretically. A good agreement was found between theory and experiment. This determination procedure was further applied to commercial products, which had average AN-contents of 40 and 48 mole%. The compositional distribution thus deduced for both the products were rather narrow, although the products must have been prepared at a very high conversion. This finding is in agreement with those reported previously by Teramachi and his colleagues[36, 38].

In the study of Wälchli et al. other interesting observations were made. One was that an upper and an immobile spot, whose shapes were a band and a circular form, respectively, were seen on chromatograms for both of the commercial products. The other was that for both products, the average composition estimated by analysis of the upper spot gave an AN-content ca. 10% lower than that given by elementary analysis. From these observations it was speculated that the commercial products might have contained either a high AN-content species or AN-homopolymer, since these species can remain immobile under this devolopment condition.

d) Cellulose Derivatives. Apart from synthetic copolymers described above, cellulose derivatives are the typical example of "polar copolymers". Since chemical modifications of cellulose are usually made via topochemical reactions, the resultant derivatives must have chemical heterogeneities to some extent, unless modification reaction does lead to 100% substitution of hydroxyl-groups attached to cellulose. In view of the fact that commercial products of cellulose acetate generally have heterogeneities both in acetyl content and molecular weight, and their complete characterization is a quite hazardous task, Kamide et al. investigated the chromatographic behavior of this derivative in TLC[39]. Experiments were made for samples with nearly the same acetyl content but different degrees of polymerization (DP) as well as for those with nearly the same DP but different acetyl contents. The range of DP and acetyl content studied were 54 ~ 334 and 54 ~ 61%, respectively. Various binary and ternary solvent systems were examined to search for suitable developers.

One of their interesting observations is that a binary methylene chloride + butanol exhibited a great resolution with respect to acetyl content at ca. 15 vol.% butanol: The developer gave R_f values, 0.9 and 0.25, for samples with acetyl content of 60.5 and 54.1%, respectively. However, this developer ceased to exhibit such a high resolution when the butanol content reached 30 vol.%, and a reversion of acetyl-content dependence of R_f was found instead. The development characteristics were explained by them in a somewhat complicated manner, but this interpretation might be given generally in terms of a "demixing effect", which has once been mentioned in Section II.1. As a conclusion, Kamide et al. recommended binaries of methylene chloride and methanol, for which the methanol content must be less than $5 \sim 10$ vol.% for the purpose of separation by the acetyl content.

Cellulose nitrate has the same characteristics as cellulose acetate in the compositional heterogeneity. An attempt was made by Kamiyama and Inagaki to investigate the heterogeneity of commercially available cellulose nitrates[15]. A concentration-gradient development was performed using a binary acetone + ethylacetate (20 : 3 by vol.) and another binary chloroform + ethylacetate (1 : 2 by vol.) as the initial and second solvents, respectively. This development condition was of a special nature from the aspect that the polarity of the initial solvent was reduced with proceeding of the development so that the migration of components having higher nitrogen content was retarded (cf. Section VI. 2.). Chromatograms thus obtained for the commercial products indicated three spots with fairly long tails, which were, however, discretely separated from one another, differing from those found for synthetic copolymers having large compositional heterogeneities. This may imply that the nitration reaction occurred topochemically.

e) **Block Copolymers of Styrene and Methyl Methacrylate.** As discussed in the foregoing subsections, every statistical copolymer is accompanied more or less by compositional heterogeneity unless it is prepared at low conversions of monomer to polymer. In contrast to this understanging, block copolymers are apt to be regarded as homogeneous even in their compositional heterogeneity, for the reason that anionic polymerizations often applied to their preparation proceed preponderantly to yield species of narrow molecular-weight distributions. As suggested by Freyss et al.[40], the molecular-weight heterogeneity of a block copolymer prepared through anionic "living" polymer technique is always smaller than that of its most polydisperse precursor and often can be even smaller than either of the two precursors. This implies that the block copolymer cannot be homogeneous in composition as likely as in the molecular weight, and rather an appreciable heterogeneity in the composition is expected to the block copolymer.

From this standpoint, Kotaka et al. established a theory for the compositional and molecular-weight distribution of block copolymers[41, 42], and studied TLC separations of styrene (S)-methyl methacrylate (M) block copolymers by composition in order to verify their theory[43]. The experiment was performed with activated silica gel as the stationary phase under a concentration gradient formed using carbon tetrachloride and 2-butanone as the initial and second solvents, respectively. In this study it was found that the addition rate of the second solvent played an important role in enhancing the resolution with respect to the composition and also in the chromatographic distinguishment of di- and tri-block, namely SM- and MSM-copolymers. The latter problem will be discussed later in Section IV.2.

All the chromatograms thus obtained indicated that the separation occurred according solely to the compositional difference. A separate experiment supported that the R_f values for the samples found under the development conditions were independent of the molecular weight. The quantitation of chromatograms was made with a TLC-scanning spectrodensitometer. As described in the foregoing subsections, the chemical heterogeneity determination by TLC requires reference samples having different known compositions, in order to establish a calibration between in R_f and

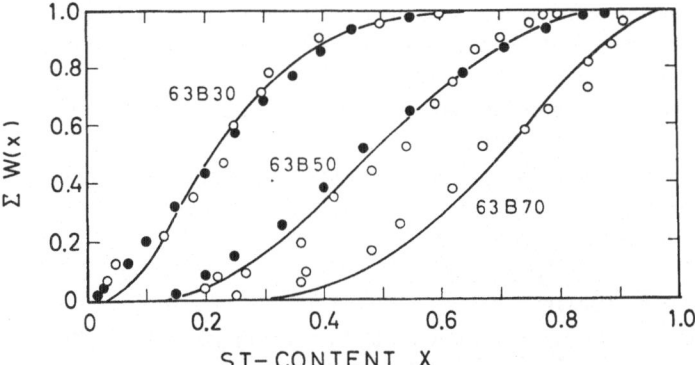

Fig. 3. Integral distribution curves obtained by TLC for ST-MMA block copolymers (reproduced from Ref. [43] by permission of the Hüthig & Wepf Verlag, Basel). For detail, see text

the composition. In this study, the investigators followed the above-mentioned method, designated as the "relative" method; but the accuracy of the determination was not satisfactory, and introduced a new method which utilized the fact that poly-styrene and polymethyl methacrylate exhibit different responses to 225 and 265 nm scanning. This method will be designated as the "absolute" method. The principle will be discussed later in Section VI.3. Figure 3 shows typical examples of the inte-gral distribution functions for the chemical composition, which were obtained for three SM-diblock samples coded as 63B30, 63B50, and 63B70, having average sty-rene mole fractions of 0.24, 0.49, and 0.69, respectively. In the figure, the circles are the TLC results, while the solid curves are the theoretical ones deduced by assuming the Schulz function of molecular-weight distribution for the precursor polymers. The compositional heterogeneities of these samples are fairly large, and this was consistent with information deduced from cross-fractionation and density-gradient ultracentrifugation for the samples[44].

f) **Copolyamides.** Last to be mentioned in this subsection is TLC application to copolyamides. The composition of copolyamides, *e.g.*, copolymers of ω-amino-caproic acid and λ-aminolauric acid, was determined with TLC by Mori and Takeuchi[45]. Development conditions to be applied to these highly polar copolymers were investigated extensively. The most important condition was that silica gel used as the stationary phase must have been deactivated completely and a closed chamber for the development was needed, which was furnished with two reservoirs for devel-oper, and another solvent, such as water or methanol, used for controlling sample migration.

III.3. Nonpolar Copolymer — Styrene-Butadiene Copolymers

A typical example belonging to this category are undoubtedly the statistical copoly-mers of ethylene and propylene. Because of their crystalline nature, however, no attempt has yet been made to apply TLC to this copolymer system. Thus, in this section, our main concern is with styrene-butadiene (SB) copolymers, which are quite important in industrial applications. Many efforts have been made toward TLC-characterization of their compositional heterogeneity[46-48].

The SB-copolymers may be classified, according to the copolymerization mechanism and technology, into several groups:

(i)   statistical copolymers obtained by emulsion polymerization technique through free radical mechanism;

(ii) random and tapered random copolymers;

(iii) tapered block and block copolymers; and

(iv) mixed radial teleblock copolymers[49].

The latter three are obtained by solution polymerization technique with alkyllithium initiator through the anionic mechanism. For these materials, the analysis of block sequences is also an interesting subject in the area of TLC application. However, because a somewhat different principle has to be applied to achieve separation by the difference in block sequences, this subject will be discussed in a subsequent section (cf. Section IV.2.).

The TLC-separation of SB-copolymers by composition presents a considerable challenge, since they are nonpolar, and moreover, the difference in polarity between

Table 2. R_f values found with single solvent developers[a]

Developer	Dielec. const.	PMMA[b]	PS	R_f-values				
				an-SBR-5[c] B:S = 52:48	an-SBR-6[c] B:S = 76:24	BR-3[d]	cis- BR-4[e]	PIB
MeOH[f]	32.60	0(N)[g]	0(N)	0(N)	0(N)	0(N)	0(N)	0(N)
Acetone	21.30	0.9–1.0	0(N)	0(N)	0(N)	0(N)	0(N)	0(N)
MEK	18.50	0.9–1.0	1.0	0(N)	0(N)	0(N)	0(N)	0(N)
EtOAc	6.02	1.0	1.0	0.7–0.9	0	0[i](N)	0[i](N)	0(N)
THF	7.42	1.0	1.0	1.0	1.0(?)	1.0(?)	1.0(?)	1.0(?)
CHCl₃	4.62	0	1.0	0.8–1.0	0.8–1.0	0.7–1.0	0.8–1.0	1.0
Toluene	2.38	0	1.0	1.0	1.0	1.0	1.0	1.0
Benzene	2.28	0	1.0	1.0	1.0	1.0	1.0	1.0
p-Xylene	2.27	0(θ)[h]	1.0	1.0	1.0	1.0	1.0	1.0
CCl₄	2.24	0(N)	0	0	0	0[i]	0[i]	1.0
Cyclohexane	2.02	0(N)	0(θ)	0	0	0	0	1.0
n-Heptane	1.90	0(N)	0(N)	0(N)	0	0	0	0.8–1.0

[a] Cited from Ref.[48] by permission of the American Chemical Society.

[b] Abbreviations: PMMA = polymethyl methacrylate; PS = polystyrene; PIB = polyiso-butylene.

[c] Commercial styrene-butadiene rubbers produced by anionic polymerization using alkyl-lithium.

[d] Commercial polybutadiene (37% cis and 55% trans).

[e] Commercial cis 1,4 polybutadiene (93% cis).

[f] For the abbreviations, refer to Table 8.

[g] (N) indicates that the polymer is insoluble.

[h] (θ) indicates theta-solvent.

[i] Upward tailing.

the comonomer units is much less than that in styreneacrylate copolymers. These two factors, *i.e.*, the low polarity and the small polarity difference, make it hazardous to adjust the polarity of developers used to achieve good separation, and this situation may be described well by the Snyder theory (cf. Section II.2.). A comparative study on this problem was made by Kotaka and White[48]. The results are summarized in Table 2 in a list of R_f values found using single solvents with different polarities for a few SB-copolymers and other homopolymers. The table implies that the same separation principles as those discussed in Chapter II are underlying, regardless of the polarity difference among the structural units of polymers: The two criteria, *i.e.*, the solubility and polarity of polymer-developer combination must be satisfied to achieve an efficient separation. Therefore, the only problem here would result from the small difference in the polarities of styrene and butadiene units, and, hence, adequate developers will be limited. However, this small difference turns out rather to be advantageous in performing the TLC analysis of compositional heterogeneity. The reason will briefly be discussed below.

As will be described in Section IV.1., the chromatographic behaviour of copolymers composed of a nonpolar and polar monomeric units, such as styrene and methacrylates, are strongly dependent on the monomeric arrangements of polymer chains (monomeric sequence), *e.g.*, triads, pentads, etc. Let us consider a block and a statistical copolymer of these comonomers having equimolar composition. It is observable that the block copolymer cannot be developed with a developer which allows migration of the statistical copolymer, whereas the block copolymer with another composition rich in styrene can be developed with this developer, giving by chance the same R_f as found for the statistical copolymer. This implies that the R_f observed for copolymer systems depends not only on the composition but also on the difference in monomeric sequences. This complicated situation often appears for copolymers composed of polar and nonpolar monomers. Thus, nobody can exclude such a case in which the heterogeneity in monomeric sequence and in composition cannot be distinguished chromatographically from one another. On the other hand, for SB-copolymers, the separation by compositional difference is possible with much less interference caused by the monomeric sequence and other sources. Rather the separation according to the block-sequence is much more problematic (cf. Section IV.2).

For the analysis of compositional heterogeneities of SB-copolymers, adoption of the adsorption mechanism in TLC was found again to be advisable. Two developer systems have been proposed to date: One is of binaries of cyclohexane + benzene reported by Tagata and Homma[47], and the other is those of carbon tetrachloride + chloroform by Kotaka and White[48]. The R_f values found with the former binary were larger with decrease in the styrene content, but a slight molecular-weight dependence of R_f was observed particularly for polystyrene- and styrene-rich copolymers. On the other hand, when the binary carbon tetrachloride + chloroform was applied for a concentration gradient development, in which the former and latter solvent were the initial and second solvents, respectively, the R_f values also increased with the styrene content from 0.18 for polystyrene to unity for polybutadiene, but were almost independent of molecular weight.

Generally speaking, when one attempts to separate a copolymer solely according to compositional difference, avoiding hazardous side effects such as those due to the difference in molecular weight, monomeric sequence, etc., it seems advisable to employ solvent pairs which have the polarity difference as much as possible (cf. Fig. 14 in Section V.3.). For this reason the binary carbon tetrachloride + chloroform was preferable over the binary cyclohexane + benzene. However, the margin of allowable composition range of the former binary was rather small: Specifically, a concentration gradient covering volume ratios from $50 : 0$ to $50 : 20$ ($CCl_4 : CHCl_3$) was enough to develop all the species from polybutadiene (large R_f) to polystyrene (small R_f). Therefore, adjusting of the concentration gradient pattern might be somewhat difficult.

It should be fair to say that the choice of binary cyclohexane + benzene by Tagata and Homma[47] was made because of some difference in motivation. They attempted to transfer TLC-data to a preparative column adsorption chromatography for the compositional heterogeneity analysis of commercial styrene-butadiene rubbers (SBR). First of all, these products are usually of low styrene content (10–30 wt.%), for which the molecular-weight dependence of R_f with the binary cyclohexane + benzene was trivial. Secondly, the fractions thus recovered were all independently characterized so that the molecular-weight dependence of R_f was no problem at all, but rather the ease in handling such solvents presented a much more preferable condition for performing the column-chromatographic analysis.

Tagata and Homma[47] analyzed in the aforementioned manner the compositional heterogeneity of two typical commercial SBR samples, E-SBR and S-SBR, which had 23.5 and 20.0 average styrene wt.%, and were produced by an emulsion polymerization and by a solution polymerization with an organometallic catalyst, respectively. The result was that the former gave a distinctly narrower compositional distribution than the latter. GPC experiments on these samples were also carried out, which indicated that the above situation was just the opposite for the molecular-weight distributions.

A qualitatively similar conclusion to the above was reported by Kotaka and White for the molecular-weight distributions of emulsion- and solution-polymerized SBR samples[48]. As to the compositional heterogeneity, the solution-polymerized samples gave broader distributions which were in agreement with the results of Tagata and Homma. However, the emulsion-polymerized sample invariably showed, on the adsorption TLC separations, broad TLC-smears, implying broad compositional heterogeneities. Particularly those polymerized at 5 °C (cold emulsion) indicated larger heterogeneities than those polymerized at 50 °C (hot emulsion). The result is in opposition to the conclusion that has been accepted so far. To solve this problem, a more thorough examination is necessary, including investigations of effects of branching, steric structure, etc. on TLC development characteristics.

When TLC developments of SB-copolymers and of other diene-containing polymers are attempted, one often encounters a crucial problem, in that a substantial amount of the polymer remains immobile at the starting point[47, 48, 50]. This could be a serious drawback in performing quantitative TLC analysis of diene-containing polymers. Tagata and Homma investigated this problem experimentally on the basis of a hypothesis that polymer species are adsorbed as they are piled up on the adsorbent surface to form a multi-layer structure, in which each layer is homogeneous in the molecular weight and composition[47]. A conclusion was drawn that the hypothesis was largely correct. This means that the composition of a portion uneluted (undeveloped) from a sample may be approximated by that of the whole sample if the recovery exceeds 70% or more.

III.4. Telechelic Prepolymers

One of telechelic prepolymers produced nowadays on an industrial scale is liquid rubber[51]. The main chain of liquid rubbers is usually composed of a nonpolar monomer, butadiene, and carries a highly polar (reactive) group at each chain end. The performance of liquid rubbers depends predominantly upon the functionality which is defined to be unity when each chain carries two polar end-groups. Average functionalities for commercially available liquid rubbers are considerably lower than unity, and this suggests that an appreciable fraction of the chains carry no end-group and/or only one end-group. Therefore, the functionality distribution is the most important factor for their characterization. Thus it is required for the quality control of liquid rubbers that a given product be separated into three components which have no functionality, the mono- and difunctionality.

To this end, several investigators have proposed different methods[52-55]. However, all of these methods were essentially based on column fractionation and/or GPC technique, although some additional devices were introduced[54, 55]. Recently Min *et al.* employed TLC and demonstrated a possibility for determining the functionality distribution of some commercial products of liquid rubbers[56]. Samples tested were 1,2-polybutadienes carrying either COOH- or OH-groups. At the outset of this study, they used conventional TLC with highly activated silica gel as adsorbent. The samples were developed using p-xylene together with 1,2-polybutadienes having no functional end-group. The chromatograms indicated that the samples were all separated into two final spots located on the solvent front and the starting point, whereas the 1,2-polybutadiene samples migrated up to the solvent front. Thus, the final spot appearing on the solvent front was assigned chromatographically to the nonfunctional component contained in the sample, while the immobile spot was assigned to a mixture of the mono- and difunctional species.

Further experiments were carried out to separate the immobile spot into two final spots which should correspond to the mono- and difunctional species. It was found that a binary carbon tetrachloride + tetrahydrofuran (100:1 by vol.) met this separation purpose. A typical example of the chromatograms is shown in Figure 4. In the figure, the upper spot is undoubtedly identified with the nonfunctional species, while the intermediate and immobile spots are, only in an intuitive manner, regarded as the mono- and difunctional species, respectively. However, the identification was proved afterwards to be correct, as described below.

For the quantitation of chromatograms, the development condition found by the conventional TLC was transferred to another TLC system which uses a thin quartz rod coated with silica gel of 75 μ thickness and may be called "thin-layer-FID chromatography". The TLC system is equipped with a flame ionization detector (FID) and commercially available as a complete set (Iatron thinchrograph model TFG-10, Iatron Co., Ltd., Tokyo). The principle of sample scanning and device for FID are almost the same as those worked out by Padley[57], Szakasits *et al.*, and

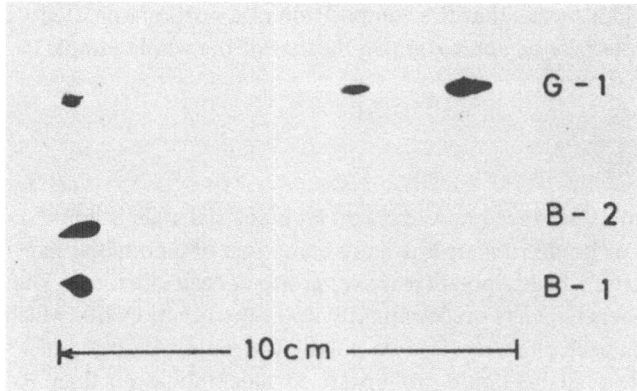

Fig. 4. Chromatograms observed for an α, ω-1,2-polybutadiene diol sample (G-1) and 1,2-poly-
butadiene samples (B-1, B-2) using a developer system CCl$_4$ + THF (100 : 1 by vol.)

others[58]. Transfer of the conventional TLC-data to the other method was successful. The chromatograms obtained with the binary carbon tetrachloride + tetrahydrofuran were scanned with the FID. Three peaks appeared, which allowed determination of each amount of the three component species. Thus, Min *et al.* confirmed that
the intermediate and immobile spots on the chromatogram were identified with the
mono- and difunctional species, respectively, by comparing the above quantitation
data with the average functionality.

In summary: When sample polymers are of low polarities and carry polar end-
groups and/or a small number of polar side-groups, TLC separation by the difference
in the number of these groups is possible by the polarity-controlled adsorption
mechanism. Since such a sample has, as a whole, only a slight affinity to adsorbent,
the development should be performed by using a "highly active" adsorbent and a
developer having a low "solvent strength". Needless to say, this condition has already
been pointed out in Section II.2. Some other separation problems presented by the
difference in end-groups will be described in Section IV.4.

IV. Characterization of Polymers with Differing Chain Architectures

IV.1. Chromatographic Distinguishment of Chain Architecture

This section deals with the effect of monomeric sequences in copolymer chains
upon TLC separation. A possibility of separating copolymers by the difference in
their chain architectures was first demonstrated by Kamiyama *et al.*[59]. For the preliminary TLC experiment they used copolymers composed of styrene and methyl
methacrylate, for the reason that this comonomer pair is endowed with the possibility of being polymerized to three different chain architectures, namely, alternating[60],
statistical, and block.

The copolymers tested preliminarily were of the statistical, azeotropic, and block type; and the separation experiment was carried out by a concentration-gradient development, for which chloroform and ethyl acetate were used as the initial and second solvents, respectively. It was found that the block copolymers could hardly be developed under the condition which allowed the statistical copolymers to migrate, despite the fact that the block copolymers had higher styrene contents, hence, lower affinities to adsorbent, than the statistical copolymers.

The above finding is quite important from the following aspect. As is well known, isolation of block or graft copolymer species from their reaction product using precipitation fractionation is often hampered by the effect that homopolymers are occluded by block or graft copolymer in solution[61]. Therefore, it has been regarded as a difficult task to extract true block or graft components, and to know the amount of occluded homopolymers. Belenkii and Gankina[11] pointed out that separation of true block or graft copolymer and attendant homopolymers was possible by the use of a two-dimensional development technique in TLC[12]. Quantitative purity tests of different graft copolymers, such as polyvinyl acetate (PVAc)-g-polystyrene, (PVAc)-g-polymethyl methacrylate, and cellulose triacetate-g-polystyrene, were made with the conventional TLC technique by Horii et al.[62]. It was confirmed that TLC allowed detection of occluded homopolymers of such small amounts as 0.5 ~ 1.0%. A similar result was reported by Inagaki et al. for a diblock copolymer of styrene and methyl methacrylate[42].

Further experiments to distinguish chromatographically the statistical copolymer with equimolar composition from the alternating copolymer were performed by Kamiyama et al.[59]. The development condition thus established was to adjust the concentration gradient lower than in the aforementioned case. To this end, a binary chloroform + ethylacetate was used as the initial solvent, in place of chloroform. Such a device is often useful to give different but still intermediate R_f values for samples having similar polarities. However, this method is applicable only to polar copolymers. The chromatographic behavior of the statistical, alternating, and block copolymers were compared with each other at equimolar composition in the foregoing manner. The result indicated that the R_f value for the alternating sample was distinctly smaller than that for the statistical sample (see Table 3). The same conclusion as above was also drawn for statistical and alternating copolymers of styrene and methyl acrylate[59].

Apparently the above results indicate that the chromatographic behavior of copolymers is different for different chain architecture. When the R_f values are compared at equimolar composition, the sample migration occurs first for the statistical copolymer and belatedly for the alternating, whereas the block copolymer remains immobile on the starting point. Such observations allow us to conclude that the mechanism of chromatographic separation of copolymers is related not only to the chemical composition but also to the chain architecture.

The above conclusion may imply that adsorbed macromolecules at liquid-solid interface take a looped chain conformation in which only a small fraction of their monomeric units are attached to adsorbent surface, as suggested by Rowland et al.[63]. Kamiyama et al. explained the difference in chromatographic behavior by chain architecture, assuming that the arrangement of monomeric sequences attached to

Table 3. Influence of monomeric arrangements of styrene-methyl methacrylate Copolymers upon R_f values[c]

Sample code	Structure	$[\eta]$ (dl/g)[a]	ST-mole %	R_f
SM7-1		3.40	71.2	0.90
SM5-10	Statistical	0.87_7	56.1	0.52
SMA2-1[b]		1.54_2	49.1	0.38
AL-1-0	Alternating	0.95_7	49.5	0.25
15B	Block	0.91_5	49.6	0
SM3-0	Statistical	1.35_9	29.6	0

[a] Values in toluene at 30 °C.

[b] Azeotropic copolymer.

[c] Reproduced from Ref. [59] by permission of Hüthig & Wepf Verlag, Basel.

adsorbent surface be responsible for the adsorption-desorption process[59]. In other words: The difference in R_f values by chain architecture was interpreted in terms of the probability for finding three dyads in a copolymer chain, *i.e.,* S-S, S-M and M-M, where S and M denote styrene and methacrylate unit, respectively. However, this theory failed in interpretation of the chromatographic behavior of block copolymers with different sequence number, as will be described in the next section, and its validity is still open for discussion.

A similar separation capability of TLC was affirmed by Donkai *et al.* for styrene-butadiene copolymers[64]. Compositions of the sample copolymers were not exactly the same and within a range of styrene contents from 28 to 33 wt.%. The TLC experiment was performed on highly activated silica gel thin-layer using a binary cyclohexane + chloroform (140 : 75 by vol.) without forming any concentration gradient in the development. It was found that the statistical copolymer gave the highest R_f value, while the tapered and block copolymers gave an intermediate and the lowest value, respectively. Further experimentation was done to investigate the compositional dependence of R_f for the statistical copolymers; and the trend in the R_f values was justified to be independent of factors other than of the chain architecture.

Before ending this subsection, another TLC application to distinguish chain architecture is to be mentioned, which was reported by Saegusa *et al.*[65]. The work concerned copolymerization products of ethylene and vinyl acetate obtained with a ternary catalyst system $ALEt_3–ZnCl_2–CCl_4$ under different conditions. As these products were oligomers having molecular weights lower than 10^3, TLC could be applied to determine the chain architecture so that the copolymerization mechanism was elucidated.

IV.2. Block Copolymers

The success in separation of copolymers by chain architecture poses the question whether or not TLC is capable of distinguishing block copolymers of a common chemical composition from one another by the difference in the number of block

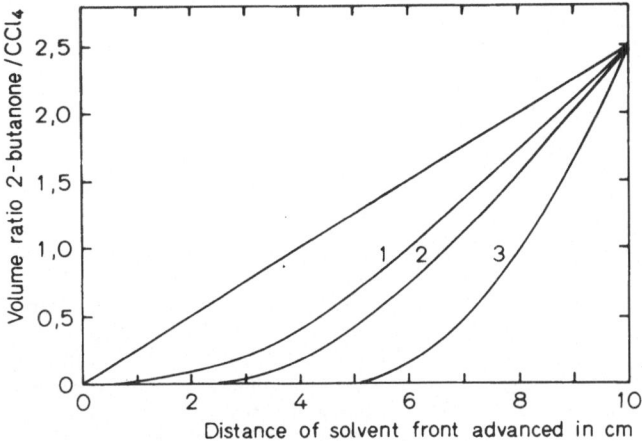

Fig. 5. Patterns (1), (2), and (3) of the second solvent addition in concentration-gradient developments for ST-MMA block and statistical copolymers (reproduced from Ref. [43]) by permission of the Hüthig & Wepf Verlag, Basel)

sequences. During the course of a study on the compositional heterogeneity of block copolymers of styrene and methyl methacrylate (SM), Kotaka *et al.* found that the triblock MSM-copolymer gave a higher R_f value than the diblock SM-copolymer on every compositional level[43]. This result was obtained on activated silica gel thin-layer by a concentration-gradient development, for which carbon tetrachloride and 2-butanone were employed as the initial and second solvents, respectively. The difference between the R_f values found on every composition level for MSM- and SM-block copolymer depended on the addition rate of the second solvent. Figure 5 illustrates patterns of the second solvent addition. Corresponding to each pattern, the chromatographic behavior of the block and statistical copolymer samples varied as shown in Figures 6a–6c. This result suggests that a good choice of developer does not present the sufficient condition for a good separation. The rate of the second solvent addition, hence, the form of concentration gradient, must be adjusted appropriately in an empirical manner.

Another result on this subject for styrene-butadiene (SB) block copolymers was reported by Donkai *et al.*[64]. Three reference samples of a tapered, di- and triblock (SBS) type were first prepared by an anionic polymerization technique proposed by Morton *et al.*[66]. The samples were subjected to concentration-gradient development in TLC with activated silica gel as adsorbent. A binary cyclohexane + chloroform (9:1 by vol.) and chloroform were used as the initial and second solvents, respectively. The chromatogram thus obtained proved that the SBS-block sample gave the lowest R_f value, while the SB-block and tapered sample gave the next higher and the highest R_f value, respectively. The observation that the SB-block sample migrated faster than the SBS-block sample is just the opposite to what occurred in the case of SM- and MSM-block copolymers. The chromatographic behavior for some commercial SB-block products was studied in a similar way, and the same trend in the R_f values as for the reference samples was confirmed.

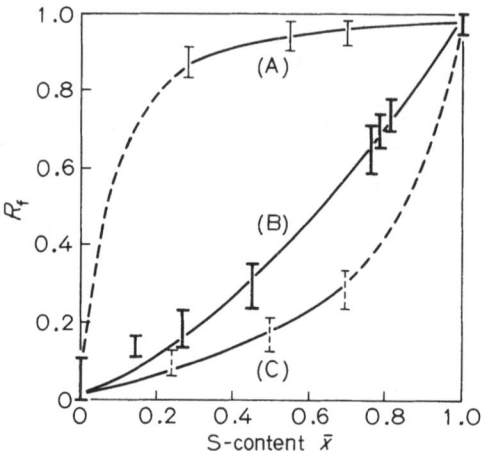

Fig. 6a. Influences of composition (ST-content) and chain architecture of ST-MMA copolymers upon R_f, observed under Pattern (1) shown in Fig. 5. Curves (A), (B) and (C) refer to statistical, tri- and diblock samples, respectively (Reproduced from Ref. [43]) by permission of the Hüthig & Wepf Verlag, Basel)

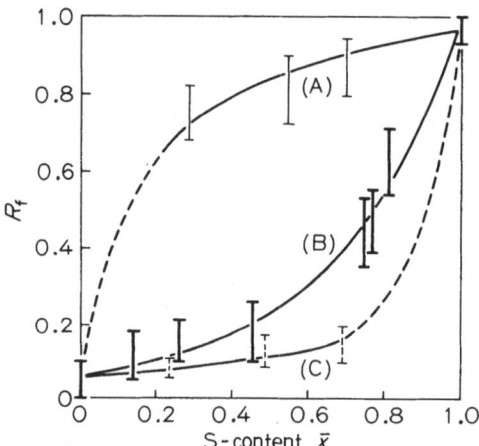

Fig. 6b. The Influences observed under Pattern (2)

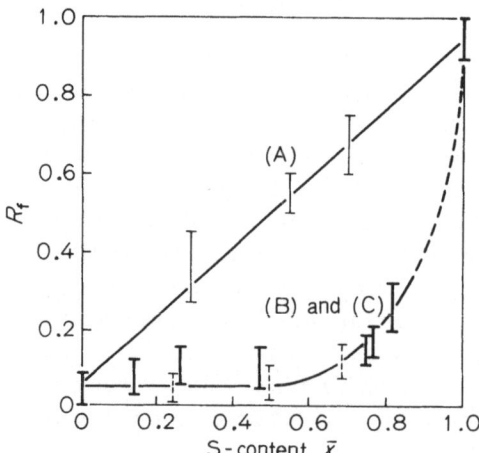

Fig. 6c. The Influences observed under Pattern (3)

No mechanism to explain the above findings has been proposed yet. However, we may point out one factor which might concern this mechanism. The factor is the great polarity difference between these copolymer systems. Before the discussion, we refer to an experimental fact reported by Thies for the adsorption behavior of mixtures of polystyrene (PS) and polymethyl methacrylate (PMMA) at silica-liquid interfaces[67]. His observation was that PS molecules, which had been adsorbed on silica gel in a nonpolar solvent, were rapidly and completely displaced by those of PMMA which were admixed afterward to the system. This implies that the PS-block in SM- as well as MSM-block copolymers do not play any role in the adsorption process. Consequently, the whole adsorptive affinity of a MSM-block copolymer would be reduced to a great extent, if either one of its PMMA-blocks is desorbed by some reason. In addition, the probability of occurrence of such desorption would be higher for the triblock than for the diblock copolymer. These might explain the difference in chromatographic behavior of tri- and di-block SM-copolymers.

In the case of SB-block copolymers, such a drastic displacement of the adsorption sites as for SM-block copolymers is quite inconceivable because of the small difference in adsorptive affinity between the PS- and PB-block, and the PB-block could exert still an additional contribution to the adsorption. Under this circumstance, it seems reasonable that the tri-block copolymer is retained on silica gel even when the di-block copolymer is desorbed. This trend might be pronounced for SB-block copolymers having lower styrene contents, which were just the case studied by Donkai *et al.* Further experiments on this problem are necessary in order to justify the above speculation.

An inherent difficulty in TLC application to the study of block copolymers should be mentioned. Kamiyama *et al.*[68] investigated the chromatographic behavior of MSM-block copolymers having equimolar composition but different molecular weight, and found that the copolymers showed an appreciably large molecular-weight dependence of R_f, even under the condition in which the adsorption mechanism in TLC was operative. The observation was argued on the basis of changes in the chain conformation caused by intramolecular phase separation between the block portions of PS and PMMA, which had been pointed out by studies on the hydrodynamic property of block copolymers[69]. Thus, such possible conformational anomalies of block copolymers should be taken into consideration when one selects the developer to be applied to TLC separation by difference in the number of block sequence.

IV.3. Homopolymers with Different Chain Microstructures

For the adsorbed state of macromolecules it has been speculated that the polymer-adsorbent interactions would be concerned not only with the overall chemical constitution but also the monomer arrangement along the chain, as described in Section IV.1. This suggests that some homopolymers may be distinguished with TLC from one another by a difference in chain microstructure, such as steric and geometrical isomerism, and stereoregularity. This section deals with this possibility, divided into

two parts, namely, separation of PMMA by difference in the steric isomerism, and
that of polybutadiene and polyisoprene by difference in the geometrical isomerism.

a) **Polymethyl Methacrylate.** As is well known, methyl methacrylate (MMA)
monomer can be polymerized to polymers with differing steric isomerism for different reactions. A first attempt to separate a blend of isotactic and atactic PMMA
into each component with TLC was made by Inagaki et al.[70]. Both PMMA were
migrated with acetone up to the solvent front but separated with ethyl acetate into
an upper and an immobile spot containing the isotactic and atactic (and syndiotactic)
PMMA, respectively. The difference in chromatographic behavior between these
stereoisomerically different polymers has not yet been elucidated clearly. As a speculation, however, it may be pointed out that the difference might be related to that
in chain conformations which these polymers assume in solution. The speculation
will be justified if the chain conformation assumed by a tactic polymer in its crystalline state is still retained in solution[71]. In this connection it should be mentioned
that Tadokoro et al. recently proposed a double-strand helix model for isotactic
PMMA[72].

The aforementioned result was applied immediately afterward by Miyamoto and
Inagaki[73] to a study on the structure of so-called stereoblock-PMMA which had been
prepared with some catalysts of Grignard's type by Fox et al.[74]. Two PMMA samples were prepared in toluene at −50 °C with n-butylmagnesium chloride and
diphenylmagnesium, and the acetone-insoluble portion of each sample was subjected
to TLC experiment using ethyl acetate as the developer. These samples were separated into two final spots which were located at the solvent front and starting point.
High-resolution NMR spectroscopic examinations of polymer species thus separated
indicated that the upper and immobile spot contained a syndiotactic-rich and an
isotactic species, respectively. Consequently, it was concluded that the acetone-
insoluble portions of the so-called stereoblock PMMA were of stereocomplex-type,
which had already been known to be formed between isotactic and syndiotactic
PMMA in solution when polar solvents like acetone were used[75]. The conclusion
was in agreement with that already drawn by Liquori et al. with another separation
technique[76].

In view of the fact that the stereoblock PMMA was merely the stereocomplex,
Miyamoto and Inagaki investigated further the mechanism of stereospecific polymerization of MMA using Grignard's type catalysts[77]. Reaction products were
recovered at different time intervals during the course of polymerization and
examined by TLC. The results suggested that the stereospecific polymerization was
closely related to the stereocomplex formation in such a way that an incipient
generation of the isotactic chains promotes preferentially the polymerization of MMA
to the syndiotactic. Further study on this problem was made by Miyamoto et al.
again using TLC[78].

On the other hand, Matsuzaki et al. studied TLC behavior of a PMMA prepared
with a phenylmagnesium bromide-tetrahyrofuran complex in toluene[79]. Good
separation into two components was achieved using either ethyl acetate singly or
a binary ethyl acetate + dioxane. However, these components could not be identified with either isotactic or syndiotactic PMMA, and the separation took place rather

largely according to the difference in heterotactic triads and molecular weight. Another study on MMA polymerization using TLC was reported recently by Buter et al.[80]. They dealt with radical polymerization products of MMA in the presence of isotactic PMMA. The polymerization product was first separated into an acetone-soluble and -insoluble portion, and the TLC behavior of the acetone-soluble portions was investigated with ethyl acetate as developer. All the samples were developed as single final spots which showed intermediate R_f values, in sharp contrast to the previous observation for the stereocomplex. For this reason, these authors presumed the samples to be of some stereoblock type.

The mechanism of stereocomplex formation was studied by the use of TLC[73, 77]. For this purpose, stock solutions were prepared by dissolving either an isotactic or a syndiotactic PMMA in chloroform (nonpolar solvent), and the stock solutions were admixed to obtain test solutions for TLC, which contained the two stereoisomeric polymers in three different ratios (isotactic/syndiotactic), i.e., 2:1, 1:1, and 1:2 (upon the admixing no visible change in the transparency of the resultant solution was observed). Acetone was used as the developer, which had been known to allow migration of both isotactic and syndiotactic PMMA to the proximity of the solvent front[70]. The chromatogram indicated that the 1:1 blend remained immobile on the starting point, suggesting formation of a 1:1 stereocomplex, whereas the other blends were separated into an upper and an immobile spot. The species involved in the immobile spot for the 2:1 mixture was assigned by NMR to the same stereoisomeric composition as the 1:1 stereocomplex. This means that the upper spot contained the isotactic species. Further NMR examination of the chromatogram obtained for the 1:2 blend proved that the blend was separated into a syndiotactic component and a 1:1 stereocomplex. However, it was pointed out by these authors[77] that the 1:1 stereocomplex was not the only possible association form if one took the other experimental results into consideration. A conclusion thus drawn by these authors was that the 1:1 stereocomplex was the primary structure, while the 1:2 was the possible secondary one. The conclusion was, however, different from that drawn by Liquori et al. through X-ray structural analysis [91].

In connection with the aforementioned study on polymerization mechanism of MMA[77, 78], Miyamoto et al. developed a preparatory method of separating blends of isotactic and syndiotactic PMMA[82]. The principle was based on a competitive adsorption of these different stereoisomeric polymers from a nonpolar solution (chloroform) onto an adsorbent surface (silica gel). The procedure was quite simple, as described below: A given polymer blend was dissolved in chloroform, in which no stereocomplex formation usually occurs, and silica gel was then dispersed in this solution for adsorptive equilibration with the polymer species. The isotactic species could be isolated as the adsorbed component. In practice, its purity was ca. 80–90%, which depended on the added amount of silica gel. By repeating the same procedure, the purity could be enhanced.

Another problem is TLC separation of atactic and syndiotactic PMMA. More exactly speaking, this concerns a question whether or not PMMA can be separated with TLC by the difference in heterotactic sequences. Inagaki and Kamiyama were engaged in this separation[32]. The TLC experiments were carried out mainly under the condition that the adsorption mechanism was operative. However, no developer system could be found for which the R_f value was dependent solely on the degree of syndiotacticity without interference of molecular weight. The difficulty in selecting the developer has been discussed later by these authors from a general standpoint of TLC (cf. Fig. 14 in Section V.3. and Refs.[15, 21]).

   b) **Polybutadiene.** This polymer has three different chain microstructures, namely, *cis*-1,4, *trans*-1,4 and 1,2-vinyl, according to the catalyst and other polymerization conditions applied. In view that the polar polymers, PMMA, had been distinguished by TLC from one another by the difference in steric isomerism, Donkai *et al.* investigated whether or not the above possibility could again be realized for less polar polymers such as polybutadiene[83]. After various tests for selecting developers, they found the following development characteristics: With carbon tetrachloride, *trans*-1,4 and 1,2-vinyl polybutadiene both migrated up to the proximity of the solvent front, whereas the *cis*-1,4 polymer remained on the starting point; on the other hand, amyl chloride allowed migration of two sample polymers other than the *trans*-1,4 polymer up to the proximity of the solvent front. The results are summarized in Table 4.

Table 4. Results on TLC developments of polybutadiene samples[c]

Polymer sample	R_f[a]	
	CCl$_4$	Amyl chloride
cis-1,4	0[b]	1
trans-1,4	0.8 ~ 0.9	0
1,2-vinyl	1	1

   [a] Highly activated silica gel thin layers were used.
   [b] Slightly tailing upward from the starting point.
   [c] Partially cited from Ref. [50] by permission of the Hüthig & Wepf Verlag, Basel.

   On the basis of this finding, a "two-step" (stepwise) development technique[12] was applied to separations of possible three binary mixtures of polybutadienes with different chain microstructure, namely, those of *cis*-1,4 + *trans*-1,4, *trans*-1,4 + 1,2-1,2-vinyl, and 1,2-vinyl + *cis*-1,4. The principle consisted of a utilization of the different development characteristics exhibited by carbon tetrachloride and amyl chloride. An example of this procedure applied by these authors[83] will be described below. A mixture *cis*-1,4 + *trans*-1,4 was developed primarily with amyl chloride until the solvent front reached a distance, *e.g.*, 10 cm from the starting point; by this development only the *cis*-1,4 polymer should have migrated up to the solvent front (cf. Table 4). In order to identify the immobile component, the chromatogram was dried in vacuum at room temperature and treated with carbon tetrachloride until the solvent front reached an intermediate distance, *e.g.*, 5 cm. It is obvious that this procedure can be alternatively used for identification of any unknown binary mixture. However, it was observed that a small amount of the sample polymer often remained immobile after the development, and this made identification difficult.

The separation possibility established so far will offer an important application of TLC to the structural characterization of so-called "equibinary" polybutadienes, reported first by Dawans and Teyssie[84]. For characterization of this polymer it is primarily necessary to ensure that the polymer is not any mixture of those having different microstructures. TLC experiments were performed by Donkai et al.[83] for (cis-1,4-1,2-vinyl)-equibinary samples obtained by Furukawa et al.[85]. The result verified that during the propagation reaction of the single polymer chain, cis-1,4 and 1,2-vinyl units were spontaneously enchained in some sequential way so that an equibinary structure was generated. However, the question, whether the microstructure was alternating or random, could not be answered by this experiment.

Table 5. Characteristics of polyisoprene samples

Sample code	$10^{-4}M^a$	Microstructure mole%[b]			
		1,4-*trans*	1,4-*cis*	1,2-vinyl	3,4-vinyl
Natural rubber					
F-5	52	0	97.4	0.2	2.4
F-6	47	0	97.4	0.2	2.4
F-7	25	0	97.4	0.2	2.4
Gutta percha (1,4-*trans*)	20.3	82.0	17.5	0	0.5
3,4-vinyl 1,4-*cis*-Syn.[c]	21	0	5.0	0	95.0
C-K	83	0	98.1	0	1.9
C-N	27	0	98.1	0	2.0
C-C	60	0	92.0	0	8.0

[a] Molecular weights determined by GPC.
[b] Estimated by IR-spectrometry (Ref. [88]).
[c] Commercial 1,4-*cis* polyisoprene products.

The separation mechanisms, which were operative in the above experiments, will be discussed below. Carbon tetrachloride is a good solvent for all types of polybutadienes, and the separation with this solvent should have proceeded according to the adsorption mechanism. On the other hand, cis-1,4 and 1,2-vinyl polybutadiene were soluble in amyl chloride even below −5 °C but trans-1,4 polymer was insoluble in this solvent below ca. 40 °C. This suggests that the separation with this solvent would have proceeded according to the solubility-controlled mechanism.

c) **Polyisoprene.** The study of organometallic compounds as polymerization catalyst made it possible to produce isoprene polymers with high cis-1,4 contents in industrial scale. However, various differences in physical properties between the synthetic and natural rubber have been exposed, which could not be interpreted in terms of the difference in their cis-1,4 contents. In light of this, Miyamoto et al.[86]

Table 6. R_f values observed for polyisoprenes using single solvent systems

Developer	Dielec. const.	$\epsilon^{\circ a}$	R_f		
			1,4-*trans*	1,4-*cis*	3,4-vinyl
Cyclohexane	2.0_2	0.04	0	0	0
CCl_4	2.2_4	0.18	$0 \sim 0.1$	$0 \sim 0.1$	1
Amyl chloride	6.6	0.26	1	1	1
p-Xylene	2.2_7	0.26	1	1	1
Benzene	2.2_8	0.32	1	1	1

[a] Solvent-strength parameters given by Snyder for alumina-adsorbent (cf. Section II.2. and Ref. [20]).

attempted to solve this problem by referring to the TLC results obtained for poly-butadiene[84]. Fractions of natural rubber with different molecular weights, gutta-percha and a (3,4-vinyl) polyisoprene prepared according to the procedure of Natta et al.[87], were used as reference samples. The molecular characteristics are listed in Table 5. In this table it should be mentioned that the *cis*-1,4 contents of natural rubber fractions were fairly lower than those to be expected. This may be attributed to a fairly low accuracy of the infrared-spectroscopic method for determining the *cis*-1,4 content, proposed by Binder and Ransaw[88].

TLC behavior of these samples observed with various single solvents on highly activated silica gel thin-layer are summarized in Table 6. From the table it is seen that carbon tetrachloride can distinguish chromatographically the 3,4-vinyl sample from either the *cis*-1,4 or the *trans*-1,4 sample. For separating the *cis*-1,4 sample from the *trans*-1,4 and 3,4-vinyl sample, binaries of (carbon tetrachloride + p-xylene) and (cyclohexane + p-xylene) were found to be effective. When the composition of these binaries was adequately adjusted, the *cis*-1,4 sample remained on the starting

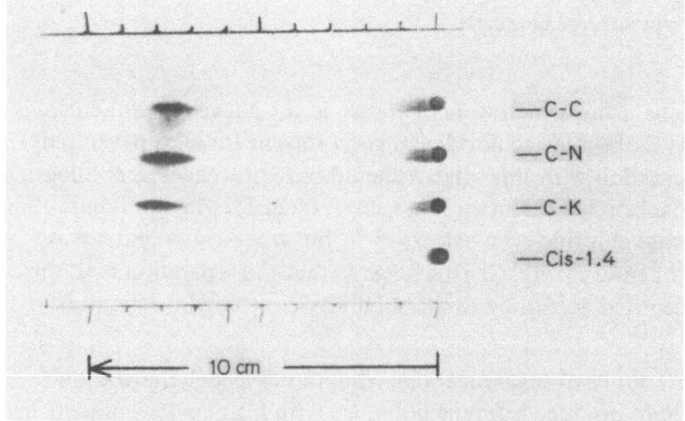

Fig. 7. Chromatograms obtained for commercial products of *cis*-1,4-polyisoprene, coded C-C, C-N and C-K, and natural rubber, coded cis-1,4. For details, see text

point, while the other samples migrated upward. This preliminary result was applied to some commercial products of (*cis*-1,4) polyisoprene (see Table 5). The developer used was a binary cyclohexane + p-xylene (20 : 80 by vol.). One of the natural rubber fractions was developed together with these products for comparison. The resultant chromatogram is shown in Figure 7, from which one sees a small amount of a component species migrated for every commercial sample. Although these authors did not further investigate the characteristics of species thus separated, the result indicates clearly some difference in microstructural features between the synthetic and natural rubber.

IV.4. Graft Copolymers

As pointed out in Section IV.1., TLC has been effectively applied to test whether or not a given graft product contained the attendant homopolymer and ungrafted mother polymer even after subjected to extraction purification[62]. A more important application of TLC to characterization of graft products was reported by Taga and Inagaki[89]. The samples were cellulose-styrene graft products prepared by a heterogeneous mutual irradiation technique with gamma rays. After the cellulose grafts had been subjected to extraction to remove the attendant homopolystyrene as much as possible, the cellulose backbone was degraded by acid hydrolysis to isolate polystyrene residue. Subsequently the residue was separated by TLC using tetrahydrofuran as developer into an upper and an immobile component, which were assigned by infrared spectroscopy to a free polystyrene fraction occluded in the cellulose matrix and a grafted polystyrene fraction that had some cellulose fragment at one of the chain ends, respectively. The separation took place according to the adsorption mechanism, and was achieved obviously owing to the end-group of polymer chains. Further endorsement for this separation possibility was given recently by Min *et al.* using well-defined polystyrene samples carrying carboxyl groups at the chain ends[90].

Recently Min *et al.* succeeded in complete TLC separation of cellulose-styrene graft products into three components, *i.e.,* a true graft copolymer, ungrafted mother polymer, and attendant homopolymer[42, 91]. The products were prepared by irradiation with gamma rays in a manner similar to that of Taga and Inagaki[39], and the cellulose backbone was subsequently modified to the triacetate derivative in order to bring the whole product into solution. The cellulose triacetate-styrene graft product was subjected to a stepwise development in which the primary development was made with a binary chloroform + dioxane (3 : 7 by vol.) up to 5 cm above the starting point, followed by a secondary development with chloroform up to 10 cm. By this procedure a chromatogram was obtained, which consists of three final spots at 0, 5, and 10 cm above the starting point, corresponding to the ungrafted mother polymer, true copolymer, and attendant homopolymer component, respectively. The above result obtained with conventional TLC was transferred to "thin-layer-FID" chromatography (cf. Section III.4. and Refs.[56-58]) in order to determine the weight ratio of the truly grafted polystyrene to the attendant polystyrene. The ratios thus estimated were in good agreement with those found for the polystyrene residues obtained by acid hydrolysis of the cellulose backbone[89]. It is thus emphasized that these applications of TLC will make it further possible to determine the true graft efficiency which has ever hardly been known by usual isolation techniques for graft copolymers.

A similar TLC application will be for the so-called "high-impact" polystyrene, which is usually produced by polymerizing styrene in the presence of diene-containing polymers. During the polymerization, occasional grafting of polystyrene to diene-prepolymer takes place, yielding a two-phase system in which polydiene phase, stabilized by polystyrene-grafted polydiene molecules, is dispersed in polystyrene matrix. For the characterization of such systems, an essential requirement is the knowledge of the true graft ratio, which is the ratio of truly grafted polystyrene to the prepolymer. However, the determination of this quantity has been complicated by the emulsifying effect of graft copolymers[61, 62], so far as the fractionating precipitation or solvent extraction method have been employed. TLC is again promising for this purpose. An obvious application of TLC is to examine the purity of the fraction to check the endpoint of the extraction or the fractionation procedure, as already pointed out. A direct estimation of the extent of grafting is also possible by TLC. Min et al.[92] developed a commercial "high-impact" polystyrene with 2-butanone continueously for 2 to 4 hours (a conventional development usually requires ca. 30 min). A solubility-based separation occurred, in which the free polystyrene migrated up to the solvent front while the graft copolymer component remained immobile. This was a solvent extraction procedure on TLC plate.

An adsorption-controlled separation of free polystyrene and graft copolymer component will also be possible, in which diene-containing components will have larger R_f values, if a common good solvent toward both polymers is properly selected as developer. In principle, this method is preferable over the solubility-controlled separation to avoid the conceivable emulsifying effect. However, in the case of diene-containing polymers a certain portion of the sample polymer often remains at the starting point[48, 50]. Such a fraction might contain graft copolymer component. If so, it might be difficult to make a clear distinction between free and graft polystyrene components. An alternative version will be to use the procedure applied for cellulose-styrene graft products[89]. Polydiene backbone of rubber-modified polystyrene can be degraded by ozonolysis or by oxidation with perbenzoic acid. Polystyrene molecules thus released would carry some polar groups, such as hydroxyl and aldehyde groups, at one of the chain ends, while free polystyrene molecules would not carry such groups. Thus, they will be readily separated by adsorption TLC.

IV.5. Branched Polymers

An appreciable progress in the fundamentals for characterization of branched polymers has been achieved to date by different investigators[93-95]. Therefore, at present, the physico-chemical behavior of branched polymers in solution as well as in bulk can be elucidated to a great extent on the molecular basis. Kurata et al. proposed a useful, iterative computer method for estimating the degree of branching

Table 7. Characteristics of tetrachain-star-shaped and linear polystyrenes

Sample code	$M_n \times 10^{-4}$ (Precursor)	$M_n \times 10^{-4}$	$M_w \times 10^{-4}$	M_w/M_n	$(M_n)_{star}/(M_n)_{prec.}$
Star-shaped samples					
S-10	2.3	8.8	9.8	1.11	3.8
S-20	6.4	24.5	24.9	1.02	3.8
S-30	9.6	33.2	33.8	1.02	3.5
S-50	12.4	49.8	52.6	1.06	4.0
Linear samples[a]					
S-AP[b]		–	2.0	–	
L-02		–	1.98	–	
L-05		–	5.10	–	
L-10			9.72	–	
L-16		–	16.0	–	
L-41		–	41.1	–	
L-67		–	67.0	–	

[a] L-series samples are those from Pressure Chemical Co.
[b] Precursor for a star-shaped sample S-A (crude sample).

and molecular-weight distribution simultaneously from a set of measurements of GPC and intrinsic viscosity[94]. However, the task remains of establishing a method for separating a mixture of linear and branched polymer species into components by difference in the degree of branching. For this separation purpose there may be three different approaches: One is to use the molecular sieving effect, and the other two are related to an aspect that branched polymers may be regarded as copolymers consisting of monomer units with and without branching. If the monomer unit having a branch changes adsorption behavior of the whole polymer, the adsorption-controlled separation mechanism in TLC will be applicable to the separation. This would be effective for polymers carrying a number of short branches. While, if it changes the solubility, the solubility-controlled separation mechanism will be applicable.

The latter possibility will be discussed later in connection with the resolution with respect to molecular weight to be attained by the solubility-controlled mechanism in TLC (see Sections V.1. and V.2.). The former possibility has been examined by Miyamoto et al.[96, 97] using tetrachain star-shaped polystyrenes, which were prepared by coupling "living" polystyrenes with 1,2,4,5-tetrachloromethyl benzene[98]. The characteristics of these samples are listed in Table 7 together with those of linear polystyrenes used as the reference sample. In this study, binaries (cyclohexane + benzene) and (carbon tetrachloride + chloroform) were tested as the developer and highly activated silica gel was used as the stationary phase. Figure 8 shows a typical chromatogram obtained with a concentration gradient development using the former binary, in which the composition was varied from 50:50 to 25:75 by volume (cyclohexane:benzene). Slightly larger R_f values were noticed for the star-shaped polymers. A similar behavior was observed for the other binary. These results imply

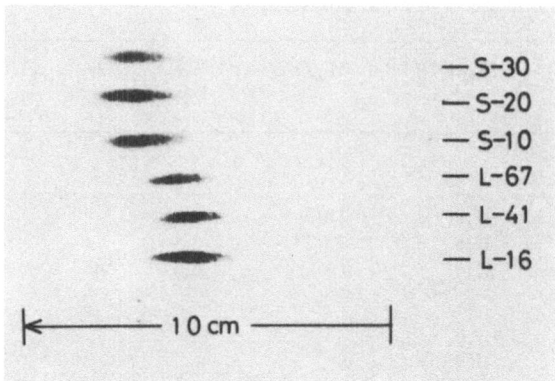

Fig. 8. Chromatograms obtained for star-shaped (S-series) and linear (L-series) polystyrene samples by a concentration gradient technique with a binary cyclohexane + benzene

that branched polymer molecules are less strongly adsorbed and/or more easily desorbed than linear polymer molecules, though the difference is quite critical.

Another approach to the TLC separation of linear and branched polystyrenes was reported by Belenkii et al.[99]. The adsorbent used was silica gel with a mean pore diameter of 80 Å (notice that no molecular sieving mechanism should be operative in this pore diameter), and the developer was composed of cyclohexane, benzene, and acetone in a composition of $12:4:\gamma$, where γ is the relative amount of acetone added. The TLC experiments were performed for different γ values ranging from 0.4 to 1.5. With increase in the γ value, the R_f values increased both for the linear and branched samples but the dependence for the linear samples was remarkably stronger than for the branched ones.

V. Separations by Solubility-Controlled Mechanism

V.1. Resolution with Respect to Molecular Weight

As described before, TLC separation based on the phase-separation mechanism causes fractionation primarily by molecular-weight differences. The resolution attained by this mechanism in TLC will be discussed below by comparing it with those attained by GPC and velocity ultracentrifugation. To this end, some findings obtained by Miyamoto et al. during the course of a study on TLC separation of star-shaped polystyrenes[96, 97] are referred to.

When a tetra-chain, star-shaped polystyrene is prepared by a coupling reaction between living polymer and coupler (e.g., 1,2,4,5-tetrachloromethyl benzene), the reaction is often carried out with the polystyryl anion in slight excess in order to avoid by-production of types of branched polystyrene other than the tetra-chain. By virtue of this reaction, the whole product is apt to contain a certain amount of precursor polystyrene. Such a crude product, coded S-A, for which the polystyryl anion was added in high excess, was used as the sample. The precursor polystyrene had a weight-average molecular weight (M_w) of 2×10^4, and 8×10^4 was expected for the molecular weight of star-shaped component.

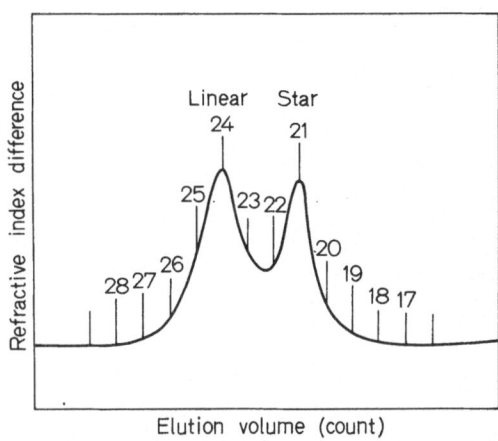

Fig. 9. GPC elution curve for a reaction product of tetra-chain star-shaped polystyrene, coded S-A. For details, see text

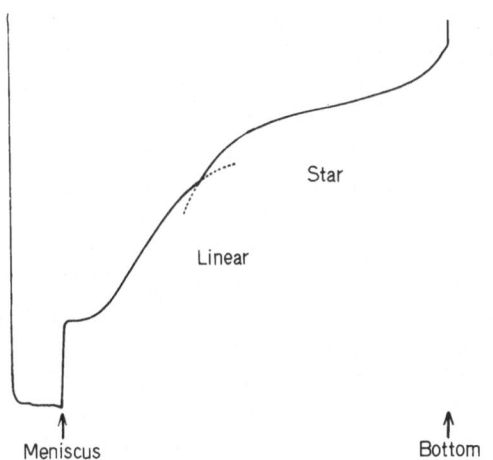

Fig. 10. Velocity ultracentrifugation pattern observed for sample S-A dissolved in cyclohexane at 0.077 g/dl. For further operation conditions, see text

First, the sample was examined by GPC, for which four columns of "styragel" of 10^6, 10^5, 10^4 and 10^3 Å nominal pore size were used. The total number of theoretical plates as determined by acetone at a flow rate of 1 ml/min was ca. 26,000. The eluent was tetrahydrofuran. The chromatogram is shown in Figure 9, which indicates two peaks at ca. 21 and 24 counts. The former may be assigned to the tetra-chain, star-shaped component, and the latter to the precursor. However, no complete separation of the two peaks was observed. For another comparison, velocity ultra-centrifugation was performed for the sample at 59,780 rpm using a θ-solvent for polystyrene, cyclohexane. The operation temperature was established at 35 °C, the θ-temperature, to minimize the concentration dependence of sedimentation velocity and other effects. A sedimentation pattern taken by UV-absorption is shown in Figure 10. It is seen that the separation of S-A sample into the two components was quite difficult even at a very low polymer concentration, 0.077 g/dl.

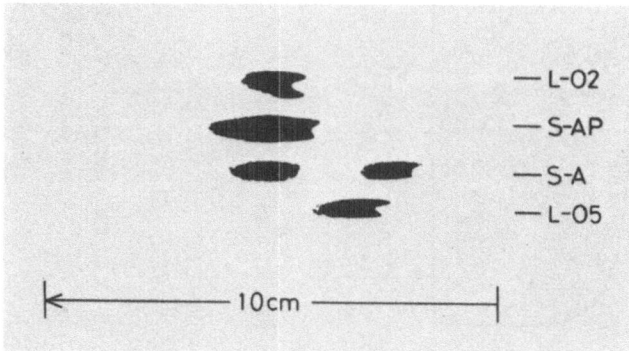

Fig. 11. Chromatograms obtained by phase-separation mechanism in TLC for sample S-A, its precursor (linear) polystyrene (S-AP), and reference polystyrene samples (L-02 and L-05). For details, see text

On the other hand, the sample was chromatographed on silica gel thin layer together with the precursor and two narrow distribution polystyrenes having $1,98 \times 10^4$ and 5.10×10^4 for M_w, coded S-AP, L-02 and L-05, respectively. A gradient development technique was applied employing a binary acetone + chloroform (10:1 by vol.) and chloroform as the initial and second solvent, respectively. By the use of acetone, the phase-separation mechanism should have been operative in this separation. The chromatogram is given in Figure 11, which shows complete separation of S-A into the precursor and star-shaped component. The molecular weight of the former component was estimated to be 2×10^4 by comparing its R_f value with that of L-02, while that of the latter fell on a value between 7×10^4 and 8×10^4, if one compared its R_f with that of L-05. Therefore, the molecular weight estimated chromatographically for the precursor was in good agreement with that determined independently. The value for the star-shaped component was also close to that expected for the tetra-chain polymer. The comparison presented here is indicative of the high resolution with respect to molecular weight to be attained by the phase-separation mechanism in TLC. The high resolution can also be affirmed by data given by Belenkii et al.[99]. They found R_f values of $<0.1, 0.19, 0.39, 0.56,$ and 0.80 for narrow distribution polystyrenes with M_w of $1.73 \times 10^5, 5.10 \times 10^4, 1.98 \times 10^4, 1.03 \times 10^4,$ and 5×10^3, respectively, when these samples were developed with the developer system cyclohexane + benzene + acetone (12:4:0.6). The high performance of TLC in its resolution was utilized by Higashimura et al. to separate cationic polymerization products of styrene into components with differing molecular weight[100].

V.2. Linear and Branched Polymers

The foregoing section dealt with TLC separation of linear and branched polymers. from the standpoint of resolutions with respect to molecular weight. In this section we summarize TLC results of chromatographic distinguishment of linear and star-shaped polystyrenes on the same molecular-weight level. It is now well known that

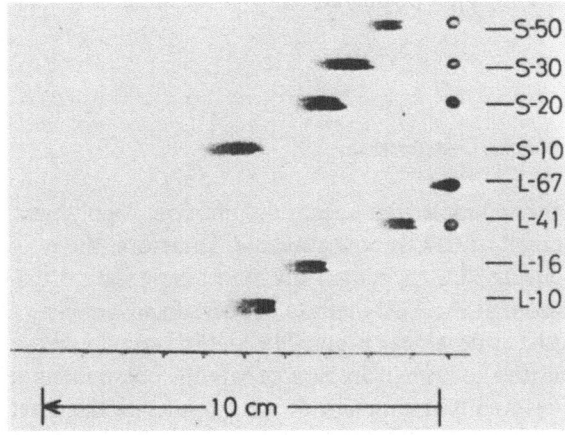

Fig. 12. Chromatograms obtained by phase-separation mechanism for star-shaped (S-series) and linear (L-series) polystyrene samples

a slight difference exists in the solubility between linear and branched polymers having the same molecular weight[95]. This allows us to expect a distinction between their TLC behavior to be observed by the phase-separation mechanism.

Miyamoto *et al.*[96] examined the possibility using tetra-chain, star-shaped polystyrene samples whose characteristics have already been given in Table 7 (see Section IV.4.). A gradient development was performed with a binary acetone + chloroform (50:9 by vol.) and chloroform as the initial and second solvents, respectively. Figure 12 illustrates a chromatogram on which the star-shaped samples were developed together with narrow distribution polystyrenes used as the linear reference sample. Figure 13 shows plots of R_f vs. $\log M_w$ for the star-shaped and linear samples. The star-shaped samples exhibit larger R_f values than the corresponding linear standards. The result may be interpreted in terms of the observation that the θ-temperature determined in cyclohexane for star-shaped polystyrenes was lower than that for the linear[95]. Unfortunately, the difference in R_f was not very large, so that

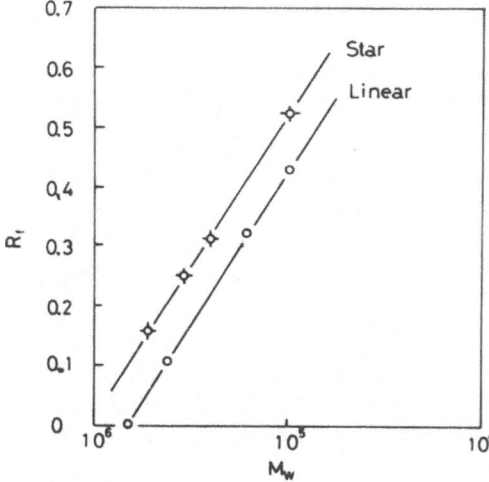

Fig. 13. Plots of R_f vs. M_w for star-shaped and linear polystyrene samples, constructed on the basis of chromatograms given in Fig. 12

no satisfactory separation of these two types of polystyrenes of the same molecular weight could be carried out.

V.3. Determination of Molecular-Weight Distribution

For the analysis of molecular weight and molecular-weight distribution of polymers, a sophisticated and convenient method of GPC is now available. Therefore, there might not be much point in introducing TLC to analyze the molecular-weight distribution. However, we may emphasize that the TLC method has certain advantages: A wider choice of solvents to be used as developer is possible, and in certain cases, the TLC method is much more sensitive to a small amount of satellite component (cf. Section V.1.). This section is devoted to a summary of some results on the determination of molecular-weight distributions by the phase-separation mechanism in TLC.

The molecular-weight distribution of a homopolymer sample can be determined by TLC if a chromatogram is obtained, on which the sample is separated into components by the difference in molecular weight together with reference samples having different known molecular weights. Using the chromatogram, a calibration curve between logarithm of molecular weight and R_f is first established, which can usually be approximated by a linear relationship. The next task is to analyze the final spot of the sample, which will be a smear extended in the development direction. Thus, the determination involves two problems: One is how to find an appropriate developer system which allows separation by molecular weight and shows only minimized tailing effects, and the other is how to analyze the chromatogram.

a) **Polystyrene**. The determination for this polymer by TLC was achieved independently by Otocka[19] and Kamiyama *et al.* [17]. Previous to the determination, Otocka and Hellman made a series of preliminary tests to select the developer[18]. It was found that a binary acetone + chloroform with a fixed composition (80:20 by vol.) gave good separation of samples in a molecular-weight span ranging from 1×10^4 to 41×10^4, but other samples of 8.6×10^5 and 1.8×10^6 remained immobile.

This result suggests two important features in application of the phase-separation mechanism in TLC. First, a polar solvent like acetone must be present in the developer in order to eliminate adsorptive interaction between adsorbent and polymer. Second, TLC developments with single solvent or solvent mixtures having fixed compositions lead to too high resolutions, so that only a limited number of different molecular-weight samples can be developed within the range of $0 \leq R_f \leq 1$, and, furthermore, to final spots which often exhibit unfavorable tailing phenomena. This situation is the same for adsorption TLC, as pointed out in Section III.1. Thus, the above authors employed a gradient development technique using aluminum oxide as the adsorbent, for which acetone was used throughout as the initial solvent, while toluene and tetrahydrofuran were alternatively used as the second solvent. Under the development condition with acetone and tetrahydrofuran, all the samples gave R_f values different from one another.

In practice, Otocka employed a binary acetone + isopropyl alcohol (96 : 4 by vol.) and chloroform as the initial and second solvents, respectively, to determine the molecular-weight distribution of a commercial polystyrene product[19]. Quantitation of the chromatogram was performed, on one side, by combining a silica-gel chromatoplate containing zinc silicate fluorescent indicator with a TLC scanning spectrodensitometer (the Schoeffel Model SD 3000, Schoeffel Instrument Co., USA), and, on the other side, by combining a chemical visualization technique using a perchloric acid solution with the above densitometer. The result was in good agreement with that obtained by GPC. Differing from the developers used by Otocka, a much more complex mixture was employed by Kamiyama et al.[17]. A mixture of acetone, benzene, and ethanol (3 : 1 : 2 by vol.) and another mixture of benzene and 2-butanone (1 : 1 by vol.) were applied as the initial and second solvents to form a gradient on silica-gel thin layer. The chromatogram thus obtained was visualized by staining with thymol blue method[12] and photographed. Then, the picture was subjected to a conventional photodensitometer for the quantitation. A good agreement was found between TLC and GPC.

b) Polymethyl Methacrylate and Polyethylene Oxide. The determination of molecular-weight distribution by TLC for the former polymer has not yet been made. However, it was found by Inagaki et al.[23] that this polymer could be separated by molecular weight using a binary chloroform + mechanol (29 : 71 by vol.). The molecular weight of samples tested ranged from 4×10^4 to some 40×10^4. In this connection it should be mentioned that two polymethyl methacrylate samples having 16.5×10^4 and 41.2×10^4, coded PMA-3 and PMA-2, respectively, showed a strong molecular-weight dependence of R_f even under the condition that adsorption mechanism was certainly operative[15, 21].

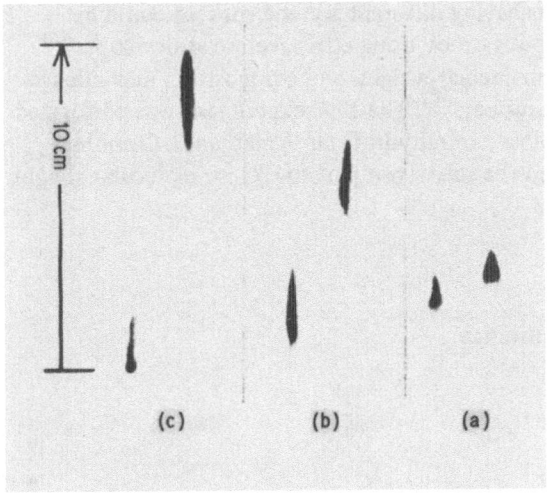

Fig. 14. Molecular-weight dependences of R_f observed for polymethyl methacrylate samples, coded PMA-3 and PMA-2 (left and right spot in each chromatogram, respectively), using three different but equi-eluotropic binaries, (a) benzene + acetone, (b) isopropyl acetate + methyl formate and, (c) ethyl acetate + methyl acetate

The above finding was obtained in a series of TLC experiments using three different binaries as developer, namely, (a) benzene + acetone (2:8 by vol.), (b) isopropyl acetate + methyl formate (100:62), and (c) ethyl acetate + methyl acetate (100:23), all of which were good solvents toward the sample polymer. The composition of each binary was so adjusted that the sample polymers might migrate intermediately on chromatoplate. Thus, these binaries were regarded as "equieluotropic developers". As shown in Figure 14, the chromatograms indicate that the strongest molecular-weight dependence of R_f was observed with the binary (c). In contrast to this result it should be recalled that a binary chloroform + methanol (95:5 by vol.) exhibited no molecular-weight dependence of R_f for polymethyl methacrylate samples[23]. On the basis of these observations it is pointed out that the R_f value can depend on molecular weight even in adsorption TLC if a binary is used, the component solvents of which resemble in their dielectric constant, just as in the case of binary (c). This trend cannot, of course, be interpreted in terms of the phase-separation mechanism.

The fractionation posssibility for another homopolymer was demonstrated by Otocka and Hellman[18]. Polyethylene oxide samples, whose molecular weights ranged from 1.5×10^3 to 1.0×10^4, were separated by molecular weight using a binary ethylene glycol + methanol (80:20 by vol.) and silica gel as developer and adsorbent, respectively. Under this separation condition, however, two samples of 1.0×10^4 and 2.8×10^4 could not be distinguished chromatographically, both showing zero for R_f, and a gradient development technique using methanol and N, N-dimethylformamide as the initial and second solvent was applied to complete separation.

c) **Styrene-butadiene Rubbers.** The solubility of copolymers is dependent on both molecular weight and composition. For example, White *et al.* were able to show that styrene-butadiene rubbers having different styrene content could be separated largely according to composition by using ethyl acetate as developer[46]. Kotaka and White investigated the molecular-weight- and composition dependence of R_f for various styrene-butadiene rubbers[48]. The TLC experiment was performed by a gradient development with a binary tetrahydrofuran + methanol. Combining these dependences, they summarized the relationship of the R_f vs. molecular weight (M) and composition (x) as

$$R_f = (A + Bx) - C_2 \log M$$

with A, B, and C_2 being positive constants.

VI. Experimental Problems

VI.1. General Remarks

As described in the preceding chapters, a TLC system designed well for a given polymer sample indicates a surprisingly high sensitivity in detecting small structural differences among sample molecules. While poorly designed experiments are misleading, in which, for example, a supposedly homogeneous sample splits into discrete final spots. Several factors for designing TLC systems well have been investigated extensively for low-molecular-weight compounds[12]. The investigations involved the effects of sample size, sample concentration in stock solution, and sample application method, etc. upon the chromatographic behavior of a sample, in addition to the major effects, *i.e.,* those of the developer, adsorbent, its degree of activation and other development conditions. From the standpoint specific to TLC of polymers, these factors have been discussed in Chapter II and also in an individual manner relevant to each polymer species. In addition, it is noted that considerations on the solubility parameter reported by Hansen[101] and by Teas[102] will be of great help in selecting developers for TLC separations of polymers. In the following selections we summarize our experiences for selecting the developer and describe some newer problems in quantitation of chromatograms.

Before going into the next section, we will discuss another important factor: the rate of sample dissolution into developer solvent, which has not yet been mentioned in this article. It is conceivable that polymeric samples are dissolved much more slowly than low-molecular-weight samples. However, the rates of dissolution determined by Ueberreiter and Asmussen for polystyrenes having different molecular weight were in an order of magnitude of some 10^{-3} g/min.[103]. This result implies that the rate of dissolution exerts no serious influence upon TLC developments because the sample size for polymer samples usually lies in a range smaller than 10 μg, but it must be taken into account that the polymer dissolution process ultimately begins after a certain swelling time. According to the above authors, the swelling times for polystyrenes depended prominently on molecular weight and were several minutes even for those having molecular weights lower than 10^4. Thus, the rate of dissolution will in some way be concerned with TLC experiments for polymers. To minimize the influence, the sample size should be diminished to as low as possible (but with the reservation that the resultant chromatogram can still be visualized). Otherwise it will be observed that a small portion of sample remains undeveloped at the starting point. A high-temperature operation of TLC will be quite useful for the purpose.

VI.2. Development Characteristics

This section deals with a comparative discussion on some features of TLC developments by adsorption mechanism, which are performed in a constant solvent composition and in a composition gradient. Let us suppose we have a sample composed of two component species having slightly different polarities. When the sample is chromatographed using a single solvent, two kinds of chromatogram results will be expected: One is that no separation occurs showing either zero or unity for R_f, while the other is that separation does occur but the components thus separated show R_f values greatly differing from one another, as already pointed out in Section III.1. However, the latter result is difficult to achieve and may only be a matter of luck, because it is extremely difficult to select a single solvent having a solvent strength which can just satisfy the migration condition of a sample described by the Snyder theory, *i.e.*, Eqs. (2) and (3) given in Section II.2. Therefore, one uses solvent mixtures, the solvent strength of which can easily be adjusted to an appropriate value by varying the composition. The development characteristics are almost the same as those observed with single solvents, except for the demixing effect (cf. Section II.1. and Ref.[20]). High resolutions with respect to the composition of sample may be expected by this type of development. In this connection it should be remembered that a TLC development by adsorption mechanism results in a distinct molecular-weight dependence of R_f, when a binary is used as developer whose constituent solvents resemble each other in their dielectric constant, as stated in Section V.3.b. (cf. Fig. 14).

When the sample is a copolymer with a large compositional heterogeneity, the development characteristics mentioned above mean that every component species of the sample cannot be assigned by R_f values, all of which are different from one another in the limited range between zero and unity. Such degeneration of R_f values is unfavorable to the determination of compositional heterogeneities. For this reason, the gradient development technique[12, 34] has been applied. In practice, a developer solvent (or solvent mixture) is placed in advance in a TLC chamber. After the solvent front reached a certain height above the starting point at which the sample was applied, a second solvent is added to form a composition gradient during the development process. Such gradients can be of a linear, concave, or convex type, and this type can be chosen by adjusting the rate of addition of second solvent.

The essential action of this technique is that the solvent strength of a developer can be varied continuously so as to satisfy the migration condition which corresponds to the polarity of each component species. Two types of gradients are generally formed, in which the solvent strength either ascends or descends with the elevation of solvent front (in the case of ascending developments). Figures 15(a) and (b) schematically illustrate separation processes of a sample having a large compositional heterogeneity, which will be operative in a constant composition (a) and a composition gradient (b), respectively. The figures were prepared, assuming that the relationship

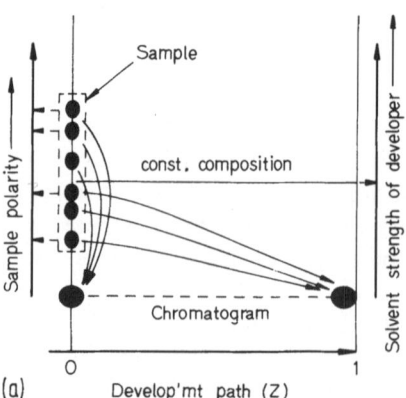 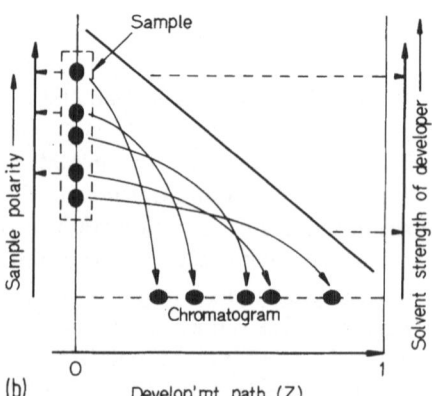

Fig. 15. Schematic representations for TLC-separation under a constant and changing composition of developer. For the details, see text

between solvent strength and development path (Z) is linear and the sample is an assembly of component species with discrete polarities.

To present the features of these two types of gradients, Table 8 is given, in which developers and development conditions for adsorption TLC of various polymers reported to date are listed. From the experimental viewpoint, the following aspects seem important. As seen in the table, the gradient development technique has generally been applied for the determination of compositional heterogeneities in copolymers. Specifically, gradient developments of descending solvent strength have been employed for the purpose. The merit is that unfavorable tailing phenomenal in final spots can be appreciably depressed by this type of development. On the other hand, those of ascending solvent strength tend to suppress predominantly the

Table 8. Developers and development conditions employed for adsorption TLC of polymers

Polymer[a] sample	Separation by difference in:	Developer[b]	Gradient (Develop'mt cond.)	Ref.
co[ST-MA]	comp. (CH)[c]	(CCl$_4$ + MeOAc) and MeOAc	yes (desc.)[d]	16)
co[ST-MA]	comp.	CHCl$_3$ and EtOAc	yes (desc.)	16)
co[ST-MA]	chain architec.	(CCl$_4$ + MeOAc) and MeOAc	yes (desc.)	59)
co[ST-MMA]	comp.	(CHCl$_3$ + Acetone) or (CHCl$_3$ + MEK)	no	11)
co[ST-MMA] (azeotropic)	comp.	(CHCl$_3$ + EtOAc)	no	33)
co[ST-MMA]	chain architec.	CHCl$_3$ and EtOAc or (CHCl$_3$ + EtOAc) and EtOAc	yes (desc.)	59)
co[ST-b-MMA]	comp. (CH)	CCl$_4$ and MEK	yes (desc.)	43)
co[ST-b-MMA]	block seq.	CCl$_4$ and MEK	no	43)
co[ST-AN]	comp.	(CHCl$_3$ + MeOAc)	no	36, 39)
co[ST-AN]	comp. (CH)	C$_2$H$_2$Cl$_4$ and EtOAc	yes (desc.)	37)
co[ST-AN]	comp.	C$_6$H$_6$ or C$_6$H$_5$Me and Acetone	yes (desc.)[e]	106)
co[ST-BD]	comp.	(C$_6$H$_{12}$ + C$_6$H$_6$)	no	47)
co[ST-BD]	comp. (CH)	CCl$_4$ and CHCl$_3$	yes (desc.)	48)
co[ST-BD]	chain architec.	(C$_6$H$_{12}$ + CHCl$_3$)	no	64)
co[ST-b-BD]	block seq.	(C$_6$H$_{12}$ + CHCl$_3$)	no	64)
LR	functionality	(CCl$_4$ + THF)	no	56)
Cell-Ac	comp. (CH)	(CH$_2$Cl$_2$ + BuOH or MeOH) (Acetone + EtOAc) and	stepwise	39)
Cell-NO$_2$	comp. (CH)	(CHCl$_3$ + EtOAc)·	yes (asc.)[f]	15)
Copolyamides	comp.	HCOOH or (HCOOH + Phenol)	no	45)
PMMA	steric isomerism	EtOAc	no	43)

[a] Abbreviations: co[X-Y] = copolymers of X and Y; co[X-b-Y] = block copolymers of poly X and poly Y; ST = styrene; MA = methyl acrylate; MMA = methyl methacrylate; AN = acrylonitrile; BD = butadiene; LR (liquid rubbers) = α, ω-polybutadiene-diols and -dicarboxylic acids; Cell-Ac = cellulose acetate; Cell-NO$_2$ = cellulose nitrate.

[b] Abbreviations: MeOAc = methyl acetate; EtOAc = ethyl acetate; MEK = 2-butanone; BuOH = butyl alcohol; MeOH = methyl alcohol; THF = tetrahydrofuran.

[c] Compositional heterogeneity (CH) was determined.

[d] Descending solvent strength with development path.

[e] Two-dimensional development.

[f] Ascending solvent strength with development path.

migration of component species for which higher R_f values are expected. Therefore, this type of development is superior to the other if a sample is composed of constituent species having discrete polarities which are greatly different from one another. A typical example is the separation of cellulose nitrate by difference in the degree of substitution, which was described in Section III.2.d and also in Ref.[15].

VI.3. Quantitation of Chromatograms

A polymer samples composed of several components can be easily separated on the TLC-plate as discrete spots, and qualitative identification of such spots is a relatively easy matter. It is somewhat a different problem, however, to perform quantitative analysis of such chromatograms, even when good separation among the spots is achieved. In this section we will describe some problems of quantitation of chromatograms. To do the quantitation, a few more steps should follow the development. The first is the visualization of chromatogram; the second is the recording of visualized chromatogram by photographing; and thirdly, the tracing is done, for example, by photodensitometry. On the other hand, several TLC-scanning spectrodensitometric devices are now commercially available[12, 15]. Use of such a device eliminates the visualization process and allows the evaluation of chromatogram directly on the TLC-plate.

For quantitative analysis of spot darkness recorded on a photofilm, one must establish a calibration of darkness intensity and polymer amount. The calibration may not be a linear function over a wide range of polymer amount, and for copolymers it may vary with the composition[16]. The absolute value of darkness intensities depends on various factors, and therefore is not well reproducible. Generally speaking, in the same series of photodensitometric tracings, a linear relationship between the darkness and polymer amount can be achieved for the spots with sample densities perhaps below 200 $\mu g/cm^2$. The tolerance was rather large in comparison with that in using a TLC scanning spectrodensitometer (TLC-scanner), as found by different authors for different polymer samples[11, 16, 19, 70, 89]. When the specific darkness, defined by darkness intensity per unit weight of sample, was compared for undeveloped spots of copolymers with differing composition in the same series of tracings, they invariably gave almost the same composition dependence[16, 37, 48].

On the other hand, use of a TLC-scanner is possible only for the samples which have specific absorptions in the ultraviolet and/or visible region. Chromatograms visualized by a staining method can, of course, be subjected to a TLC-scanner, instead of photographing and densitometric tracing. In the measurements, an incident light beam of a wavelength specific to the sample is projected to a given chromatogram, and usually the response of the reflected beam is recorded as a function of the scanning distance, *i.e.*, the development distance. The specific response is highly sensitive to the sample amount, if the sample contains styrene units, but rapidly levels off with increasing sample amount (perhaps 100 $\mu g/cm^2$).

An advantage in the use of TLC-scanner was pointed out by Kotaka et al.[43]. When one wants to determine the compositional heterogeneity of a given binary copolymer, a series of reference samples with different composition are required to establish a calibration between the composition and R_f. If the comonomers have specific light absorptions at different wavelengths, the use of a TLC-scanner eliminates the necessity of reference samples, because the point-by-point composition of each separated component can be determined by two scanning runs of the chromatogram with different wavelength beams. This method was applied to the compositional heterogeneity determination of styrene-methacrylate block copolymers, and the result was compared with that deduced from the other method using reference samples (cf. Section III.2.e).

In spite of the above advantage, the quantitation by TLC-scanner is accompanied by various difficulties. The most direct difficulty is caused by hypochromic effects in the ultraviolet spectra of copolymers, especially those containing styrene units. For such statistical copolymers, the UV-extinction coefficient depends not only on the composition, but also on the monomeric arrangement along the polymer chain[104]. The other difficulties are somewhat indirect and connected rather with the development conditions; but they are serious from the experimental standpoint, as described below.

It has sometimes been observed that a sample applied to chromatoplate migrated in the development toward the glass-adsorbent interface rather than the adsorbent surface, the final spot thus having become hardly visible from the adsorbent side[48]. Of course, such a final spot cannot respond to the scanning of reflection mode. On the contrary, Hezel pointed out that when the sample size was comparatively high, the response of the TLC-scanner from a final spot tended to increase with increase in the migration distance of the spot[105]. This result was interpreted in such a way as schematically illustrated in Figure 16. The sample molecules, just after applied, will be distributed not only on the adsorbent surface but also within the adsorbent layer, and the molecules present within the layer will gradually come up to the surface with elevation of the solvent front. Thus, such a wandering of samples molecules may be responsible for the result if the scanning is done in the reflection mode.

The visualization-and-photodensitometric-tracing method is also liable to produce such a result. However, an outcome of this kind appears to be less significant

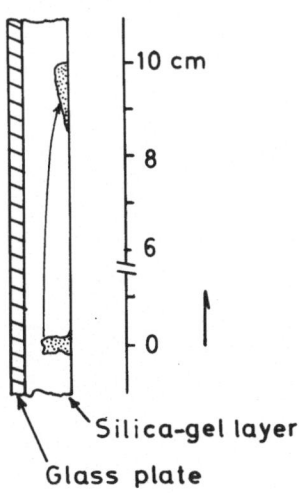

Silica-gel layer

Glass plate

Fig. 16. A schematic representation for distribution of sample molecules within a thin layer

because of the large tolerance of this method to the spot density, as was mentioned already. It must be emphasized that this method is really superior to the method by TLC-scanner for the compositional heterogeneity determination of copolymers if one does not mind somewhat hazardous experimental procedures. When a given sample is composed of several discrete component species, as in the case of liquid rubbers, the "thin-layer-FID" chromatographic technique will be quite useful for quantitation of chromatograms (cf. Section III.4. and Ref.[56]).

VII. References

1) Pauling, L.: College chemistry, 3rd Edit., W. H. Freeman and Co. 1964, p. 13.
2) Kollinsky, F., Markert, G.: Makromol. Chem. *121*, 117 (1969).
3) Teramachi, S., Kato, Y.: J. Macromol. Sci. Chem. *A4*, 1785 (1970); Macromolecules *4*, 54 (1971); Teramachi, S., Fukao, T.: Polymer J. *6*, 532 (1974).
4) v. Tavel, P., Bollinger, W.: Helv. Chim. Acta *51*, 278 (1968); v. Tavel, P., Bieri, V.: Makromol. Chem. *149*, 63 (1971);
   Wälchli, J., v. Tavel, P.: Bull. Inst. Chem. Res., Kyoto Univ. *53*, 424 (1975).
5) Kuhn, R.: Makromol. Chem. *177*, 1525 (1976).
6) See, *e.g.*, Ellerstein, S., Ullman, R.: J. Polym. Sci. *55*, 123 (1961).
7) Kern, W., Rauterkus, K. J., Weber, W.: Makromol. Chem. *43*, 98 (1961); Konishi, K., Yamaguchi, S.: Anal. Chem. *38*, 1755 (1966).
8) Braun, D.: Kunststoffe *52*, 2 (1962).
9) Nagasawa, K., Ohta, K.: Japan Analyst. *16*, 1285 (1967).
10) Langford, W. J., Vaughan, D. J.: Nature *184*, 116 (1959); J. Chromatog. *2*, 564 (1959).
11) Belenkii, B. G., Gankina, E. S.: Dokl. Akad. Nauk SSSR *186*, 857 (1969); J. Chromatog. *53*, 3 (1970).
12) Stahl, E.: Dünnschicht-Chromatographie, 2nd Edit. Berlin-Heidelberg-New York: Springer 1967.
13) Randerath, K.: Thin-layer chromatography. New York-London Academic Press 1968.
14) Shellard, E. J.: Quantitative paper and thin-layer chromatography. Academic Press New York-London: 1968.
15) Inagaki, H.: Thin-layer chromatography. In: Fractionation of synthetic polymers. Tung, H. L. (ed.). New York: Marcel Dekker (in press).
16) Inagaki, H., Matsuda, H., Kamiyama, F.: Macromolecules *1*, 520 (1968).
17) Kamiyama, F., Matsuda, H., Inagaki, H.: Polymer J. *1*, 518 (1970).
18) Otocka, E. P., Hellman, M. Y.: Macromolecules *3*, 362 (1970).
19) Otocka, E. P.: Macromolecules *3*, 691 (1970).
20) Snyder, L. R.: Principles of adsorption chromatography. New York: Marcel Dekker 1968.
21) Kamiyama, F., Inagaki, H.: Bull. Inst. Chem. Res., Kyoto Univ. *52*, 393 (1974).
22) See *e.g.*, Flory, P. J.: Principles of polymer chemistry, Chapter 13, Cornell Univ. Press 1953.
23) Inagaki, H., Kamiyama, F., Yagi, T.: Macromolecules *4*, 133 (1971).
24) Kamiyama, F., Inagaki, H.: Bull. Inst. Chem. Res., Kyoto Univ. *49*, 53 (1971).
25) Otocka, E. P., Hellman, M. Y., Muglia, P. M.: Macromolecules *5*, 227 (1972).
26) Johansson, B. G., Rymo, L.: Acta Chem. Scand. *16*, 2067 (1962); *18*, 217 (1964).
27) See *e.g.*, Determann, H.: Gel chromatography, 2nd Edit. Berlin-Heidelberg-New York: Springer 1969.
28) Halpaap, H., Klatyk, K.: J. Chromatog. *33*, 80 (1968).
29) Donkai, N., Inagaki, H.: J. Chromatog. *71*, 473 (1972).
30) White, J. L., Kingry, G. W.: J. Appl. Polym. Sci. *14*, 2723 (1970).
31) Belenkii, B. G.: Private communication, Nov. 1974.
32) Inagaki, H., Kamiyama, F.: Macromolecules *6*, 107 (1973).
33) Donkai, N., Inagaki, H.: Unpublished experiments in 1973.
34) Rybicka, S. M.: Chem. Ind. (London) *1962*, 308.
35) Inagaki, H.: Bull. Inst. Chem. Res., Kyoto Univ. *47*, 196 (1969).
36) Teramachi, S., Esaki, H.: Polymer J. *7*, 593 (1975).
37) Wälchli, J., Miyamoto, T., Inagaki, H.: To be published in Polymer J.
38) Teramachi, S., Fukao, T.: Polymer J. *6*, 532 (1974).
39) Kamide, K., Manabe, S., Osafune, E.: Makromol. Chem. *168*, 173 (1973).
40) Freyss, D., Rempp, P., Benoit, H.: J. Polym. Sci. *B2*, 217 (1964).
41) Kotaka, T., Donkai, N., Min, T. I.: Bull. Inst. Chem. Res., Kyoto Univ., *52*, 332 (1974).
42) Inagaki, H., Kotaka, T., Min, T. I.: Pure & Appl. Chem., *46*, 61 (1976).
43) Kotaka, T., Uda, T., Tanaka, T., Inagaki, H.: Makromol. Chem. *176*, 1273 (1975).
44) Nakazawa, A., Kotaka, T., Inagaki, H.: Brit. Polym. J., in press.

45) Mori, S., Takeuchi, T.: Kobunshi Kagaku (in Japanese) 29, 383 (1972).
46) Quisenberry, D. O.: Tenn. Engr. (Oct.), 1971, 5; White, J. L., Salladay, D. G., Quisenberry, D. O., MacLean, D. L.: J. Appl. Polym. Sci. 16, 2811 (1972).
47) Tagata, N., Homma, T.: J. Chem. Soc., Japan (Nippon Kagaku Zasshi), 1972, 1330.
48) Kotaka, T., White, J. L.: Macromolecules 7, 106 (1974).
49) See, e.g., Haws, J. R., Middlebrook, T. C.: Rubber World 167 (4), 27 (1973).
50) Donkai, N., Murayama, N., Miyamoto, T., Inagaki, H.: Makromol. Chem. 175, 187 (1974).
51) See, e.g., Consaga, J. P.: J. Appl. Polym. Sci. 14, 2157 (1970).
52) Muenker, A. H., Hudson Jr., B. E.: J. Macromol. Sci. Chem. A3, 1465 (1969).
53) Law, R. D.: J. Polym. Sci. A-1, 9, 589 (1971).
54) Anderson, J. N., Baczek, S. K., Adams, H. E., Vescelius, L. E.: J. Appl. Polym. Sci. 19, 2255 (1975).
55) Baczek, S. K., Anderson, J. N., Adams, H. E.: J. Appl. Polym. Sci. 19, 2269 (1975).
56) Min, T. I., Miyamoto, T., Inagaki, H.: To be published in: Rubber Chem. Technol., in press.
57) Padley, F. B.: J. Chromatog. 39, 37 (1969).
58) Szakasits, J. J., Peurifoy, P. V., Woods, L. A.: Anal. Chem. 42, 351 (1970); Mukherjee, K. D., Spaans, H., Haahti, E.: J. Chromatogr. 61, 317 (1971).
59) Kamiyama, F., Matsuda, H., Inagaki, H.: Makromol. Chem. 125, 286 (1969).
60) Hirooka, M., Yabuuchi, H., Iseki, J., Nakai, Y.: J. Polym. Sci. A-1, 6, 1381 (1968).
61) Molau, G. E.: Characterization of macromolecular structure, McIntire, D. (ed.). Washington, D. C.: National Academy of Sciences 1968, p. 245; Ikada, Y., Horii, F., Sakurada, I.: J. Polym. Sci., Polym. Chem. Ed. 11, 27 (1973).
62) Horii, F., Ikada, Y., Sakurada, I.: J. Polym. Sci., Polym. Chem. Ed. 13, 755 (1975).
63) Rowland, F., Bulas, R., Rothstein, E., Eirich, F. R.: Ind. Eng. Chem. 57, No. 9, 46 (1965).
64) Donkai, N., Miyamoto, T., Inagaki, H.: Polymer J. 7, 577 (1975).
65) Saegusa, T., Yatsu, T., Miyaji, S., Fujii, J.: Polymer J. 1, 7 (1970).
66) Morton, M., McGrath, J. E., Juliano, P. C.: J. Polym. Sci., Part C 26, 99 (1969).
67) Thies, S.: J. Phys. Chem. 70, 3783 (1966).
68) Kamiyama, F., Inagaki, H., Kotaka, T.: Polymer J. 3, 470 (1972).
69) See, e.g., Ohnuma, H., Kotaka, T., Inagaki, H.: Polymer J. 1, 716 (1970).
70) Inagaki, H., Miyamoto, T., Kamiyama, F.: J. Polym. Sci. B-7, 329 (1969).
71) See, e.g., Inagaki, H., Miyamoto, T., Ohta, S.: J. Phys. Chem. 70, 3420 (1966).
72) Kusanagi, H., Tadokoro, H., Chatani, Y.: Macromolecules, 9, 531 (1976).
73) Miyamoto, T., Inagaki, H.: Macromolecules 2, 554 (1969).
74) Fox, T. G., Garrett, B. S., Goode, W. E., Gratch, S., Kincaid, J. F., Spell, A., Stroupe, J. D.: J. Am. Chem. Soc. 80, 1768 (1958).
75) Watanabe, W. H., Ryan, C. F., Fleisher, Jr., P. C., Garrett, B. S.: J. Phys. Chem. 65, 896 (1961); Ryan, C. F., Fleischer, Jr.; P. C., J. Phys. Chem. 69, 3384 (1965).
76) Liquori, A. M., Anzuino, G., D'Alagni, M., Vitagliano, V., Costantino, L.: J. Polym. Sci. A-2, 6, 509 (1968).
77) Miyamoto, T., Inagaki, H.: Polymer J. 1, 46 (1970).
78) Miyamoto, T., Tomoshige, S., Inagaki, H.: Makromol. Chem. 176, 3035 (1975).
79) Matsuzaki, K., Kanai, T., Kono, Y., Yoshida, T.: J. Fac. Engng. Univ. Tokyo A-10, 1 (1972).
80) Buter, R., Tan, Y. Y., Challa, G.: Polymer 14, 171 (1973).
81) Liquori, A. M., Anzuino, G., Coiro, V. M., D'Alagni, M., De Santis, P., Savino, M.: Nature 206, 358 (1965).
82) Miyamoto, T., Tomoshige, S., Inagaki, H.: Polymer J. 6, 564 (1974).
83) Donkai, N., Murayama, N., Miyamoto, T., Inagaki, H.: Makromol. Chem. 175, 187 (1974).
84) Dawans, F., Teyssie, Ph.: Makromol. Chem. 109, 68 (1967).
85) Furukawa, J., Kobayashi, E., Kawagoe, T.: Polymer J. 5, 231 (1973); Furukawa, J., Kobayashi, E., Kawagoe, T., Katsuki, N.: Makromol. Chem. 175, 237 (1974).
86) Miyamoto, T., Tomoshige, S., Inagaki, H.: Unpublished experiments in 1973.
87) Natta, G., Porri, L, Carbonaro, A.: Makromol. Chem. 77, 126 (1964).
88) Binder, J. L., Ransaw, H. C.: Anal. Chem. 29, 503 (1957).
89) Taga, T., Inagaki, H.: Angew. Makromol. Chem. 33, 129 (1973).

90) Min, T. I., Miyamoto, T., Inagaki, H.: Bull. Inst. Chem. Res. Kyoto Univ. *53*, 381 (1975).
91) Min, T. I., Kotaka, T., Inagaki, H.: To be published in J. Appl. Polym. Sci.
92) Min, T. I.: Unpublished experiments in 1974.
93) See, *e.g.*, Zimm, B. H., Kilb, R. W.: J. Polym. Sci. *37*, 19 (1959);
    Kilb, R. W.: J. Polym. Sci. *38*, 403 (1959);
    Grubisic, Z., Rempp, P., Benoit, H.: J. Polym. Sci. *B5*, 753 (1967);
    Kurata, M., Abe, M., Iwama, M., Matsushita, M.: Polymer J. *3*, 729 (1972).
94) Kurata, M., Okamoto, H., Iwama, M., Abe, M., Homma, T.: Polymer J. *3*, 739 (1972).
95) Candau, F., Rempp, P., Benoit, H.: Macromolecules *5*, 627 (1972).
96) Miyamoto, T., Donkai, N., Takimi, N., Inagaki, H.: Annual Reports of Res. Inst. Chem.
    Fibers, Japan (in Japanese) *30*, 27 (1973).
97) Inagaki, H.: A. Final Report to the International Institute of Synthetic Rubber Producers,
    Inc., Oct. 1975.
98) See, *e.g.*, Yen, S. S.: Makromol. Chem. *81*, 152 (1965); Altares, Jr., T.,
    Wyman, D. P., Allen, V. R., Meyersen, K.: J. Polym. Sci. *A3*, 4131 (1965).
99) Belenkii, B. G., Gankina, E. S., Nefedov, P. P., Kuzentsova, M. A., Valchikhina, M. D.:
    J. Chromatog. *77*, 209 (1973).
100) Higashimura, T., Kishiro, O., Matsuzaki, K., Uryu, T.: J. Polym. Sci., Polym. Chem. Ed.
    13, 1393 (1975).
101) Hansen, C. M.: J. Paint Technol. *39*, 104, 505 (1967).
102) Teas, J. P.: J. Paint Technol. *40*, 19 (1968).
103) Ueberreiter, K., Asmussen, F.: J. Polym. Sci. *23*, 75 (1957).
104) See, *e.g.*, Stützel, B., Miyamoto, T., Cantow, H.-J.: Polymer J. *8*, 247 (1976).
105) Hezel, U.: Angew. Chem. *85*, 334 (1973).
106) Glöckner, G.: Plaste u. Kautschuk *23*, 338 (1976).

Received September 21, 1976

Author Index Volumes 1-24

PLASTICS AND HIGH POLYMERS

New Series

Polymers/Properties and Applications

Editors: H.-J. Cantow, Freiburg; H. J. Harwood, Akron/ USA; J. P. Kennedy, Akron/USA; S. Okamura, Kyoto

Vol. 1
B. RÅNBY; J. RABEK
ESR Spectroscopy in Polymer Research

The fundamental principles of ESR spectroscopy are first outlined, the experimental methods including computer application are described in more detail, and the main emphasis is on the application of ESR methods to polymer problems.

Scheduled to come out in summer 1977.

B. VOLLMERT
Polymer Chemistry

Springer Study Edition.
Translated from the German by Immergut, E. H.
630 figures, XVII, 652 pages. 1973.

A comprehensive and fundamental treatment of the formation of synthetic and natural macromolecules, their chemical reactions and their structural characteristics. Special consideration is given to the relationship between synthetics, structure and properties.

ADVANCES IN POLYMER SCIENCE

Editors: H.-J. Cantow; G. Dall'Asta; J. D. Ferry; H. Fujita; M. Gordon; W. Kern; G. Natta; S. Okamura; C. G. Overbörger; T. Saegusa; G. V. Schulz; W. P. Slichter; A. J. Staverman; J. K. Stille

Vol. 17
Polymerization

32 figures, III, 103 pages. 1975.
ISBN 3-540-07111-3

A. Casale; R. S. Porter: Mechanical Synthesis of Block and Graft Copolymers (112 references)
W. H. Sharkey: Polymerization through the Carbon-Sulfur Double Bond (70 references)

Vol. 18
Macroconformation of Polymers

45 figures, V, 149 pages. 1975.
ISBN 3-540-07252-7

P. A. Small: Long-Chain Branching in Polymers. The effects of long-chain branching on the molecular dimension and related properties of polymers, and on their molecular-weight distributions, are reviewed. Available information on the degree of branching in free-radical polymers is summarized (208 references).
A. Teramoto; H. Fujita: Conformation-dependent properties of synthetic Polypeptides in the Helix-Coil Transition Region. (127 references)

Springer Verlag
Berlin
Heidelberg
New York

ADVANCES IN POLYMER SCIENCE

Vol. 19
Polymerization Reactions

37 figures, 24 tables, III, 146 pages. 1975.
ISBN 3–540–07460–0

A. Ledwith, D. C. Sherrington: Stable Organic Cation Salts: Ion Pair Equilibria and Use in Cationic Polymerization (167 references).
J. P. Kennedy, J. E. Johnston: The Cationic Isomerization Polymerization of 3-Methyl-1-butene and 4-Methyl-1-pentene (20 references).
E. B. Mano, F. M. B. Coutinho: Grafling on Polyamides (194 references).
F. Millich: Rigid Rods and the Characterization of Polyisocyanides (37 references).

Vol. 20
New Scientific Aspects

95 figures, IV, 227 pages. 1976.
ISBN 3–540–07361–X

Y. Imanishi: Syntheses. Conformation and Reactions of Cyclic Peptides (166 references).
E. Ilizuka: Properties of Liquid Crystals of Polypeptides with Stress on the Electromagnetic Orientation (93 references).
J. Sohma, M. Sakaguchi: ESR Studies on Polymer Radicals produced by Mechanical Destruction and their Reactivity (104 references).
T. Kunitake, Y. Okahata: Catalytic Hydrolysis by Synthetic Polymers (146 references).

Vol. 21
Mechanisms of Polyreactions — Polymer Characterization

68 figures, III, 151 pages. 1976.
ISBN 3–540–07727–8

J. P. Kennedy, T. Chou: Poly (isobutylene-co-β-Pinene): A New Sulfur Vulcanizable. Ozone Resistant Elastomer by Cationic Isomerization Copolymerization (55 references).
J. A. Semlyen: Ring-Chain Equilibria and the Conformations of Polymer Chains (224 references).
S. Inoue: Asymmetric Reactions of Synthetic Polypeptides (67 references).
J.-M. Braun, J. E. Guillet: Study of Polymers by Inverse Gas Chromatography (108 references).

Springer-Verlag
Berlin
Heidelberg
New York

Vol. 22
Physical Chemistry

77 figures, IV, 154 pages.
ISBN 3–540–07942–4

Y. S. Lipatov: Relaxation and Viscoelastic Properties of Heterogenous Polymeric Compositions (171 references).
B. R. Jennings: Electro-Optic Methods for Characterizing Macromolecules in Dilute Solution (32 references).
A. M. Basedow, K. Ebert: Ultrasonic Degredation of Polymers in Solution (100 references).

Scheduled to come out in 1977.